the
UNIVERSITY
of
GREENWICH

Tropical fruits
Second edition

TROPICAL AGRICULTURE SERIES

The Tropical Agriculture Series, of which this volume forms part, is published under the editorship of Gordon Wrigley

ALREADY PUBLISHED

Tobacco   *B. C. Akehurst*
Sugar-cane   *F. Blackburn*
Tropical Grassland Husbandry   *L. V. Crowder and H. R. Chheda*
Sorghum   *H. Doggett*
Sheep Production in the Tropics and Sub-Tropics
      Ruth M. Gatenby
Rice   *D. H. Grist*
The Oil Palm   *C. W. S. Hartley*
Cattle Production in the Tropics Volume 1   *W. J. A. Payne*
Spices Vols 1 & 2   *J. W. Purseglove* et al.
Tropical Fruits   *J. A. Samson*
Bananas   *N. W. Simmonds*
Agriculture in the Tropics   *C. C. Webster and P. N. Wilson*
Tropical Oilseed Crops   *E. A. Weiss*
An Introduction to Animal Husbandry in the Tropics
      G. Williamson and W.J.A. Payne
Cocoa   *G. A. R. Wood and R. A. Lass*

Tropical fruits
Second edition

J. A. Samson

Longman
Scientific &
Technical

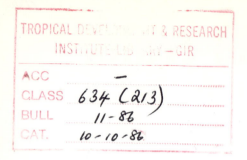
Longman Scientific & Technical,
Longman Group UK Limited,
Longman House, Burnt Mill, Harlow,
Essex CM20 2JE, England
and Associated Companies throughout the world.

Published in the United States of America
by Longman Inc., New York

First published 1980
Third impression 1984
Second edition 1986

British Library Cataloguing in Publication Data
Samson, J. A.
   Tropical fruits. – 2nd ed. – (Tropical
   agriculture series)
   1. Tropical fruit
   I. Title   II. Series
   634′.6   SB359

ISBN 0-582-40409-6

Library of Congress Cataloging-in-Publication Data
Samson, J. A. (Jules A.)
   Tropical fruits.

   (Tropical agriculture series)
   Includes bibliographies and index.
   1. Tropical fruit. I. Title. II. Series.
SB359.S25 1986   634′.6   85-24235
ISBN 0-582-40409-6

Set in Linotron 202 10/11pt Times
Produced by Longman Singapore Publishers (Pte) Ltd.
Printed in Singapore

Contents

Foreword

I am glad to have the privilege of writing a foreword to Jules Samson's authoritative book *Tropical Fruits*. Longman's Tropical Agriculture Series is a comprehensive family of well prepared, well illustrated, up-to-date books covering various aspects of agriculture in the tropics.

The author has prepared it with students of agriculture in colleges and universities, as well as growers, in mind. Whereas recently published monographs on citrus fruits, banana, pineapple, mango and other fruit crops of major importance are available for purchase in bookstores or reference in libraries, no new book with general coverage has been forthcoming in more than 20 years. Perhaps the last was the 1958 edition of W. H. Chandler's *Evergreen Orchards* which is now out-of-print. Prior to publication of *Evergreen Orchards* Wilson Popenoe's *Manual of Tropical and Subtropical Fruits* (1920) was the tropical horticulturists' 'bible', although it did not include banana, pineapple and citrus fruits.

An excellent work is John W. Purseglove's *Tropical Crops* consisting of four volumes and 1,326 pages published between 1968 and 1972. However, it tends to be encyclopedic in that it covers all crops, consequently the treatments for many are limited to the barest essential. An earlier work, *Tropical and Subtropical Agriculture* (1961) by J. J. Ochse, M. J. Soule, M. J. Dijkman, and C. Wehlburg, which consists of two volumes with 1,446 pages suffers from similar deficiencies in the eyes of persons interested in fruit culture.

The existing situation with regard to a more generalized book from the standpoint of both teacher and student is well stated by Franklin W. Martin in an article titled 'Texts in tropical horticulture: a difficult choice' (*Hortscience* 5, 145–6, 1970). It is not only a matter of choice, however, but also of availability. The lack of a good up-to-date text has become increasingly apparent in recent years. One needs only to reflect on the facts that 40 per cent of the world's surface is in the tropics, virtually all of the newly emerging and undeveloped nations are in the tropics, and students from these nations are going in ever-increasing numbers to European and American universities for instruction in agriculture that they cannot obtain at home. It is reassuring, therefore, to see a new book on the subject appear and fill the void.

Jules Samson is truly qualified to write an authoritative book of this kind. He was born in Surinam, and grew up surrounded by the crops of which he writes. He received college instruction at the prestigious Agricultural University in Wageningen, The Netherlands. He has carried out a programme of research on citrus fruits and various tropical fruits for more than 20 years. He has travelled extensively and observed agricultural practices in the tropical and subtropical regions of Europe and Africa and the subtropical regions of the United States of America.

For the last 12 years he has been teaching courses and directing the studies and researches of graduate students as a member of the faculty of the Department of Tropical Crops of The Agricultural University, Wageningen.

His compendium of facts on the tropics and the fruit crops growing therein, much of it from first-hand knowledge and experience, constitutes a timely and valuable fund of information for everyone with an interest in tropical horticulture.

William B. Storey
Professor of Horticulture (Emeritus), University of California, Riverside

Preface to the second edition

The first edition of this book was kindly received by the reviewers. However, they all agreed that the last two chapters were far too short. As I was of the same opinion, I was glad to get the opportunity to revise the text completely for a new edition.

The Department of Tropical Crop Science and the Main Library of the University of Wageningen provided the facilities to find the required literature quickly. This was incorporated into the book up to January 1985. All chapters were thoroughly revised, a task I could take ample time for, since I retired from the University in September 1981.

Chapter 8 (on mango, avocado and papaya) was completely rewritten and split into three separate chapters. Chapter 11 (formerly 9), on the minor fruits, was much extended and special attention was paid to such 'new' crops as feijoa, kiwi, passion fruit and litchi. However, the emphasis on tropical conditions has remained unchanged.

More than 200 new references were added to the book and the text was weeded of factual and grammatical errors to the best of my knowledge. Three figures of the first edition were scrapped, one was improved and thirty-four new ones were added. As far as possible, all tables were brought up-to-date till 1983 and their number was increased by eleven. A list of families and genera of fruit crops became Appendix 1 and to Appendix 2 (formerly 1) authors' names were attached.

Acknowledgements

Thanks are due to the Department of Tropical Crop Science for permitting me to work on the first edition of this book. Within this department I want to thank my colleague Jan Bink for his comments on the first four chapters, Roel Boekelman for doing the drawings, Jack van Zee for his work on the photographs and Fien van Dijk for library help. A one-time guest lecturer in this department, Dr W. B. Storey of Riverside, California, kindly wrote the Foreword. Mrs Gerda Rossel and several students helped to collect literature for (what are now) the last four chapters.

Thanks also go to my wife who typed part of the manuscript and to Dr J. Soule, University of Florida; Dr S. J. Wellensiek, retired professor of Horticulture, Wageningen; Gerard Verhoeven and Peter Gobets, for their helpful remarks and suggestions. I am grateful to the publisher and his editors for their careful scrutiny of my manuscript and to the following persons and institutions for allowing me to use photographs and/or drawings:
— Agricultural Experiment Station and Department of Agriculture of Surinam: figs 2.4, 3.5, 5.5, 5.9, 5.10, 5.14, 5.15, 5.16, 6.7, 6.12, 6.18, 8.1, 8.5, 8.6, 10.6, 11.2, 11.3, 11.4 and 11.12
— Department of Plant Taxonomy, AU Wageningen: fig. 11.15
— Fruits (d'outre mer): fig. 7.2
— IMAG, Wageningen: fig. 4.5
— KWH Whirlwind BV, Holland: fig. 4.6
— Longman: figs 2.3, 4.1 and 6.5
— Maisonneuve et Larose: figs 6.9, 6.11, 6.14, 7.5 and 7.7
— McGraw-Hill: figs 4.7, 4.8 and 4.9
— Mr H. Miller, Kibbutz Beth Haemek, Israel: fig. 9.7
— Ochse, J.J. estate: figs 8.4, 11.5, 11.9, 11.13, 11.16, 11.17, 11.19, 11.22, 11.27, 11.29 and 11.31
— Royal Botanical Society of the Netherlands: fig. 3.4
— Springer-Verlag, Berlin: fig. 2.2
— Surland NV, Surinam: figs 6.6, 6.19, 6.20, 6.21 and 6.22
— Times Atlas, comprehensive edition 1968: fig. 2.1
All other figures come from the Department of Tropical Crop Science, AU Wageningen, or from the author.

Chapter 1

Introduction

Definitions

This book deals with tropical fruits and, more particularly, with tropical fruit growing. Tropical, of course, refers to the region between the 'tropics' of Cancer and Capricorn: that part of the earth that lies between 23° 27' latitudes North and South. Contrary to popular belief, the equator is not the hottest region on earth. A band of high temperatures moves back and forth with the seasons, though the highest temperatures generally occur in the Sahara and the North Mexican desert.

What characterizes the tropics is not so much the heat, as the equable warm temperatures throughout the year. The average temperature is about 27 °C and the warmest month is only a few degrees warmer than the coldest; indeed the difference between day and night is greater than between winter and summer. Daylength varies little throughout the year and the longest day is less than 13 hours long. Most fruit crops show little response to daylength, with the exception of pineapple, as we shall see later.

This book will often refer to the subtropics, the region between the two tropics and about 40° latitude. Here the summer is hotter and the winter cooler than in the tropics. Humidity is generally lower and the difference in daylength is greater. A better definition, in environmental terms, is that the subtropics are bounded by the isotherms of 10 °C average temperature for the coldest month.

The word 'fruit' can be variously defined. Botanically, a fruit is the developed ovary of a flower, after fertilization has taken place. Some fruits develop without fertilization and do not contain seeds. A fruit, in the horticultural sense, is something which is eaten fresh and out of hand. Thus, apples, oranges and bananas are fruits; plantain and tomatoes are 'fruit vegetables'; peanuts and coconuts are 'oilseeds'.

Fruit growing can be defined in various ways. In Holland, for instance, it has been defined as 'the cultivation of edible fruits on woody plants'. This definition excludes the melon and the strawberry. In the tropics it would be even less appropriate as pineapple and banana are not woody.

If we accept that fruits are eaten out of hand, then table grapes would be fruit, but wine grapes would not; similarly with citrus fruits and pineapple. The term 'evergreen' would misinterpret the contents of this book, as attention is paid to some deciduous fruit crops too. Although subtropical crops and conditions are not neglected, the emphasis will be on the tropics. I therefore propose as a definition of fruit growing: 'the cultivation of edible fruits that are consumed either fresh or processed'. It is a simple matter to derive a definition of tropical fruit growing from this.

The present state of tropical fruit growing

If size of production is regarded as the most important criterion, we can roughly distinguish four groups of fruit crops in the world:

(a) Those having a production of more than 10 million tonnes per year, namely grape, citrus, banana, apple, plantain and mango.
(b) Those producing more than 1 million tonnes, but less than 10 million tonnes, namely pear, avocado, papaya, peach, plum, pineapple, date, fig and strawberry.
(c) Those with production figures of more than 100,000 tonnes, but less than 1 million, such as cashew nut.
(d) The rest for which there are no reliable statistics, such as guava, Brazil nut, litchi, macadamia and soursop.

The order of treatment chosen for this book is more or less based on this sequence.

Statistics are not always available. The best sources are the FAO Production Yearbooks and the French periodical *Fruits d'outre mer*, or briefly *Fruits*. From these sources we have gleaned Table 1.1.

Production figures for grape, pome and stone fruits have doubled since 1950, while those of banana and citrus tripled and have increased six-fold for pineapple. The production of mango and cashew nut seems to be stationary. A fast development is taking place in avocado, kiwi and litchi, but it is too early for predictions. Cadillat predicted in 1974 that citrus would reach a level of 55 million tonnes by 1980. This was realized, but somewhat differently from what he expected. An increase in production is not always accompanied by a rise in consumption. If consumption continues to lag behind for many years, then a relative overproduction is caused which will in turn lead to a diminishing production; however, fruit growing is a long-term process and can change only gradually.

Table 1.1   *Production of the major fruit crops (10⁶ tonnes)*

	1960	1970	1980	1981	1982	1983	1984
(Sub)tropical							
Citrus	21	38	56	55	54	57	56
Banana	15	31	39	40	41	41	41
Plantain	—	—	22*	22	23	20	20
Mango	10	12	13	13	13.5	14	14
Pineapple	2	4.2	7.8	8.6	8.9	8.7	8.8
Date	1.4	1.3	2.7	2.7	2.6	2.8	2.4
Papaya	—	—	1.8	1.9	1.9	2	2.1
Avocado	—	0.9	1.6	1.5	1.5	1.6	1.6
Cashew nut	0.4	0.5	0.5	0.5	0.5	0.5	0.4
Temperate							
Grape	45	60	67	61	71	65	64
Apple	15	24	34	33	39	37	40
Pear	4.2	7.2	8.5	8.7	8.9	9.5	9.1
Peach	3.8	5.5	7.3	7.3	7.1	7.4	7.7
Plum	4.5	3.9	4.6	5	6.2	6.2	6.1

— no data available   tonne = 1,000 kg
* Formerly partly included in banana
Source: FAO Production Yearbooks

The consumption of tropical fruits varies considerably between countries. The average consumption in Western Europe and the United States is about 10 kg/head/year for citrus and banana, and below 1 kg for the other tropical fruits. It is much lower in Eastern Europe, but growing steadily (Pieniazek, 1977).

An important development has been the rise in consumption and export of processed fruit products, mainly juices. Between 1960 and 1970 this nearly tripled for citrus fruits, so that it is now responsible for a third of the market; this occurred especially in Florida and Brazil.

Fruit prices fluctuate strongly between and within years. In general the highest prices are obtained from October to November, according to Fajac (1974) and Naville (1975). The countries of the southern hemisphere such as Brazil, South Africa and Australia thus have a decided price advantage, since they can deliver their fruit in the autumn of the northern hemisphere. Of course, other factors also regulate the prices, especially quality.

It is only in the twentieth century that tropical fruits have come to the fore. Before 1900 many of them were completely unknown in Europe and the United States. Furthermore, those that were known could only be shipped during certain months and then became entirely unavailable for the rest of the year. What has brought about this enormous expansion? Storey (1969) puts forward several causes, amongst which are:

(a) better opportunities for travel, so that people have had the chance to become familiar with these exotic fruits;
(b) better and faster ships (with refrigeration) taking them to the market in less time and with less spoilage;
(c) better processing methods to convert tender fruits into products that would not spoil;
(d) a better distribution system;
(e) appropriate propaganda and information;
(f) research into cultivation methods, crop protection, technology, etc.

Much research has taken place in the subtropics, but interest has also awakened in tropical countries.

Nutritional considerations

To what extent can tropical fruits meet dietary needs? In parts of East, Central and West Africa banana or plantain are an important part of the diet and the daily consumption may exceed 2 kg per head. Similarly with breadfruit on some islands of the Pacific, the West Indies or Sao Tomé. The 'nuts' have high contents of protein and fat; avocado is also high in fat. But in general fruits do not play a major role in nutrition as far as calories and proteins are concerned. The picture changes when we consider vitamins and minerals. Then the fruits, as well as the vegetables, become indispensible, as Table 1.2 shows.

We see from Table 1.2 that Brazil nut, cashew nut, date and fig are high in calories, proteins, calcium and iron. Banana and

Table 1.2   *Calories and nutrients per 100 g edible portion*

Fruit	Cal.	protein (g)	Ca (mg)	Fe (mg)	Vit. A (IU)	thiamine (mg)	Vit. C (mg)
Orange	53	0.8	22	0.5	—	0.05	40
Banana	116	1.0	7	0.5	100	0.05	10
Mango	63	0.5	10	0.5	600	0.03	30
Pineapple	57	0.4	20	0.5	100	0.08	30
Date	303	2	70	2	50	0.07	—
Fig	269	4	200	4	100	0.10	0
Avocado	165	1.5	10	1	200	0.07	15
Guava	58	1	15	1	200	0.05	200
Papaya	39	0.6	20	0.5	1,000	0.03	50
Cashew nut	590	20	50	5	—	0.60	0
Brazil nut	688	14.5	180	3	—	1.00	10

— = at, or near zero   1 cal. = 4.187 kJ
Source: Platt (1962)

avocado are also fairly high in calories. Papaya and mango are rich in vitamin A, the nuts in thiamine and guava in vitamin C (vitamin A is no longer measured in international units (IU) but in mg; 1,000 IU equals 0.3 mg). Some fruits, not shown in Table 1.2, contain far more vitamin C than the guava: e.g. West Indian cherry jaboticaba and emblic (see Appendix 2 for their botanical names).

Man cannot live by fruit alone, but it should definitely be part of his diet. Nutritionists advise a daily intake of at least 100 g, and as much variety as the season permits. Total fruit production in the world is 250 million tonnes per year, about 50 kg/person/year or 137 g/day, but this is spread very unevenly. In many tropical countries there is a serious shortage of fresh fruit, at least during part of the year.

Not all fruits are harmless. In the akee, the aril, an outgrowth of the seed, is edible but should be eaten from naturally-opened fruits only; the unripe aril and the pink vein that attaches it to the seed are highly poisonous (Purseglove, 1968). In carambola and bilimbi the fruits contain from 1 to 6 per cent oxalic acid, which can cause calcium deficiency and kidney stones. Plantains contain serotonin, which may cause high blood pressure when consumed in large quantities, as in East Africa.

Social and economic factors

Why grow tropical fruits? A desire to provide fruit for the family could be one motive. As our body cannot store vitamin C we must eat fresh fruit (and/or vegetables) every day. There are many different fruits, varying in flavour and in seasonal availability. It is

Table 1.3 *Fruit trees for one family (five persons)*

Fruit	Spacing (m)	Number	Space (m²)
Mango	9 × 9	1	81
Avocado	7 × 7	2	98
Grapefruit	7 × 7	1	49
Lime	7 × 7	1	49
Mandarin	7 × 7	1	49
Sw. orange	6 × 6	5*	180
Guava	6 × 5	1	30
Soursop	6 × 5	1	30
W.I. cherry	6 × 5	2	60
Papaya	4 × 2.5	4	40
Banana	3 × 2	6	36
Plantain	3 × 2	6	36
Total space			738

* two early, one mid-season, two late cvs

therefore sensible to grow as many different kinds as one can. Consider the case of a family consisting of five persons; say we want to supply each member with 200 g of fresh fruit (edible portion) for every day of the year. Assuming that on average 50 per cent of the fruit is edible we need 2 kg of fruit a day, or 730 kg/year. Reckoning with a modest yield of 10 tonnes/ha we need 730 m² of garden space. This could be used as shown in Table 1.3.

The area could be interplanted with shade-loving species, such as salak, duku, carambola and bilimbi. No doubt, further space will be found for interplanting with pineapples, some hedges of passion fruit and quite a few more bananas and plantains than have been

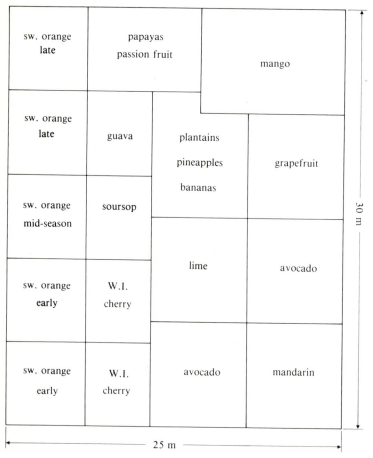

Fig. 1.1 Plan for a family plot

indicated in the table. Well-timed pruning would coax the bananas into yielding one or two bunches a month. Papaya would be harvested every week and the citrus cultivars would provide fruit for a succession of maybe six months.

So far we have spoken of pure stands of fruit trees, but generally they occur in combination with other crops. These *home gardens*, as described by Terra (1954) contain not only fruits and vegetables but also bamboo, firewood, medicinal herbs, spices, flowers etc. Altogether the farmer spends just a few days per year on his home garden and the yield (in calories and in cash) is not very high, but the benefits are spread out over the year. The system has been widely adopted in South-East Asia, Polynesia and the West Indies, but is not popular in Africa. Its significance is generally overlooked. On the island of Java, for instance, 90 per cent of all fruit and 79 per cent of all vegetables are grown in home gardens; these provide 15 per cent of the national income: as much as rice! (Dhahiyat *et al.*, 1975).

Some fruits are gathered in the forest, such as durian in Malaysia, sapucaia nuts in the Guianas and Brazil and several palm fruits in South and Central America. Mixed cropping systems are quite common e.g. bananas under coconut or as temporary shade with cacao or coffee; pineapples and papaya as a catch crop with rubber or oil palm etc.

Let us now consider the production of fruit for the local market,

Fig. 1.2 A fruit market in Abidjan (Ivory Coast)

a system which may expand to become production for export. This may take place on several small farms or on a few very large estates. An intermediate form is also possible: the nucleus estate, surrounded by many small farms.

Exact data on labour productivity are scarce, especially where fruit crops are concerned. Andreae (1972) puts the labour requirement at 1–3 man-years/ha for banana and at 0.8 man-year/ha for citrus.

Quite elaborate data are furnished by INEAC (1958). A nursery, which produced citrus plants for 20 ha for three to four years, used 667 man-days or 33.5/ha. Land clearing cost 120 man-days and other work 50 man-days; this includes a cover crop of *Pueraria*. The non-bearing orchard required 468 man-days in 5 years, or 94 man-days/year. For the bearing orchard 174 man-days were needed for maintenance and 12.5 man-days for harvesting 10 tonnes of fruit (or 25 man-days for 20 tonnes).

Some of these requirements now seem excessive: 12 man-days for fruit-thinning, 8 man-days for gathering fallen fruit and 22 man-days for pruning. For banana, nursery work came to 50 man-days, land clearing to 305–420 man-days, planting to 26 man-days and maintenance of the non-bearing orchard to 205 man-days/ha; harvesting took 30 man-days.

The INEAC figures appear high when compared with data from Surinam (Samson, 1961). However comparison is difficult because of differences between Zaire (Belgian Congo) and Surinam in climate, soil, social and economic conditions. Besides, in the Surinam data the labour requirement for land clearing, planting and transport of fruits have not been taken into account. From these (see Table 5.21) we may conclude that one man can take care of 4 ha of citrus, not including clearing, planting and transport of crops. In Israel a banana plantation requires 120 man-days/ha (including a harvest of 30 tonnes): i.e. irrigation 10 man-days, desuckering 15 man-days, weed control 10 man-days, removal of dead leaves 10 man-days, binding, propping and bagging 30 man-days, cutting, transportation and packing 30 man-days, miscellaneous 15 man-days/ha (Ticho, 1970).

In the Jalapa region (20° N) of Mexico, Marten and Sancholuz (1982) investigated the land-use systems in several climatic zones, from sea level up to 4,000 m high. Some of their findings are presented in Table 1.4. The survey was made in 1978, when wages were about 3 US dollars a day. It appears from these data that it pays to use irrigation and other modern techniques. It is also clear that fruit trees give far more protection against soil erosion than field crops.

There is little information on the labour cost for setting up fruit plantations in the tropics. Zwaardemaker (1963) calculated a labour

Table 1.4   *Fruit and field crops in Jalapa (Mexico); data for 1 hectare*

Crop	Land use	Labour (days)	Yield (tonnes)	Net profit (100 pesos)*	Erosion (tonne. ha^{-1} $year^{-1}$)
Orange	RP	5	5	30	2
Orange	RM	54	26	72	10
Banana	RM	20	12	76	5
Banana	IM	103	40	199	5
Pineapple	IM	168	60	349	15
Mango	RP	23	5	115	5
Mango	RM	90	13	222	10
Mango	IM	114	16	267	10
Papaya	RP	32	15	140	15
Papaya	RM	42	20	166	15
Papaya	IP	67	27	217	15
Papaya	IM	92	40	328	15
Sapodilla	RM	54	10	175	10
Rice	IM	95	7	105	35
Sugarcane	IM	100	120	149	18
Cassava	RM	75	29	180	40

Source: Marten and Sancholuz (1982)
* 100 pesos (Mex.) was equivalent to 4 US dollars in 1978
I = irrigated, R = rainfed
M = modern technology, P = present technology

requirement of 58 days/ha for a guava and papaya plantation in Surinam. In harvesting papaya, production was 143 to 370 kg/man-hour with a small tractor and 103 to 167 kg without the tractor. De Willigen (1968), also in Surinam, calculated picking costs of $2\frac{1}{2}$ US cents/kg for guava, 4 cents for West Indian cherry and 7 cents for the yellow passion fruit.

How long does a fruit plantation last? The economic life of a citrus orchard, from first to last harvest, is put at 40 years in California and 33 years in Florida. Of course, individual trees live longer but the orchard, as an economic unit, is written off at $2\frac{1}{2}$ or 3 per cent a year. In the tropics an economic life of from 20 to 25 years seems more appropriate; in other words, the orchard should be written off at 4 or 5 per cent a year.

Under favourable circumstances a banana plantation might last for two or more decades, but generally hurricanes and nematodes keep its economic life down to four or five years; then it has to be replanted on disinfected land – the life of a pineapple plantation is similar. Papaya lasts even less time: after three years the trees are so high that harvesting becomes too expensive. Diseases may force a change; the problem is particularly great in passion fruit where a heavy investment has to be made in poles and wires.

Another matter requiring much consideration is the minimum area of planting. Banana for export must be delivered every seven to ten days, and at a certain stage of maturity. Suppose 36 times a year at least 1,000 tonnes have to be delivered shipside; if we put average production at 24 tonnes/ha, then a minimum area of 1,500 ha is required. A pineapple cannery needs a much greater area to back it up, perhaps 10,000 ha. Even if pineapple is grown for export as fresh fruit the minimum area would have to be something like 1,000 ha, if regular shipments are to be made. A cashew processing factory has a capacity of more than 50 tonnes a day, or 15,000 tonnes a year. We may therefore put the minimum area at 15,000 ha.

The term 'Agroforestry' has been coined only recently but the idea is centuries old. It embraces all forms of land use in which trees are present, in pure or mixed stands, with or without intercropping and animals. This system helps to preserve soil fertility, while food, forage and fuel are provided. Kumar (1981) gives an example of a beach protected by screw pine (*Pandanus*), behind which wind screens of *Casuarina* trees border plantations of coconut and cashew. Another illustration comes from Bavappa and Jacob (1982): coconut and areca nut with pepper using them as support, growing together with coffee, lime, banana, papaya and other fruit trees. Preto (1983) mentions *Ceratonia*, *Tamarindus*, *Zizyphus* and many other fruit spp. as useful in arid regions and *Dacryodes* as such for the humid zone.

Is fruit growing a profitable business? There is no single answer to that question: in some places profits are made, in others losses will occur. One may also ask *who* profits from these activities. Local labour is assured of regular work, which is, of course, desirable, but whether or not the labourers get a reasonable share of the profits is often open to question.

We can only refer to social problems in passing. What effect does theft of agricultural produce have on the inclination to grow fruit? In many places fruit trees are regarded as communal property. Population density, public health, land tenure, religious beliefs etc. are also likely to influence the establishment and development of a fruit industry in many countries. This is a field which social scientists might profitably investigate.

Centres of tropical fruit research

Samson (1977) pointed out that the best-known centres of citrus research are found in the subtropics (Riverside, Lake Alfred, Campinas, Rehovot, Nelspruit, etc.), but that the French Institut de Recherches sur les Fruits et Agrumes or IRFA (formerly IFAC) has many small experimental stations in the French-speaking countries of Africa. Their monthly periodical, *Fruits* (*d'outre mer*), now in its 40th volume (1985), is a major contribution to international fruit research. IRFA is not only active in citrus research, but also in banana, pineapple, mango, cashew nut and other fruit crops. Other periodicals of international interest to the fruit grower are: *Horticultural Abstracts* and *Citrograph*. Recently *Cultivos tropicales* (Cuba) and *Frutas tropicais* (Campinas, Brazil) started publication; extensive reviews on avocado, cashew, guava and mango have already appeared in the last named. A data base called FAIREC has been set up in Valbonne, France (Anon., 1982); facts on eight fruit groups (citrus, banana etc.) are stored in twenty-five categories, e.g. ecology and rootstocks.

No really international body for research into tropical fruits has so far been founded. Yet there is a great need for one. It could do for fruits what IRRI has done for rice, or CIAT for cassava. Nevertheless, fruit research is being carried out in many countries. The big banana and pineapple companies in Central and South America, Hawaii, and some other countries spend a great deal of time and money on research, but very little is published. This is probably due to fierce competition between these companies, forcing them to secrecy in findings on cultivars, cultivation practices and so on. However, pests and diseases have no respect for international boundaries or company property and so cooperation is sometimes forced. It is to be hoped that it will spread to other areas of research.

A stimulus has been provided by organizations such as the International Society for Horticultural Science (ISHS), the American Society for Horticultural Science (ASHS), the International Organization of Citrus Virologists (IOCV) and the International Society of Citriculture (ISC). All regularly have congresses, symposia and conferences. Their publications include proceedings and periodicals such as *Hortscience* (ASHS), *Chronica Horticulturae* and *Acta Horticulturae* (ISHS), all of which are indispensable to those interested in tropical fruit.

Large collections of tropical fruit crops can be found in Homestead, Florida (Mowry *et al.*, 1967), Mayaguez, Puerto Rico (Kennard and Winters, 1960; Martin, 1976), Cuba (Cañizares, 1969) and in places such as Bangalore, India; Peradeniya, Sri Lanka; Bogor, Indonesia. Here again international cooperation is desirable for exchange of seed lists, plant materials and technical knowledge.

References

Andreae, B. (1972) *Landwirtschaftliche Betriebsformen in den Tropen*, Parey, Hamburg.

Anon. (1982) 'A data base for tropical and subtropical fruits (FAIREC)', *Fruits*, **37**, 275–6.

Bavappa, K. V. A. and **Jacob, J. V.** (1982) 'High intensity multi-species cropping', *World crops*, **34**(2), 47–50.

Caddillat, R. M. (1974) 'La situation mondiale agrumicole', *Fruits*, **29**, 831–43.

Cañizares, J. (1969) *Banco de germoplasma de frutales tropicales y subtropicales*, Ciencia Tecnica, La Habana, Cuba.

Dhahiyat, Y. *et al.* (1975) *Education, Research and Public Service project on soil erosion control in the Jatiluhur area in West Java*, Inst. Ecology, Padjadjaran Univ., Bandung, Indonesia.

Fajac, F. (1974) 'Le marché des fruits tropicaux et subtropicaux en France en 1973', *Fruits*, **29**, 155–62.

FAO *Production Yearbooks*, Rome.

INEAC (1958) *Normes de main-d'oeuvre pour les travaux agricoles au Congo Belge*, Hors Série.

Kennard, W. C. and **Winters, H. F.** (1960) *Some fruits and nuts for the tropics*, ARS, USDA Misc. Publ. 801, Washington.

Kumar, P. H. (1981) 'Problems and prospects of establishing a plantation forestry with *Casuarina*, cashew and coconut in the coastal belt of India', *Riv. Agric. subtrop. e trop.*, **75**, 317–23.

Marten, G. G. and **Sancholuz, L. A.** (1982) 'Ecological land-use planning and carrying capacity evaluation in the Jalapa region (Vera Cruz, Mexico), *Agro-ecosystems*, **8**, 83–124.

Martin, F. W. (1976) 'Introduction and evaluation of new fruits in Puerto Rico', 105–10, in *First Int. Symp. on tropical and subtropical fruits, Lima, Acta Hort.*, **57**, ISHS.

Mowry, H. *et al.* (1967) *Miscellaneous tropical and subtropical Florida fruits*, Bull. 156A, Agr. Ext. Serv., Gainesville.

Naville, R. (1975) 'Le marché francais des fruits tropicaux et subtropicaux en 1974', *Fruits*, **30**, 359–66.

Pieniazek, S. A. (1977) 'Eastern Europe – a potential market for tropical fruits', *Acta Hortic.*, **53**, 293–6 (Fourth Africa Symp. Hortic. crops).

Platt, B. S. (1962) *Tables of representative values of foods commonly used in tropical countries*, HMSO, London.

Preto, G. (1983) 'Importanza e prospettive dell'agro-selvicoltura tropicale', *Riv. Agric. subtrop. e trop.*, **77**, 319–41.

Purseglove, J. W. (1968) *Tropical crops, Dicotyledons*, Longman.

Samson, J. A. (1961) *Handleiding voor de citruscultuur in Suriname*, Mededeling 24, Landb. Proefst. Suriname.

Samson, J. A. (1977) 'Problems of citrus cultivation in the tropics', *Span*, **20**, 127–9.

Storey, W. B. (1969) 'Recent developments in tropical fruit crops', *Proc. Fla. State Hort. Soc.*, **82**, 333–9, Miami.

Terra, G. J. A. (1954) 'Mixed garden horticulture in Java', *Malayan J. Trop. Geogr.*, **3**, 33–43.

Ticho, R. J. (1970) *The banana industry in Israel*, Min. Agr. Israel.

WHO (1974) *Handbook on human nutritional requirements*, Geneva.

Willigen, P. de (1968) 'Kostprijsonderzoek vruchtenproefbedrijf Boma', *Surin. Landb.*, **16**, 99–109.

Zwaardemaker, J. R. (1963) *Enkele bedrijfseconomische en technische aspecten van de cultuur en het oogsten van enkele vruchtencultures*. Landb. Proefst. Suriname. mimeogr. report.

Chapter 2

Environment

Climate in relation to tropical fruit growing

The weather changes from day to day but climate has a more permanent character. It may be defined as 'average weather' or as 'the whole of average atmospheric phenomena for a certain region, calculated for a period of thirty years'. These phenomena are generally taken to be light, heat, water and air

Light is necessary for photosynthesis, growth and development of plants. Generally, fruit trees need much light and must be grown in a sunny climate, but some, such as the banana, tolerate shade. Others, for example young mangosteen, need shade during part of their development. A third group requires permanent shade; this group includes the salak palm, duku and carambola (Terra, 1949).

Daylength, that is the time elapsing between dawn and dusk, may exert considerable influence on flowering. It is customary to distinguish between short-day, long-day and day-neutral plants, according to their demands on the duration of the light period. However, most tropical fruit trees show only slight reactions, or none at all, to the photoperiod which is always between 11 and 13 hours in the tropics. The pineapple is an exception.

The mean temperature at sea level near the equator is 26–27 °C and the range is small; it usually amounts to 2–3 °C between months and 6–10 °C between day and night. Farther away from the equator the range increases. Altitude has the effect of lowering the temperature by 5–6 °C for every 1,000 metres.

The growth rate of plants depends primarily on temperature. This means that plants which grow optimally at sea level, will grow more slowly in the mountains. In Jamaica, for instance, it was shown that the banana cultivar Lacatan had a growth cycle of 13 months at sea level and one month more for each 100 metres of altitude. 'Heat index' is a term which indicates the sum of average daily temperatures during the growth cycle of the crop. For perennial crops, however, average temperature is not a good criterion as the plant stops growing when it is too cold (below the minimum temperature) or too hot (above the maximum). It is therefore better to use only *effective* temperatures, those between the two extremes for growth. Examples will be given later on.

The apple and its relatives need low temperatures in order to break the dormancy of the buds. For this reason pome and stone fruits such as apple, pear, peach and cherry cannot be grown in the tropical lowlands. Depending on the cultivar, 250–1,000 hours with temperatures below 7 °C are necessary for flowering to occur (Ruck, 1975). Other crops that need chilling are mentioned in Chapter 11.

Frost kills most tropical fruit trees, citrus and date are exceptions. Even a temperature several degrees above zero may be harmful. Banana fruit turns brown when exposed to a temperature below 12 °C because latex in the skin coagulates. Similar 'chilling' damage occurs in soursop, sapodilla, etc. On the other hand, the extremely high temperatures found in arid regions cause wilting, 'sun scald', necrotic spots and even the death of plants. The date is resistant, up to about 50 °C.

Water is essential for the growth of plants. About 2,000 mm/year of rain falls on either side of the equator, up to about 10° latitude. Farther away there is less rain, especially from 20–40°N. and from 20–30°S. Mountain ranges and deserts change the picture considerably. More important than total rainfall is the distribution of rain over the year. The number of dry months, in which evapotranspiration exceeds precipitation, is of great significance.

Terra (1949) investigated this for the island of Java, Indonesia (about 6°S.) and reached the following conclusions. Mangosteen, kapulasan, banana and papaya grow best when more than 100 mm of rain falls in every month, but little harm results if one or two months should fall to a level below 60 mm. Rambutan and durian also need plenty of rain, but can tolerate two or three dry months. On the other hand, mango and cashew nut should have at least three dry months and not more than seven wet months, otherwise bloom and fruit setting will be disturbed. Citrus occupies a position somewhere between those extremes.

The yearly course of rainfall follows the apparent movements of the sun. Consequently, there are two wet and two dry seasons close to the equator. Nearer the 23rd parallel there is only one wet and one dry season. The closer one gets to that parallel, the longer the dry period lasts. Perennial crops need irrigation if they are cultivated here. An example is provided in Table 2.1.

As Table 2.1 shows, Yangambi and Bombay with practically the same yearly rainfall have entirely different patterns of distribution. In Yangambi most months receive from 100 to 200 mm of rain and even 'dry' months are only slightly below 100 mm; it is not likely

Table 2.1 *Number of dry months (< 100 mm) for six locations*

	Lat.°	Alt. (m)	Rain (mm)	Dry months	
Colombo, Sri Lanka	7	7	2,370	1	II
Bombay, India	19	11	1,813	6	XI–IV
Yangambi, Zaire	1	487	1,964	2	I, VI
Freetown, Sierra Leone	8	11	3,510	5	XII–IV
Paramaribo, Surinam	6	4	2,297	2	IX–X
Port of Spain, Trinidad	10	40	1,609	3	II–IV

I = January, II = February etc.
Source: Walter *et al.* (1975)

that fruit crops will ever be subjected to drought here. Bombay, on the other hand, has a long dry spell from November to April and fruit growing is impossible without irrigation. In Freetown, which has much more rain than Bombay, irrigation is also required for fruit growing.

A rainfall below 100 mm/month is considered low in the tropics as evapotranspiration usually lies between 120 and 150 mm. On the other hand, more than 300 mm/month is not useful as the surplus cannot be stored in the soil; it is even harmful as it promotes erosion and leaching. Distribution of rainfall within the month is important: one cloudburst of 100 mm has an effect entirely different from ten well-spaced showers of 10 mm.

The best way to evaluate the usefulness of rain for a crop is by the concepts: 'potential evapotranspiration' and 'water supply in the rooted zone'. Evaporation is a physical process, transpiration a physiological process. Evapotranspiration (E) comprises both the upward movement of water or vapour from the soil and the transpiration of the plant. Potential $E(E_p)$ refers to a closed leaf canopy that is well supplied with water. It is difficult to measure E_p directly but it can be estimated from E_w, the evaporation from a free water surface. It can also be calculated from Penman's formula which takes into account the increase of E_p caused by bright sunlight, high temperature, low humidity and wind. Under average tropical conditions E_p reaches a value of 4–5 mm a day, but it may well exceed 10 mm in a sunny climate. Maximum E for a fast growing crop such as banana may be 50–60 per cent higher than E_w. In order to calculate the total needed for irrigation, the water lost by run-off from the field and that lost by deep percolation must be added to E_p.

The water supply in the rooted zone is generally taken to be equivalent to 100 mm of rain, so that it is easily exhausted in one rainless month. In very good soils it may exceed 200 mm. Some crops can tolerate drought for months, others (banana) have to be irrigated in even a short dry season.

Using these two concepts we can now calculate the water balance of the soil for a certain crop. Frémond *et al.* (1966) gave the following figures for coconut in Port Bouet, Ivory Coast. Table 2.2 shows that 'effective' evapotranspiration is equal to E_p from May to August and from October to December; in other words: when rainfall and soil reserve permit it. In the other months the effective E is lower than E_p and the excessive rainfall of May and June serves no useful purpose. For a discussion of yield responses to water, see Doorenbos *et al.* (1979).

Table 2.2   *Water balance of the soil in mm, Port Bouet*

Month	Rain	E_p	Reserve	Deficit	Excess	Effective E
J	30	147	58	59	—	88
F	50	153	—	103	—	50
M	108	182	—	74	—	108
A	160	178	—	18	—	160
M	382	164	100	—	118	164
J	570	132	100	—	438	132
J	177	124	100	—	53	124
A	35	107	28	—	—	107
S	61	114	—	53	—	88
O	207	138	69	—	—	138
N	188	152	100	—	5	152
D	113	155	58	—	—	155
Total	2,081	1,746		307	614	1,466

Source: Frémond (1966)

Water is also present in the atmosphere as vapour. Atmospheric humidity influences growth and development of plants: low humidity has a drying effect, while high humidity creates favourable conditions for fungi, causing diseases. When humidity is high by day dew is formed at night, this promotes infection and diseases such as banana leaf spot.

Our last climatic factor is air. Pollution may occur, even in the tropics; for example, when gasses escape from a factory. But we are more concerned with wind. The so-called trade winds blow from ENE in the northern and from ESE in the southern hemisphere; they are noted for constancy in speed and direction (Lockwood, 1974). In the subtropics the wind is more often westerly. A hot desert wind like the notorious Harmattan of West and Central Africa can do a great deal of damage to crops. Storms, with wind speeds over 50 km/h, and hurricanes (wind speed over 100 km/h) may ruin a banana crop completely and do much damage to woody crops such as avocado, citrus and mango. Most tropical storms originate between 8 and 15 degrees latitude and move away from the equator.

Many attempts have been made to classify the climates of the world. One of the most popular, by Köppen, recognizes five main groups:

A   rainy climates, coolest month above 18 °C
B   dry climates, either warm or cold
C   rainy climates, with a mild winter
D   rainy climates, with a cold winter
E   polar climates.

We shall ignore D and E, as the winters are too cold for tropical fruit crops. In Köppen's system, as modified by Trewartha (1954), the following subdivisions are added:

a   warmest month over 22 °C
b   warmest month below 22 °C
f   'feucht' (humid), no dry season
h   hot (B-group), all months above 0 °C
m   monsoon, short dry season
s   dry season in summer
w   dry season in winter
S   'Steppe', savanna
W   'Wüste', desert.

Symbols c, d, k, and n have been disregarded here as they refer to cold or misty conditions. The system is surveyed and some examples are given in Table 2.3.

Table 2.3   *(Sub)tropical climates, Köppen-Trewartha*

Symbol	T	Tw	Tc	Pd (mm)	Examples
Af			>18	>60	Surinam, Yangambi (Zaire), Malaysia
Am			>18	<60	Puerto Rico, Sierra Leone, Bombay
Aw			>18	<60	Maracaibo, Guinea, Calcutta
Caf		>22	0–18	<30	Louisiana, Hong Kong, Sydney
Caw		>22	0–18	<30	Florida, Allahabad
Csa		>22	0–18		Los Angeles, Valencia
Cb		<22	0–18		Western Europe, tropical mountains
BShs	>18				Tunis, Tehran
BShw	>18				Sahel zone
BWh	>18				Arizona, Sahara, Arabia

T = Average temperature, *Tw* = average temperature warmest month,
Tc = average temperature coldest month, *Pd* = precipitation dryest month

For a map, giving many examples of climatic types, see *The Times Atlas of the World*, comprehensive edition 1978. Figure 2.1 shows the climates of Africa and the Mediterranean region.
The Köppen-Trewartha classification is not always satisfactory

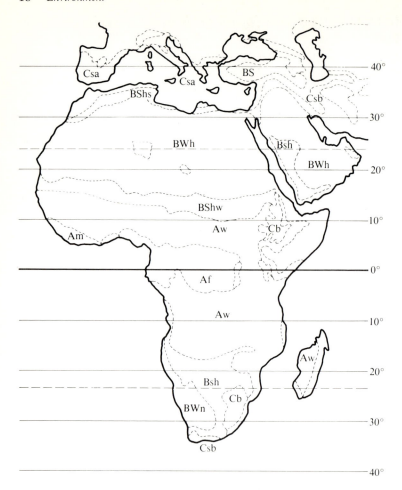

Fig. 2.1 Climate types of Africa and the Mediterranean region, according to Köppen-Trewartha (1954). For explanation of symbols see text

for our purpose. For instance, the Cb-type stretches all the way from Bergen, Norway to Bordeaux, France but also comprises Bogota, Colombia and Addis Ababa, Ethiopia. In the latter two places, however, the temperature range is far smaller and all days are short.

How do fruit trees fare in these climatic types? Table 2.4 attempts to show this; but it only gives a rough idea and its value is limited. Variations in climate or cultivar may admit or exclude a fruit species and cultural measures, such as irrigation and protection

Table 2.4   *Relation of climate to fruit species*

Crop/climate	Af	Am	Aw	Caf	Caw	Csa	Cb	BShs	BShw	BWh
Citrus	+	+	+	+	+	+	?	?	?	?
Banana	+	+	+	?	?	—	—	—	—	—
Mango	?	+	+	—	+	?	—	—	+	—
Cashew nut	?	+	+	—	+	—	—	—	+	—
Pineapple	?	+	+	—	—	—	—	?	+	—
Avocado	+	+	+	—	+	?	—	?	+	—
Date	—	—	—	—	—	—	—	+	?	+
Grape	—	—	?	—	—	+	+	+	—	—
Apple	—	—	—	—	?	—	+	—	—	—

+ = fit, — = unfit, ? = doubtful

against frost or heat, may make the difference between success and failure.

There are many other classifications of climate, but we cannot discuss them here. We shall only mention Thornthwaite's system (Thornthwaite and Mather, 1957) because it contains elements that are useful to the crop specialist such as indices for thermal efficiency, aridity, humidity and E. Unfortunately, it is not (yet) widely used. A simple rule by Holdridge (1959) also deserves to be mentioned: $E_p = 59T$. In other words, if the average temperature is 25 °C, E_p will amount to $59 \times 25 = 1,475$ mm. Walter *et al.* have constructed climate diagrams that show at a glance many important details of interest to the crop specialist. They are based on an equivalence of 10 °C with 20 mm of rain on the vertical axis. On the horizontal axis the months are set out, with the warmest always in the middle. Thus in the northern hemisphere it runs from January to December and on the southern from July to June. The temperature curve serves as a rough indication of E_p. Therefore, the intersections of the curves for rainfall and temperature indicate transitions from wet to dry periods and *vice versa*. The fields between those lines are coarsely dotted for dry seasons and finely tinted for wet periods; very wet periods have a medium tint. The diagram also records elevation, average temperature, total rainfall per year and the number of years during which records were taken. Some examples are presented in Fig. 2.2.

Soils and fertilizers

In this section we will indicate topics which are of particular interest to the tropical fruit grower. More information will be found in specialized books as Russell (1973) Sanchez (1976) and Buringh (1980).

Soil is defined as a three-phase system in which plants grow. All three phases: solid, liquid and gas, are essential. The solid part is

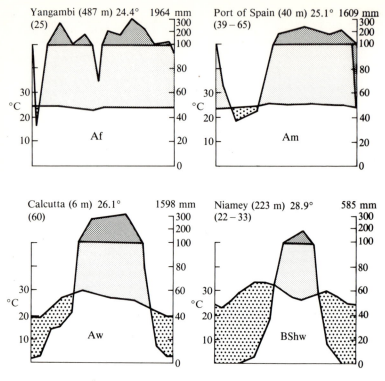

Fig. 2.2   Climate diagrams of four tropical towns, according to Walter *et al.* (1975)

the frame that provides space for the other two; it also contains minerals and organic matter. Texture depends on the size of particles: gravel, coarse and fine sand, silt and clay. Soils are classified according to the relative contribution of these particles (Fig. 2.3). Structure describes the arrangement of soil particles into bigger units. A crumb structure is considered the best. It assures the presence of large and small pores, so that sufficient water and air are available for root growth.

Soil is a living system, it teems with organisms that break down organic matter and build it up again; it is constantly mixed with mineral particles. Under favourable conditions a stable compound called humus results, which helps to maintain soil structure and soil fertility. However, it is not stable at high temperatures. Above 25 °C humus is broken down faster than it is built up. In the tropics accumulation of organic matter can therefore only take place under water. When such peaty soils are reclaimed without burning, the organic matter decays gradually and confers great fertility to the

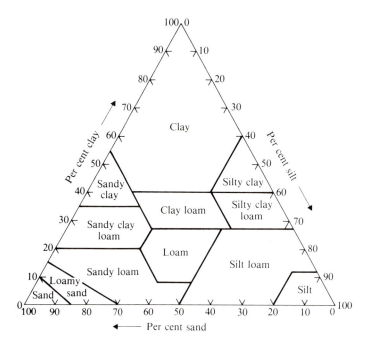

Fig. 2.3 Soil texture classes used by US Soil Survey. (Sand 2–0.05 mm, silt 0.05–0.002 mm, clay below 0.002 mm). *Source*: Russell (1973)

soil. The modern banana industry of Surinam is based on such soils (Samson, 1968).

Water is needed for plant growth but too much water in the soil, means too little air. The amount of water present after rain or irrigation, when no more water is seeping down, is called the field capacity (FC) of that particular soil (pF 2). The roots can only remove water from large pores, not from the capillaries. When all water has been sucked from the larger pores the permanent wilting point (PWP) is reached (pF 4.2). The difference between FC and PWP determines the water-holding capacity of a soil, as is shown by the pF curve in Fig. 2.4. Lines drawn at pF 2.0 and 4.2 intersect with the clay curve at 61 and 29 volume per cent soil moisture; this clay thus holds 61 − 29 = 32 per cent available water. Likewise the fine sandy clay holds only 52 − 35 = 17 per cent water the roots can use.

Good soil must be permeable. Sometimes a crust is formed on the surface, causing rain water to run off to ditches where it is lost. Run-off may also cause erosion. Hardpans prevent the vertical

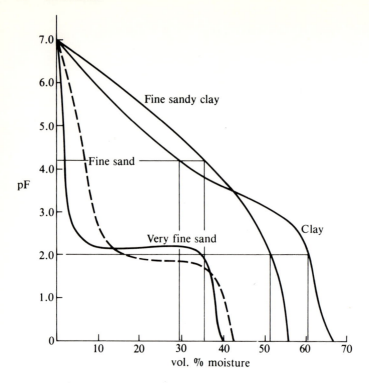

Fig. 2.4 pF curves for four soil types in Surinam
Source: Agr. Expt. Sta. Surinam, Bull. 91 (1973)

movement of water through the soil. The fruit grower must identify these impediments to good drainage.

In soil air, the content of carbon dioxide is about 0.2 per cent, six times as high as in the atmosphere. This air must be changed constantly by diffusion to prevent CO_2 rising to noxious levels. The fruit grower should choose a planting site with the best possible physical properties. He must, where necessary, improve drainage and soil structure by adding organic matter to his soil and by cultivation.

Defects in chemical fertility are generally easier to remedy. The main constituent elements of plants: oxygen, hydrogen and carbon, occur in huge quantities in water and air. Nitrogen, in order to be absorbed by plant roots, must be presented in the form of nitrate (NO_3^-) or ammonia (NH_4^+). Some bacteria can convert N_2 into NH_2^- if an energy source is available. The 'nitrogen-fixing bacteria' live freely in the soil, or form 'nodules' on the roots of a leguminous plant. The nitrogen thus fixed becomes available to the host plant; either directly, or after old nodules have been cast off and disintegrate (Cobley and Steele, 1976). Leguminous plants are therefore extremely useful as cover crops and green manures.

The other 'major' elements are sulphur (S), phosphorus (P), calcium (Ca), potassium (K) and magnesium (Mg). The 'minor' or trace elements occur in far smaller amounts in plants, but are essential: iron (Fe), manganese (Mn), zinc (Zn), copper (Cu) and boron (B). In some cases sodium (Na), molybdenum (Mo) and a few others are also essential.

Sizable amounts of nutrients are withdrawn from the soil when fruit is harvested. According to Montagut and Martin-Prével (1965) 40 tonnes of bananas remove 80 kg N, 8 kg P, 200 kg K, 10 kg Ca and 10 kg Mg from the soil. Such losses have to be made good, but one must also supply nutrients for initial growth; this amounts to about five times the yearly loss.

Before deciding on a planting site, one should have the soil examined. Samples of soil layers are taken and analysed for texture, organic matter, nutrients and pH. At pH 7 the soil solution is neutral, below that it is acid. The most favourable pH for nearly all fruit crops is about 6, when all nutrients are easily absorbed by the roots. At a pH below 5 it may be necessary to apply lime or dolomite. At a pH over 7 deficiencies of iron and zinc will cause trouble.

Soil processes are complicated and soil analysis alone gives only an indication of the potential to feed the plant. Therefore, the crop too must be analysed. Generally leaves are used; they should be of uniform age and position (e.g. all from non-fruiting branches). Furthermore, the grower has to inspect his crops regularly and look for signs of deficiency or excess. The symptoms vary. Usually a lack of N is shown by a light green colour of the leaves, of P by their dark green colour. Potassium-deficiency is recognized by marginal scorch and lack of Mg by loss of chlorophyll. Leaves deficient in Fe show a fine network of green veins against a lighter background. Zinc deficiency is characterized by narrow leaves with yellow bands between the veins; if Mn is lacking we see light green bands between the veins, but the leaf size is normal. Shortages of boron, copper and molybdenum are less prevalent, those of calcium and sulphur are rare.

Organic fertilizers: farmyard manure, compost and green manure, improve structure. However, they are bulky and not easy to handle. Compound fertilizers contain several major elements, e.g. 12.10.18.2 contains (per cent) 12N, 10P_2O_5, 18K_2O and 2MgO. Generally it is not advisable to use compound fertilizers. They rarely correspond to the peculiarities of soil and crop. However, the well-informed grower can make his own mixtures. One saves money by using highly concentrated fertilizers, which are lower in price per unit and cheaper to transport.

To sum up, fruit crops should be grown on soils with good physical properties; in general a sandy loam is preferable. Chemical fertility is desirable, but not essential. In most cases the pH should be near 6.

The principal nitrogen fertilizers are ammonium sulphate (21% N), ammonium nitrate (33%), calcium ammonium nitrate (21%) and urea (45%). Long-term use of ammonium sulphate acidifies the soil. Urea is relatively cheap and can be sprayed on leaves, but this may cause damage if it contains more than a trace of biuret.

Phosphate fertilizers, with the P_2O_5 content in parentheses ( ), come in the form of superphosphate (16–20%), double = triple super (36–48%), basic slag (14–18%) and rock phosphate (30%). In super and double super the P is water soluble and fast acting. It dissolves slowly in the other forms and they should not be used at a pH higher than 6. As P does not disperse easily through the soil profile, care must be taken to apply it to deeper layers too. Fruit growers achieve this by putting rock phosphate at the bottom of planting holes.

The main potassium fertilizers (K_2O content in parentheses) are KCl or muriate of potash (50 or 60%), sulphate of potash (48–52%) and sulphate of potash and magnesia ('patentkali', 'sulpomag') which contains 26–30 per cent K_2O and 9–12 per cent MgO. KCl should not be used on crops sensitive to chlorine.

Lime ($CaCO_3$) and dolomite ($CaCO_3.MgCO_3$) are used to adjust the pH; the rate depends on pH, organic matter and clay content of the soil. The liming factor is the quantity of lime needed per ha to raise pH one point, from 5 to 6. At pH 4 ten times as much lime is required. Fertilizer is deposited in a circular area about 50 cm wide under the outer branches of the tree and never closer than 20 cm to the stem; for full-grown trees it is broadcast or laid in a broad band between the rows.

Micronutrients are usually applied as sulphates (Zn, Mn, Cu), but these are not available to the roots at high pH. It is easier to spray them on the trees. Table 2.5 gives the concentrations recommended by Samson (1966). These amounts are based on the assumption that one hectare receives 2,000 litres of a dilute solution (high volume spraying) or 100 litres of a concentrated solution (low volume spraying). However, an antagonism exists between zinc and

Table 2.5 *Amounts of spray material in g/100 litre*

	High Volume	*Low Volume*
Zinc sulphate	500	5,000
Manganese sulphate	250	2,500
Hydrated lime	250	2,500
Copper-oxy-chloride	250	2,500

manganese, so it is better not to mix them (Mann and Takkar, 1983). According to Kotur (1984) a 0.1 per cent solution of zinc oxide gave better results than zinc sulphate.

Iron is best applied in the form of chelates: complex organic compounds such as Fe-EDTA or Fe-EDDHA. This is especially necessary on alkaline soils. Micronutrients should be applied carefully as they may tie up other elements such as P and may become toxic. The range between deficiency and toxicity is rather small, particularly in boron.

The mineral content of farmyard manure is rather low, about 0.45 per cent N, 0.2 per cent P_2O_5 and 0.4 per cent K_2O but 'it has a beneficial effect on the soil, promotes microbial processes, improves structure, aeration and water holding capacity' (Jacob and Uexküll, 1960). An application of 20 tonnes/ha/year would deliver 90 kg N, 40 kg P_2O_5 and 80 kg K_2O. Compare this to the amounts of 80, 8 and 200 taken up by a good banana crop (Montagut and Martin-Prével, 1965) and we see that only an additional amount of K would be required.

Compost can be made from all kinds of waste products such as town rubbish, weeds, factory refuse, sawdust etc. It is not used to a great extent in the tropics, except in South-East Asia. Yet it would be worthwhile to make better use of the opportunities presented, of the rubbish dumps of big cities in Africa for example.

A green manure is an auxiliary crop grown for the sake of a main crop that follows it. Leguminous plants are often chosen for this purpose, e.g. *Crotalaria*, *Tephrosia* and *Sesbania* spp. They are turned under to enrich the soil before they set seed and become woody, or before the dry season when they would compete for water with the main crop. They usually need an application of phosphate and potassium fertilizer. Pigeon pea (*Cajanus cajan*), cowpea (*Vigna unguiculata*) and other pulses provide food as well as being a soil amendment.

The same plants may be used for mulching. They are planted in rows, two metres apart, and regularly cut back; the branches and leaves are spread over the soil. Grass, straw, other plant residues and even paper can be used as mulch. This protects the soil from heavy rain and direct sunlight and thus from loss of structure, humus and moisture and from high temperature in the upper layers.

Some mulches are rich in carbohydrates and poor in protein, their C/N ratio being over twenty; bacteria and fungi will then use soil ammonia or nitrate for growth and thus cause a temporary N-deficiency. It is therefore necessary to apply some N fertilizer together with a straw or sawdust mulch. The C/N ratio of humus is around ten.

Cover crops differ from green manures in not being ploughed under. They are generally herbaceous creeping legumes such as

Pueraria phaseoloides (tropical kudzu), *Centrosema pubescens* and *Calopogonium mucunoides*. Mixtures are often used, in which *Calopogonium* grows fastest at first; after about a year *Pueraria* outgrows the others, but when shade of the main crop becomes noticeable *Centrosema* or *Vigna hosei* will gradually take over. The ability to fix nitrogen has been estimated at 200 to 400 kg/ha/year for *Pueraria*, which makes it unnecessary to apply a N-fertilizer; in fact that would only cause the cover crop to become 'lazy' and fix less nitrogen. However, other elements (P, K, Mg and trace elements, particularly Mo) must be provided. Deficiency symptoms of tropical kudzu have been described by Dirven and Ehrencron (1961). A good non-creeping cover crop is *Moghania macrophylla* (syn. *Flemingia congesta*).

Some disadvantages of cover crops must be mentioned. They have a tendency to climb and smother trees, particularly *Pueraria*; this is easily overcome by a few slashes of the machete (cutlass) during monthly inspections. Secondly, they compete for water with the main crop in the dry season and might even become a fire hazard. This can be averted by cutting the cover crop at the start of the dry period. Thirdly, seed is sometimes hard to get. This is particularly the case with tropical kudzu. However, once that cover crop is established it can be propagated from layers; these are rooted in baskets and set out at a density of 50/ha at the beginning of the wet season.

One final remark on cover crops. Where the dry season lasts a long time, say more than three months, it is undesirable to maintain a permanent cover crop. It is better to resow each year at the start of the rainy season.

We shall conclude with a few remarks on salt. This problem arises in arid or sub-arid regions where E_p exceeds rainfall. Water containing salts is drawn to the surface and evaporates, leaving the salt behind. Gypsum and lime are not too troublesome, but the more soluble chlorides, sulphates, nitrates, carbonates, bicarbonates and borates may cause a lot of trouble (Buringh, 1968).

The amount of salt in the soil solution is measured by electrical conductivity (EC) which is expressed in milli-mho/cm. Mho is the reverse of ohm which measures resistance. At an EC of 0–2 mmho/cm all crops grow well, but at EC 2–4 mmho/cm sensitive crops begin to show symptoms. Between EC 4–8 mmho/cm most crops suffer and from EC 8–16 mmho/cm only a few can function normally, date palm and some citrus and avocado rootstocks among them. EC can also be used to assess the quality of irrigation water. This is said to be low in salinity if EC is below 0.25 mmho/cm and highly saline if it is above 2.25 mmho/cm.

Redress to salt can be found by washing out the Na^+ ions in salty soils and by treating with gypsum in alkali soils (containing carbonates) with the effect that Na^+ is replaced by Ca^{2+} ions.

Tropical vegetation

The grower who wants to set up an orchard cannot always obtain reliable data on the climate and soils of that region. In that case vegetation may serve as an indication of the agricultural potential. Near the equator the tropical rain forest is the climax vegetation. Its luxurious growth, caused by a constant recycling of nutrients, gives a mistaken impression of high soil fertility. As soon as this is checked by clearing and burning large losses occur. The ash, which temporarily enriches the soil, is leached by the first rainstorm (Webster and Wilson, 1980). Other formations recognized by these authors are seasonal, dry evergreen, montane, swamp and seasonal swamp.

References

Buringh, P. (1968) *Introduction to the study of soils in tropical and subtropical regions*, Wageningen; 3rd edn, 1980.

Cobley, L. S. and **Steele, W. M.** (1976) *An introduction to the botany of tropical crops*, Longman.

Dirven, J. G. P. and **Ehrencron, V. K. R.** (1961) 'Deficiency symptoms in tropical kudzu', *Sur. Landb.*, **9**, 41–7.

Doorenbos, J. *et al.* (1979) *Yield response to water*, FAO Irrigation and Drainage paper **33**, Rome.

Frémond, Y. *et al.* (1966) *Le cocotier*, Paris.

Holdridge, L. R. (1959) *Ecological indications of the need for a new approach to tropical land use*, Turrialba.

Jacob, A. and **Uexküll, H. von** (1960) *Fertilizer use*. Verlagsgesellschaft für Ackerbau, Hannover.

Kotur, S. C. (1984) 'A comparison of zinc oxide, unneutralized and neutralized zinc sulphate as foliar sprays in Coorg mandarin', *Indian J. Agric. Sc.*, **54**, 186–8.

Lockwood, J. G. (1974) *World climatology: an environmental approach*, London.

Mann, M. S. and **Takkar, P. N.** (1983) 'Antagonism of micronutrient cations on sweet orange leaves', *Scientia Hortic.*, **20**, 259–65.

Montagut, G. and **Martin-Prével, P.** (1965) 'Besoins en engrais des bananeraies antillaises', *Fruits*, **20**, 265–73.

Ruck, H. C. (1975) *Deciduous fruit tree cultivars for tropical and sub-tropical regions*, Farnham Royal.

Russell, E. W. (1973) *Soil conditions and plant growth* (10th edition), Longman.

Samson, J. A. (1966) *Handleiding voor de citruscultuur in Suriname*, Mededeling 39, Landbouwproefst. Suriname.

Samson, J. A. (1968) 'Citrus cultivation in Surinam', *Neth. J. Agric. Sc.*, **16**, 186–96.

Sanchez, P. A. (1976) *Properties and management of soils in the tropics*, Wiley-Interscience.

Terra, G. J. A. (1949) *De tuinbouw in Indonesië*,'s Gravenhage.

Thornthwaite, C. W. and **Mather, J. R.** (1957) *Instructions and tables for computing potential evapotranspiration and the water balance*, Centertown.

Trewartha, G. T. (1954) *An introduction to climate*, McGraw-Hill, New York.

Walter, H. *et al.* (1975) *Climate diagram maps of the individual continents and the ecological climatic regions of the earth*, Berlin (with 9 maps).

Webster, C. C. and **Wilson, P. N.** (1966) *Agriculture in the tropics*, Longman; 2nd edn, 1980.

Chapter 3

Botany of tropical fruits

The successful fruit grower should be familiar with certain fundamental facts of botany; some of particular interest are discussed briefly here. For more details the interested reader should consult general textbooks and books such as Loveless (1983), Cobley and Steele (1976) and Abercrombie (1968).

Taxonomy

This branch of botany describes, classifies and attaches names to plants. In view of the enormous number of plant species, more than half a million, a knowledge of taxonomy is indispensable for anyone wishing to study plants.

Two classes of higher (seed-bearing) plants are recognized: dicotyledons and monocotyledons. Banana, pineapple and palms belong to the latter group; most other fruit plants to the former. The most important taxonomic groups below the rank of class are family, genus and species. Each species (sp.) bears a double name, of which the first is the name of the genus. Specific names should be followed by an author's name or initials, e.g. L. for Linné; these names are given in Appendix 2.

The term variety was formerly used for both botanical and horticultural varieties. It has been decided, see Gilmour (1969), that a variety (var.) must be based on distinct differences in form. Where agronomic properties, such as colour, taste, ripening time, are concerned the term cultivar (cv.) should be used. A fruit cv. is often a clone, i.e. a group of plants vegetatively propagated from one single plant. All members of a clone have the same genotype. An example of a variety is the yellow passion fruit *Passiflora edulis* var. *flavicarpa*, whereas the Valencia orange is an example of a cultivar: *Citrus sinensis*, cv. Valencia. We may also write 'Valencia' (note: with single inverted commas), in which case the indication cv. is omitted. For a list of families and genera of fruit crops, see

Appendix 1. Common and matching botanical names are found in Appendix 2.

Morphology

Herbaceous plants usually blossom once and then die, having set fruit and seed; they are called annuals. Some perennials, like banana and pineapple, also flower once and then die: they are monocarpic. Our woody fruit crops are polycarpic, they bloom year after year. Some trees, e.g. apple, drop their leaves in winter; they are deciduous. Others, like citrus, mango, avocado, retain their leaves for two or more years: evergreen trees.

The fruit grower must pay attention to the flower; flower parts are shown in Fig. 3.1. If the pistil is missing the flower is male, if stamens are missing it is female. A flower with both pistil and stamens is hermaphrodite. A plant bearing male and female flowers is monoecious; in dioecious plants they are found on separate plants. The date palm and most papaya plants are dioecious.

The ovary contains an outer layer (nucellus) which is diploid and an inner layer (embryo sac) where the diploid sexual embryo and the triploid endosperm are formed. After this 'double fertilization' the seed and fruit develop. The fruit develops from an ovary, but other parts of the plant may also grow larger. In the cashew nut the swollen fruit stalk forms an 'apple' which is bigger than the real fruit (the nut). A fig is a fleshy receptacle containing fruitlets. In pineapple, *Annona* spp. and breadfruit many fruitlets combine to produce a compound fruit.

A berry is a fleshy fruit with seeds lying free in the pulp, as in avocado, banana, papaya. A true nut contains only one seed: cashew nut and macadamia. A Brazil nut, however, has woody capsules containing many seeds. Mango, West Indian cherry and date have drupes or stone fruits. Seeds are covered by a seed coat and attached to the fruit by a stalk. This stalk becomes fleshy and edible in passion fruit, litchi and rambutan; it is called an aril.

Next to embryo and endosperm a seed may also contain nucellar tissue, which is wholly descended from the mother plant. Consequently, if buds appear in the nucellus, as in mango and citrus, they are all of the same genotype as the mother plant. This phenomenon is called nucellar embryony (see Figs. 3.2 and 3.3). Its significance will be explained in later chapters.

Flowers may stand alone, or in groups called inflorescences. The exact point where they appear is of interest: at the end of branches (terminal), or on the side (lateral). Citrus and avocado have lateral flower groups and can therefore be planted fairly close, without

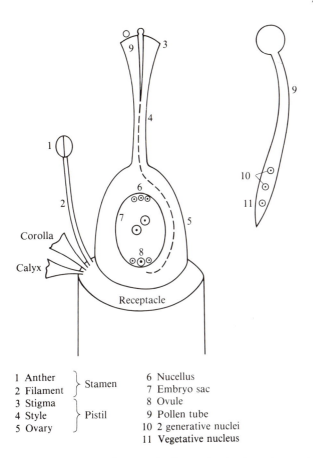

1 Anther	⎫ Stamen	6 Nucellus	
2 Filament	⎭	7 Embryo sac	
3 Stigma	⎫	8 Ovule	
4 Style	⎬ Pistil	9 Pollen tube	
5 Ovary	⎭	10 2 generative nuclei	
		11 Vegetative nucleus	

Fig. 3.1 Schematic diagram of a flower showing its various parts

crowding. Trees with terminal flowers, like cashew nut and mango, need more room.

Pollination is not always followed by fertilization, yet a fruit may still grow out without seed being set; this is called parthenocarpy. It occurs in banana and some citrus cvs. If pollen can fertilize a pistil of the same flower (tree or clone) we call that cv. self-fertile or self-compatible. However, nature quite often favours cross-pollination. One of the ways this may be brought about is protogyny, in which the pistil ripens before the stamens, as in avocado. In such cases the grower must ensure cross-pollination by planting two or more cross-compatible clones. Pollination may be effected by wind, birds or animals, but usually insects are responsible, especially bees. It

Fig. 3.2   Polyembryony of citrus

is therefore a good idea to keep hives in the orchard, certainly during bloom time. Hand pollination is sometimes necessary, as in date palm and some *Annona* spp. A very remarkable case is the fig, where one highly specialized wasp species acts as pollinator.

Fig. 3.3   Polyembryony of mango

Physiology

Photosynthesis is the basic process of assimilation in nature. It takes place in light by means of chlorophyll and other pigments. There are three distinct chains of reaction known as C_3 (the general case), C_4 (mainly in tropical grasses) and CAM (as in the pineapple).

This process is reversed in respiration: sugar and oxygen deliver carbon dioxide, water and energy which is used by the plant for the manufacture of essential compounds such as proteins. Respiration goes on day and night, photosynthesis can only take place in light. Clearly, the rate of assimilation must be greater than that of dissimilation in a growing plant.

For every activity three cardinal points are in force: minimum, optimum and maximum. The optimum temperature for photosynthesis is generally around 25 °C, whereas the optimum for respiration is frequently over 30 °C. It follows that very high temperature, especially at night, is a distinct disadvantage.

LAI (leaf area index) is the ratio between leaf and ground surface; it is less than one in young crops, then increases to about four. It may rise to nearly ten in a densely planted crop such as pineapple.

Growth in plants is coordinated by hormones (growth regulators). They are produced at the growing points and transported from there. Their action depends on the concentration at a certain point and on the target. For instance, auxin stimulates a stem into growth at 0.1 ppm but inhibits growth of buds and roots at that concentration. At one thousandth of that value it causes buds to shoot, but promotion of root growth takes place at even lower concentrations (Fig. 3.4).

NAA   0 M 10^{-7} M 10^{-6} M 10^{-5} M 10^{-4} M 10^{-3} M

Fig. 3.4 Effect of NAA concentration on root growth of cassava stem segments; the optimum is 10^{-5}M. *Source*: Eskes *et al.* (1974)

The top (apex) of a growing stem produces auxin. When transported downwards, it inhibits the growth of side branches. This phenomenon is called apical dominance. If we cut the top off, the dominance is broken temporarily and some buds will shoot. Soon, however, one of the shoots becomes the new top and dominance is restored. Thus the effect of pruning may be adversely influenced. A more effective way to overcome apical dominance is bending: the dominance is broken on the upper side of the branch and buds can sprout.

Well-known synthetic auxins are IBA, NAA and IAA. Related compounds, such as 2,4-D, MCPA and 2,4,5-T are used as weed-killers. Other activities primed by auxin are: cambium division, differentiation of xylem and phloem, flower formation, parthenocarpy, fruit setting and fruit growth. There are many applications in horticulture.

The gibberellins (GA) induce abnormal extension of stems and leaves. Their synthesis is inhibited by growth retardants, e.g. CCC and Alar. The cytokinins originate in the roots and move up into the stem where they induce buds to shoot. Ethylene induces ripening in fruits and flowering in pineapple. Abscissic acid (ABA) promotes leaf fall and aging. It is inactivated by low temperature and drought. Every activity connected with growth in the plant is influenced by the interaction of these hormones. The reader should consult Luckwill (1981) and Wareing and Phillips (1978) for additional information.

Plants propagated from cuttings or grafts flower early. On the other hand, a plant produced from seed has to pass through a juvenile stage which takes many years. During that time the tree cannot bloom; furthermore it exhibits thorns and upright growth habit. However, juvenility is not spread evenly through the tree. Near the centre and the ground it lasts years, while the outer and higher parts are entering the generative phase. This should be remembered when budwood is being gathered.

In temperate regions flower induction and bloom are seasonal processes, but in the tropics flowering is possible at any time of year. Some induction by periods of drought may be involved. This is the case in most *Citrus* spp., avocado and rambutan; they flower shortly after the first good rain that follows a dry season. Mango and cashew nut flower during the dry season. In lemon, papaya, banana, guava and soursop, growth and bloom continue throughout the year, as long as there is sufficient moisture in the soil.

Alternate bearing occurs in many perennial crops; during the 'on-year' very many small fruits are set whose development exhausts the tree. In one or more subsequent 'off-years' there is little or no harvest. Certain cvs of mango (Table 3.1) and mandarin exhibit this strongly. 'Dashehari' is a typical alternate bearer, the other cvs are fairly regular.

Table 3.1 *Number of fruits per tree in four mango cultivars (Singh, 1971)*

Cultivar	Year	1963	1964	1965
Dashehari		0	247	0
Chausa		38	29	46
Jonathan Pasand		43	35	42
Totapari Red Small		78	170	65

Alternate bearing can be overcome by stripping flowers or thinning fruitlets in an on-year and by spraying with chemicals such as NAA or DNOC. Application of fertilizers may also help, particularly if they contain Mg, Zn and Mn. The tree may be partly girdled, as is sometimes done in avocado. Each year a different branch is ringed and will fruit heavily. In this way the alternation is shifted from within the orchard to within the tree. It reminds me of an adage of the late Professor A. M. Sprenger (who taught me horticulture): 'a tree is a federation of branches'. Efforts are made in every fruit crop to select regular bearers and to discard alternate bearing cultivars. A comprehensive review is presented in *Hortic. Reviews*, vol. 4 (1982).

Replanting crops on the same soil is a constant source of troubles. Bad growth is caused by root parasites, mainly fungi and nematodes. Disinfection of the soil works well in banana and pineapple growing. However, in woody crops such as apple, citrus and guava toxic substances, usually phenolics, were found in the soil (Brown *et al.* 1983; Burger and Small, 1983). It is therefore better not to replant immediately with the same crop, but to wait several years. Nurseries should never be established in such a place.

Breeding, selection and propagation

Plant breeding and selection aim at making, or choosing crop plants that are genetically most able to satisfy our needs; propagation supplies the means of multiplying these plants. The wild ancestors of our crops are found in certain well-defined areas, the 'centres of origin' or 'gene centres'. Here they show a rich diversity of forms and types, so they are also called 'centres of diversity'. The most important as far as tropical fruits are concerned, are:
1. China; citrus, litchi.
2. South-East Asia; citrus, banana, mango, durian, mangosteen, rambutan, salak palm, bilimbi, carambola, *Syzygium* spp.
3. Central Asia; apple, pear, apricot, almond.
4. Middle East; grape, fig, date palm, pistachio, pomegranate, apple, pear.
5. South America; pineapple, cashew nut, guava, pejibaye palm, Brazilnut, sapucaia nut, West Indian cherry, *Eugenia* spp. *Annona* spp.
6. Central America; avocado, papaya, *Passiflora* spp., *Inga* spp., sapodilla, star apple, mamey sapote, mammey apple.
7. Africa; tamarind, akee, melon and water-melon.
8. Australia; macadamia, *Spondias cytherea*.

As can be seen from this list, South-East Asia, South and Central America have made the greatest contribution by far. But there must be hundreds more potential fruit crops. For Central and South America alone, Fouqué (1977) has described and portrayed 908 species.

Many wild plants are in danger of extinction. All over the world forests are cut down and grasslands are burnt at a frightening rate. More than 20,000 plant species are now in need of protection. Thus, much potential breeding material may get lost. It is therefore necessary to set up habitat reserves, field collections and botanic gardens to preserve these genetic resources. The International Board for the Protection of Genetic Resources, IBPGR, is doing this work. The Board also publishes a periodical under the name IBPGR.

The plant breeder tries to achieve higher yields and better quality, aims at regular bearing, earliness, resistance to disease and pests, lower cost of cultivation and better processing properties in his cultivars. Quality is not only a matter of taste and appearance but also of good holding and transport properties.

We must discern between genotype and phenotype. A ripe 'Valencia' is a dark orange colour in California, light orange in Florida, light green to yellow in Surinam. The membranes are soft in the first case and fairly tough in the latter. These are differences in phenotype, the genotype has remained the same.

Sudden changes in genotype are called mutations. They occur regularly, e.g. once in two million plants per year in banana, but are not always noticed. The fruit grower should be aware of the possibility, for while he wants to propagate a useful 'sport', he must avoid propagating bad ones. In grafting, or as a result of a local mutation, two different tissues may lie next to or around each other. Such chimaeras are seen now and then in citrus. More information is given in general textbooks such as Simmonds (1976), Janick and Moore (1975) and Ferwerda and Wit (1969).

Propagation is done sexually by seed or vegetatively by stems, buds, and other plant parts. Many tropical fruit crops are still grown from seed, which in most cases leads to a very diverse progeny. This is not so in most *Citrus* spp. and in certain mangoes thanks to nucellar embryony. Carambola usually comes true to type from seed and papaya is always grown from seed.

Germination power is quickly lost in durian, *Artocarpus* and *Eugenia* spp. (Teng and Hor, 1977; Kaul and Zentsch, 1979). Other spp. have seeds that only germinate after weeks or months of stratification: they must be mixed with coarse sand in layers and be exposed to winter rains, e.g. date, kaki, pistachio and *Zizyphus*. Macadamia seeds have to be scratched and soaked in water.

In most fruit crops, better results are achieved with vegetative propagation. This takes two forms: propagation on the plant's own roots as in cuttings, layers and air-layers, and propagation on a rootstock as in grafting and budding. As a rule vegetative propagation in fruit crops is cheaper and works faster than seed propagation; moreover, it results in a uniform crop and the juvenile phase is short. There are also disadvantages: virus diseases spread more quickly and shallow rooting in cuttings and layers makes them susceptible to wind damage. On the other hand, shallow rooting is an advantage on soils with a high water table.

A cutting develops from an incomplete plant part, either a stem, root or leaf, after it has been cut off. In a stem cutting new roots have to be formed. Where root primordia are already present, as in citron and the cover crop *Pueraria*, they will grow out naturally; otherwise rooting can be promoted by the application of hormones. In passion fruit and West Indian cherry a high percentage of rooting takes place, but in guava the take is low in most clones. A root cutting must be able to generate buds as happens in sour orange, apple and seedless breadfruit.

Layering as such is hardly practised in the tropics, but a variant called 'stooling' is sometimes used for mango and guava: branching is stimulated by topping and roots are formed on these shoots after earthing-up. The stool can be used many times. Air-layering is very popular in South-East Asia for propagating mango and mandarin.

Grafting is the most frequently used method. In ordinary grafting the scion (i.e. the upper part) consists of a short branch with a few buds, which is brought together with the rootstock (i.e. the lower part) in such a way that they can unite and grow as one individual. Grafting is especially useful when the rootstock is immune to a disease that attacks the clone on its own roots, e.g. footrot in citrus. Bud grafting, budding for short, is a special case of grafting with the scion reduced to one bud.

There are many grafting techniques (Garner, 1967). It suffices here to mention the whip-and-tongue graft which is effective when stock and scion are of more or less the same size, and the side graft and crown graft which are used when the rootstock is bigger than the scion.

Budding is also done in several ways. In citrus shield budding is practically always used. Many fruit trees in Indonesia are propagated by modified Forkert, a method that works well with rubber and other latex-bearing plants such as sapodilla and mamey sapote. A rectangular patch bud is used when the diameter of the stock is 12–30 mm, a shield bud when it is less than 12 mm.

Success or failure in budding depends on climate and weather, age of stock and scion, and clonal properties. For instance, it is customary in India to bud mango onto one-year-old rootstocks, but in Surinam the best take is obtained on stocks less than four months old.

In general, stock and scion must belong to the same botanical species in order to be graft-compatible, but many exceptions are known. *Citrus* spp. are easily grafted on each other and on closely related genera such as *Poncirus* and *Fortunella*, but also on distant

Fig. 3.5 Bending of budded rootstocks in Surinam

relatives such as *Citropsis* and *Swinglea*. Mangosteen and sapodilla can also be grafted on related genera. In some cases a short-lived union induces rooting on a scion that would otherwise not root. On the other hand, cases of graft-incompatibility between close relatives have been reported; all these were the result of translocation of a virus or a harmful substance such as a glucoside.

Stock and scion are known to exert considerable influence on each other. The dwarfing effect of some stocks in citrus can be ascribed mainly to a virus, but even on virus-free combinations sizes differ. Absorption of ions from the soil, number and quality of fruits, number of seeds and tolerance to diseases, pests, cold, salt and wind are among the influences of rootstock on scion. In its turn the scion can cause differences in the extension of the root system and in tolerence to cold and foot root (Samson, 1970). A wealth of information is found in Soule (1979).

Techniques of making cuttings, grafts, buddings, layers, etc. will not be discussed here; they are extensively dealt with in Hartmann and Kester (1975), Garner (1967), Garner *et al.* (1976) and Pennock (1970). However, it should be noted that the textbooks are based mainly on the experience and conditions of the subtropics. An attempt to adapt budding practices to the tropical environment has been made by Samson and Bink (1976), who show that methods known as lopping, bending and half-girdling give better results than topping, in forcing the bud to grow out. The chapter on citrus will contain more information on this subject.

An interesting new method of propagation has been developed by John Maurice of Hazorea nurseries, Israel. He stops the growth of the taproot in young seedlings to promote a well-branched fibrous root system. A thin layer of rooting medium, careful watering and high humidity are essential. The seedling rootstock is grafted as soon as possible with a softwood piece with one or more leaves. His mini-trees are packed in rolls of 50, total weight 2 kg, and can withstand seven days of travel without watering, if properly packed. At their destination they are raised to normal size for use as mother trees. Thus high-grade plant material can be established in hardly accessible places (Maurice, 1969). The Netherlands Government supports development projects using such mini-trees in Indonesia, Egypt and Tanzania (Verhey, 1983).

Natural propagules are used in banana (rhizomes or suckers), pineapple (shoots, slips, crown), date palm and salak palm (offshoots). Finally we mention *in-vitro* culture, a technique in which plants or plant parts are propagated on artificial media under sterile conditions. This is particularly important for making virus-free selections. A technique called shoot-tip grafting has been described by Roistacher *et al.* (1976). Su-Shien Ma and Shu-o Wang (1977) have reported on rapid multiplication of pineapple from crown buds. A certainly incomplete 'world list' (Winton, 1978) mentions eight culturists of whom five work on citrus. Avocado, banana, cashew, date, *Artocarpus* and *Caryocar* are also named. More detail is provided by Rao *et al.* (1981), Withers (1981), McComb (1982) and Napolione (1984). A data-base has been set up in Nottingham, England.

Nursery plants can be delivered for field planting in several ways: with a clod, in a container or bare-root, pruned and stripped of leaves or with leafy branches. The method preferred depends on climate, weather and other conditions. Transport, of course, is far cheaper with bare-root plants, but they run the risk of drying out; therefore this can only be done in a wet season, over short distances and when the plants can be set out immediately. With bare-root and leafless plants the risk of transporting disease, nematodes and scales is decreased. In most cases pruning is needed in the nursery, in order to start a framework of branches.

References

Abercrombie, M. *et al.* (1968) *A dictionary of Biology*, Penguin.
Brown, R. L. *et al.* (1983) 'Growth inhibition from guava root exudates', *Hortsci.* **18**, 316–8.
Burger, W. P. and **Small, J. G. C.** (1983) 'Allelopathy in citrus orchards', *Scientia Hortic.* **20**, 361–75.
Cobley, L. S. and **Steele, W. M.** (1976) *An introduction to the botany of tropical crops*, Longman.
Eskes, A. B. *et al.* (1974) 'Callus growth and rooting of cassava stem segments cultured in vitro', *Acta Botanica*, **23**, 315–20.
Evans, L. T. (1975) *Crop physiology; some case histories*, Cambridge.
Ferwerda, F. P. and **Wit, F.** (eds. 1969) *Outlines of perennial crop breeding in the tropics*, Wageningen.
Fouqué, A. (1977) *Espèces fruitières d'Amérique tropicale*, Paris.
Garner, R. J. (1967) *The grafter's handbook*, London.
Garner, R. J. *et al.* (1976) *The propagation of tropical fruit trees*, Farnham Royal.
Gilmour, J. S. L. *et al.* (1969) *International code of nomenclature of cultivated plants*, Utrecht.
Hartmann, H. T. and **Kester, D. E.** (1975) *Plant propagation*, Englewood Cliffs.
Janick, J. and **Moore J. N.** (eds.) (1975) *Advances in fruit breeding*, Purdue Univ. Press.
Kaul, M. L. H. and **Zentsch, W.** (1979) 'The life span of some Indian forest seeds', *Beiträge zur trop. Landwirtschft u. Veter.* **17**, 283–6.
Loveless, A. R. (1983) *Principles of plant biology for the tropics*, Longman.
Luckwill, L. L. (1981) *Growth regulators in crop production*, Studies in biology 129, Arnold.
Maurice, J. (1969) 'Suction grafting and rooting tropical trees', *World crops* **21**, 265–7.
McComb, J. A. (1982) 'Micropropagation techniques for fruit trees and vines', *Austral. Hortic.* **80(8)**, 61–6.
Napolione, I. (1984), 'Results of micropropagation of subtropical trees and shrubs' *Rivista di agric. subtrop. e trop.* **78**, 139–51.
Pennock, W. (1970) *Plant grafting techniques for tropical horticulture*, Bull. 221, Univ. P. Rico.
Purseglove, J. W. (1968, 1972) *Tropical crops: Dicotyledons, Monocotyledons*, Longman.
Rao, A. N. *et al.* (1981) 'Cotyledon tissue culture of some tropical fruits', In: *Tissue culture of economically important plants* (ed. A. N. Rao), Proc. Internat. Symp., Singapore.
Roistacher, C. N. *et al.* (1976) 'Recovery of *Citrus* selections free of several viruses, exocortis viroid and *Spiroplasma citri* by shoot-tip grafting in vitro', *Proc. Seventh IOCV Conference*, 186–93.
Samson, J. A. (1970) 'Rootstocks for tropical fruit trees', *Tr. Abstr.*, **25**, 145–51.
Samson, J. A. and **Bink J. P. M.** (1976) 'Citrus budding in the tropics: towards an explanation of the favourable results of lopping', *Proc. Seventh IOCV Conference*, 213–6.
Simmonds, N. W. (ed.) (1976) *Evolution of crop plants*, Longman.

Singh, R. N. (1971) *Biennial bearing in fruit trees*, ICAR, New Delhi.

Soule, J. (1979) *Rootstock-scion relationships*, HOS 6361 (HSC 626), Dept. Fruit Crops, Univ. Florida, Gainesville.

Su-shien Ma and **Shu-o Wang** (1977) 'Tissue culture propagation of pineapple', *J. Chin. Soc. Hort. Sci.* **23**, 107–13.

Teng, Y.T. and **Hor, Y. L.** (1977) 'Storage of tropical fruit seeds', *In: Seed technology in the tropics* (ed. H. F. Chin), Univ. Malaysia, Serdang, Selangor.

Verhey, E. W. M. (1983) 'Minute nursery trees, a break-through for the tropics?', *Chronica Hortic.* **22(1)**, 1–2.

Wareing, P. F. and **Phillips, I. D. J.** (1970) *The control of growth and differentiation in plants*, Oxford; 2nd ed. 1978.

Winton, L. (1978) 'List of world tree tissue culturists', *Newsletter Intern. Ass. for Plant Tissue culture*, **26**, 22–31.

Withers, L. A. (1981) *Institutes working on tissue culture for genetic conservation*, IBPGR 81/30, Rome.

Chapter 4

Crop husbandry

Cultural operations

Land clearing

This takes place either manually or mechanically. If a dense vegetation like the rain forest is involved the undergrowth must first be removed by slashing. At the start of the dry season the big trees are felled and left to dry out. This takes six to eight weeks in a wet climate or four weeks in drier areas. Meanwhile, brushwood is piled up around the large trees so that they burn well. After burning, stumps are removed wherever practicable and a cover crop is sown to protect the soil against erosion by heavy rain (Webster and Wilson; 1966, 1980).

As labour costs rise, it is likely that mechanical clearing will replace manual labour. This is done by bulldozers with chains and balls, treedozers and brushcutters. All these machines have to be used with the utmost skill and caution as they will otherwise compact the soil, disturb the topsoil and leave deep ruts, where water will collect.

Burning causes the destruction of most of the organic matter in the vegetation. Not only are carbon, hydrogen and oxygen lost, but much nitrogen and some sulphur will also escape into the air. The remaining ash enriches the soil with minerals, especially potassium, but not for long. A large part may be lost by run-off, or by leaching. It is therefore highly desirable to avoid burning completely, or at least to keep it to the minimum. In Central America and Cameroon the forest is underbrushed, the land is drained and banana suckers are planted; subsequently the bigger trees are felled, but not burned (Simmonds, 1966). This presumably involves softwood trees which disintegrate quickly; hardwood trees take much longer to decompose.

Trees can also be killed by ringing or girdling, which interrupts the connection between leaves and roots; the roots then starve and die. However, it may take a long time before the whole tree is dead.

Sodium arsenite and 2,4,5-T can speed up the process (but arsenite is highly toxic to men and beast and its use should be avoided). The tree is first 'frilled' with downward slashes all around, then a 3 per cent solution of 2,4,5-T in oil is sprayed on the frill. It will take 6–12 months before the tree is dead, but it is then completely disintegrated, including the roots.

It has been proposed (Samson, 1959) that the 2,4,5-T treatment be combined with the previous planting of a *Pueraria* cover crop, which will smother secondary growth. A disadvantage of this method is the waiting period of a year before the work can be continued. On the other hand, it is cheaper than felling and burning and it prevents erosion and leaching. Dioxin, a by-product of 2,4,5-T, is exceedingly harmful. For more information on 2,4,5-T and dioxin, see Lee and Orr (1980).

Drainage

Drainage is of the utmost importance in the wet tropics. Nor should it be neglected in drier climates. Restricted root development and salt problems may result from poor drainage. Drains are usually left open in the tropics. On heavy soils 'cambered beds' are made, which are usually 6–10 m wide (see Fig. 4.1), but in Surinam it was possible to increase the width to 30 or 40 m on certain permeable clay soils (van Amson, 1964).

Propagation

Particulars for each crop will be given in the relevant chapters; here we limit ourselves to general remarks. Nurseries should always be located on the best possible land: virgin soil or soil on which the same crop or a relative has not been grown previously. Water

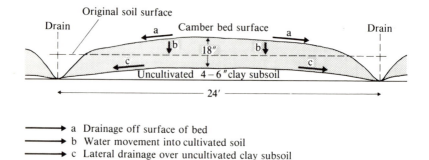

Fig. 4.1   Cambered bed, diagrammatic section. *Source*: Webster and Wilson (1980)

Fig. 4.2 A citrus nursery under *Sesbania* shade in Indonesia

should be available throughout the year. Sheds must be built to keep tools, fertilizers, pesticides etc. out of the rain and locked up. A lath house is needed for young seedlings, cuttings etc. Manure must be kept on a cemented floor with a roof over it; a compost heap is also useful. Small trees of *Leucaena*, *Sesbania* or shrubs like *Cajanus* (pigeon pea), *Tephrosia* and *Crotalaria* provide a light shade which is beneficial to most nursery plants; they also supply mulch and serve as green manure.

Planting

This is carried out at the beginning of the rainy season, holes having been made during the previous dry season. Their size is not important, provided the soil is permeable. It is not worth the trouble (and cost) of making them bigger than is necessary to accommodate the roots of the planting material. Top soil should be kept apart to be replaced on top after filling the hole. The subsoil is mixed with compost, farmyard manure and some phosphate.

A plant should never be set lower in the field than it was in the nursery. As the soil has to settle, it is sensible to plant about 10 cm higher than ground level. To avoid air pockets, which might kill the roots, it is necessary to tamp the soil around the roots and to water well. Sizaret (1983) presents a valuable exposition on the subject. Some points he emphasizes: keep soil layers apart, put them back in place, add extra topsoil, hold plants vertical, plant near the valley side on terraces.

Plant density and spacing depend very much on the nature of the crop: thus a mango needs more room than an orange tree and a pineapple plant can do with far less space than a banana. Soil and climate have to be considered too: where water and nutrients are in short supply, more root space per tree is necessary. Trees with terminal inflorescences must not be crowded.

There has been a tendency to closer planting since about 1950. Rising rates of interest on capital necessitate higher and earlier yields. Better clones and rootstocks and more efficient agronomic methods have made this possible. The attendant higher cost of plants and planting is generally not an important cost factor.

Planting is done on the square, unless there are good reasons for deviating from that system. Right angles are set out easily with a rope measuring 3 + 4 + 5 m (or dm, or feet) in accordance with Pythagoras' theorem. A tree planted at 5 m on the square occupies a space of 5 × 5 = 25 m². One hectare, therefore, contains 10,000:25 = 400 trees. We can also say that the planting density is 400 (trees/ha).

Trees with lateral inflorescences, such as citrus and avocado, can just as well be set out in a rectangular planting system. A spacing of 6 × 8 m instead of 7 × 7 m will give us more room for manoeuvring between the rows and for intercropping during the initial years; the planting density will hardly change: from 204 to 208. Double, or hedgerow planting is also practised in citrus; for instance 3 × 8 m. Usually the intention is to remove half the trees in each row when crowding sets in, but this is often neglected as the grower cannot find it in his heart to do so. Quite rightly too, as it sometimes turns out. The most extreme example I have seen was in Spain in a 5-year-old 'Satsuma' orchard, planted at 1 × 3 m (3,300 trees per ha!). It produced 50 tonne/ha, which the owner was selling at a good price; no wonder he did not consider thinning out.

Trees with circular crowns, such as palms, are best planted in an equilateral triangular system. This allows cultivation in three directions and accommodates some 16 per cent more trees per ha. If the distance between trees is d, then a tree takes up $\frac{1}{2}d^2 \sqrt{3}$ m². Let us put d at 10 m. On the square 100 trees/ha can be planted. In an equilateral triangle a tree occupies $\frac{1}{2} \times 100 \times 1.73 = 86.5$ m², so the density is 10,000:86.5 = 116. However, date palms are more often planted on the square, presumably because it allows intercropping.

Triangular planting, not equilateral, can be achieved by staggering, i.e., moving alternate rows half a space. With pineapples, for instance, two rows are set staggered at 30 × 60 cm on a bed, the next bed being 1 m distant. A simple way to calculate the planting density is as follows: spacing is 30 cm in the row and (60 + 100):2 = 80 cm between rows. Each plant occupies 0.24 m² and the density is therefore 10,000:0.24, equals 41,667 plants/ha.

Row direction must be considered too. In the subtropics preference is given to a north-south direction to take maximum advantage of sunlight (Platt, 1973). In the tropics, where the sun is never far from directly overhead, the row direction would make little difference and other considerations carry more weight. On slopes of 4 per cent or more, it is best to follow the contour lines; on steep slopes (10 per cent or more) terraces should be built. Otherwise, planting rows perpendicular to the prevailing wind is preferred, as it facilitates spraying.

Where strong winds occur the orchard must be protected by shelter belts or windbreaks. These are set at right angles to the prevailing wind in order to reduce the speed of strong winds and the resulting damage to crops. A dense belt sets up turbulence and is therefore undesirable. The effect of a good windbreak can be felt as far away as 15 to 20 times its height; this means that the belts have to be set about 200 m apart.

Walls and other artificial belts are sometimes built, e.g. in date culture, but generally living trees are used such as *Eucalyptus* spp., *Grevillea robusta*, *Casuarina equisetifolia*, and *Hakea saligna*. Leguminous trees are more often used in the wet tropics: *Albizzia*, *Erythrina*, *Gliricidia*, *Inga* and *Leucaena* spp.; between them hedges of tall grasses may be maintained.

Soil management

Management aims at maintaining the soil in good condition, or improving the condition if necessary. This includes protection from direct sunlight (which would raise the soil temperature too much) and from the impact of rainfall, which may destroy soil structure. It is usually several years after planting before a tree will form such an extensive canopy that it can provide adequate protection to the soil; with bananas this takes only three to four months. We shall discuss briefly the following methods of soil management: clean weeding, intercropping, slashing, mowing, chemical weed control and a permanent cover crop.

Weeds compete with the crop for light, water and nutrients; so the grower must try to get rid of them. However, clean weeding leads to loss of organic matter in the soil and to erosion, even on flat land. Cultivation with ploughs, discs, harrows and other implements is therefore not advisable in fruit farming except, perhaps, just before planting. Even so, it will seem inconceivable to most small-holders not to use good land whenever possible; they usually grow a food crop or cash crop for a few years between the rows of their fruit trees. Preferably, this should be an annual crop that provides good cover and improves the soil with its root nodules, such as cowpea (*Vigna unguiculata*).

Where labour costs are low, weeds can be slashed periodically with machetes or cutlasses. A good man does about 800 m² a day,

Fig. 4.3   Banana intercropped with cocoyam (*Xanthosoma*) in Nigeria

so it takes 12½ man-days/ha. In the wet tropics three to five rounds a year are necessary. As the cut weeds are left in place, it results in a form of mulching. However, the required labour may be temporarily otherwise engaged, for instance when rice is planted or harvested.

Cutlassing can be done in many ways. Koenraadt (1962) distinguishes short weeding (close to the ground), round weeding (around trees) and long weeding (to a height of 15 cm). Short and round weeding causes much erosion, long weeding does not; furthermore it is considerably cheaper, as is clear from his data:

	man-days/ha/year
short weeding (×2), round weeding (×2)	24.55
short weeding (×4), no round weeding	15.93
long weeding (×8)	9.60
machine weeding (×8 plus ditches)	13.17

Koenraadt, who based his observations on 110 discharges of water during rainy seasons, reports that long weeding should be undertaken as soon as the weed is about 40 cm high. No erosion, better weeds (more *Commelina*), better soil structure and lower costs were the result on this 150 ha coffee–citrus–cacao estate on heavy soil in an equatorial climate.

The next step is to enlist the help of mowing machines; first on a part-time basis, later full-time. It is a difficult decision to make: on the one hand the work must be done, on the other hand many people may lose their jobs. The fast rising cost of machinery and fuel has also to be considered. Tractor-drawn rotary mowers generally cut a swath 2 m wide. Regular mowing favours the grasses and a sort of lawn will develop. This is an advantage because it facilitates transport, but there are disadvantages (competition for nitrogen and compaction of the soil).

Chemical weed control is hazardous in the wet tropics. If heavy rain occurs shortly after spraying, massive erosion is likely to result. Furthermore, residual effects are possible. Some crops are extremely susceptible to weedkillers, e.g. the banana to 2,4-D. Nevertheless, a combination of chemical weed control (in strips along the tree rows) and machine mowing (between the rows) seems possible; this would decrease the danger of erosion. Ditches can be kept clean with total weed killers. The effect of five herbicides on six weeds is shown in Table 4.1.

Table 4.1 *Resistance and susceptibility to herbicides*

Weed	Herbicide				
	1	*2*	*3*	*4*	*5*
Bermuda grass *Cynodon dactylon*	R	R	S	R	S
Guinea grass *Panicum maximum*	R	I	I	S	S
Para grass *Brachiaria* sp.	R	I	S	R	S
Nuts edge *Cyperus rotundus*	R	R	S	R	R
Dog fennel *Eupatorium* sp.	S	S	S	I	R
Redroot pigweed *Amaranthus* sp.	I	S	I	S	R

R = resistant, S = susceptible, I = intermediate
1 = simazine, *2* = diuron, *3* = bromacil, *4* = paraquat, *5* = dalapon
Source: Lange *et al.*, 1975

We infer from this table that no single herbicide is adequate under all circumstances; several have to be used, either in combination or in succession. Soil properties have to be considered too; bromacil is an efficient weedkiller for pineapple on sandy soils of the Ivory Coast at 1.5–2 kg/ha, while on volcanic soils in Cameroon 6 kg/ha is needed (Gaillard and Haury, 1974). For more information, see Kasasian (1971) and Holm (1977).

One of the most obnoxious weeds in the tropics is lalang, *Imperata cylindrica*. It can be killed by dalapon or glyphosate, but a cheaper and more elegant method is available: replacing it by a cover crop of *Mimosa invisa*. This thorny plant can then be rolled up on sticks, taking the lalang with it. Under high rainfall Para grass and most other weeds are suppressed by *Pueraria phaseoloides* (kudzu, Fig. 4.4). A permanent cover crop is a good solution for soil management problems in humid areas; in drier regions green manuring and mulching are more attractive.

Fig. 4.4   A cover crop of kudzu, *Pueraria phaseoloides*

Soil conservation

This aims at effective control of soil erosion. Very steep slopes should be left under forest or permanent pasture; gentler slopes may be used for fruit trees and other crops that provide some protection to the soil. Soil-conserving methods include: tree rows along the contours (not up and down the slope), cover crops, terraces and silt pits. The work involves three phases: a survey for mapping and classification, mechanical measures and agronomic practices.

As an example we may cite the experiment by Wang *et al.* (1977) in Taiwan, in a high rainfall area where banana was planted on a 24 per cent slope on clay loam soil. Treatments, repeated four times, were: clean cultivation, terracing, vegetative barriers, a cover crop and mulching with Bahia grass. If soil loss for clean cultivation is put at 100, then it was only 0.23 for the cover crop and mulching; yields were lowest in clean cultivated plots and 13–26 per cent higher in the other treatments. Also see Table 1.4.

Water conservation furthers infiltration of rainfall into the soil and minimizes run-off. Dams and reservoirs have to be built for storage of water (Webster and Wilson, 1966, 1980).

Irrigation

Irrigation of fruit crops is necessary in arid and semi-arid regions where total rainfall is insufficient to supply the water needs of the crop; in the humid tropics seasonal deficits may occur. It depends on the crop and the extent of the deficit whether irrigation will be needed or not. Even two weeks of drought may have serious consequences for the banana, while citrus is not harmed by two dry months; in fact, it will benefit as a concentrated bloom will follow the rains. Mango, cashew nut and date palm need several dry months for proper blooming and fruiting.

In Chapter 2 we saw that the water balance of a soil may be calculated if one possesses the data on rainfall, evapotranspiration and available moisture in the rooted zone. Alternatively, one may determine the state of soil moisture with an auger and decide to irrigate when it drops below a certain percentage of field capacity. Tensiometers are also used; they are calibrated in centibars: zero indicates a very wet soil, 80 an extremely dry one. At a soil suction of 40 centibars most crops will need irrigation.

Furthermore, trees have to be watched constantly in time of drought. On warm days leaves tend to wilt during the hot hours, but they will recover their turgor later in the day. If the drought continues, wilting will start earlier and last longer each day. This is a sign that irrigation is necessary. However, this does not apply to fruit-bearing citrus, as the leaves draw water from the fruits. These shrink and shrivel and they, not the leaves, have to be checked in order to determine the need for irrigation.

Before irrigation facilities can be made available, data on climate and hydrology of a region have to be assembled and analysed; social, economic and financial aspects have to be considered. The designing and planning stage comes next and is followed by construction of the primary and secondary canals, reservoirs and pumping stations. To make the scheme fully operational management, maintenance, extension services, credit facilities and agricultural supplies have to be taken care of (des Bouvries and Rydzewski, 1977).

When the water is finally flowing in the tertiary canals (to the farm), irrigation may take place in two ways: by gravity and under pressure. Gravity is especially useful on land with a gentle slope and smooth topography; the major methods are the flooding or border method, basin flooding and furrow irrigation. In the border method the whole area is wetted, in the second case basins are constructed containing from one to five trees and in the furrow method water is led along two sides of each tree. In all these methods capital investment is low, but labour cost is high.

Irrigation by pressure is used in installations that are portable, semi-portable or permanent; labour costs decrease in that order, but capital costs increase. In the first instance perforated aluminium pipes are used, in the other cases sprinklers are fixed onto the pipes, or rain guns are used.

A special case is drip or trickle irrigation, where water is applied under low pressure to the rooted zone of the crop only; consequently a great saving of water is achieved and weed growth around the tree is restricted (Davidson ben David, 1975). There is no run-off; even sewer water or water with a high salt content can be used. Only 30 per cent of the soil volume needs to be wetted for a normal yield (Black, 1976).

Pruning

The purpose of pruning is to establish a balance between vegetative growth and fruit bearing. A certain minimum leaf area has to be maintained for each fruit. On the other hand too dense a canopy will put lower leaves in the shade. The pruner should have an image in his mind of the ideal form of his tree. This depends on the natural growth habit of the tree, but may deviate from it. For instance, a 'profit line' may be maintained, above which harvest is not economic. The problem is how to achieve or approach the ideal form with the least possible interference to the tree. When pruning a branch, select a bud that is pointing the way you want the new twig to grow; cut about $\frac{1}{2}$ cm above this spot. The main rule of pruning should be: when in doubt, don't! It is easy to cut off a branch but it takes years to put it back on the tree. Knowledge of growing and flowering patterns is also necessary: are inflorescences terminal or lateral? Do they bloom on young or older wood? Is there one bud per axil or two (as in grape and macadamia)? Does the plant produce a good reaction to severe pruning, or bad as in avocado which has few dormant buds for rejuvenation?

Spacing also has some influence on pruning, especially in the system called 'hedging' in which fruit trees are planted closely and pruned mechanically every year on either the left, right, or top of the row. The pruned one third part of the tree will not bear much fruit for one year, but will make up for it in the other two years. Picking costs are much reduced by this non-selective system, as the trees remain small.

Some trees need much more pruning than others. The grapefruit may be left virtually unpruned, but the lemon must be pruned every year while the pruning requirements of orange and mandarin lie between those two. It seems likely that trees with terminal inflorescences, such as mango and cashew, may profit from regular pruning as this allows closer spacing. Vine crops need special care (see Chapter 11).

Three types of pruning are recognized: frame, maintenance and rejuvenation pruning. A framework is best formed in the nursery; it usually consists of a single stem split up in four main branches, each occupying a sector. A similar division of space takes place with the branches of the second and third order. However, seedling trees and juvenile grafted trees retain a dominant main stem for a long time. They should not be forced into a growth pattern that is foreign to them; bending some branches is more appropriate than heavy pruning.

Maintenance pruning aims at the preservation of the status quo. It is often remarked that 'water sprouts' have to be suppressed. They arise from dormant buds on older wood. All the same, many grow out into full light, lose their wild habit and start to behave like other branches. There is little sense in cutting them off, unless they grow straight up or obliquely through the tree.

Rejuvenation pruning is meant to bring trees in decline back into production. The tree is severely cut back, except for one 'nursing branch', so that dormant buds may develop; the best-placed branches are then retained. It is necessary to protect the bare stem against bright sunlight by whitewashing.

Pruning is an entirely different matter in monocotyledons. In the banana it means allowing a limited number of shoots of different ages to grow simultaneously on one stool. In the date palm older leaves are cut to make the tree more accessible for harvest, or to create a less humid atmosphere for the ripening fruit.

Deciduous trees in moderate climates are generally pruned in winter, when the leaves have fallen; the tree is dormant and little damage is done. In the humid tropics pruning may take place at almost any time of the year. It is important to use good equipment. Cuts should be small, smooth and sloping, so that no water can collect on the wound. Large wounds must be treated with a covering material, but not on the same day; the wound must have a chance to sweat. The pruner must also remove shoots of the rootstock, dead wood, parasites (such as many Loranthaceae), epiphytes (e.g. ferns), climbing vines and nests of bees, wasps, ants and termites.

Crop protection

Crop protection comprises all activities, agronomic and legal, to protect the crop against injury. Injury is every change leading to lower quality of the desired product; damage results from injury and loss is the social and economic result. The term plant pathology refers to the science of diseases caused by fungi, bacteria and viruses. A 'pest' is any animal or agent causing damage to crops. (Zadoks, 1975; Hill, 1975; Hill and Waller, 1982).

Causes of injury are biotic (living) or abiotic (non-living). Among the latter we find: drought and inundation, storms, deficiency and excess of elements, salinity, chemicals etc. However, we shall chiefly deal with the following biotic causes: fungi and bacteria;

viruses and mycoplasmas; insects and mites; nematodes (roundworms).

Other biotic causes are rats, birds, snails, weeds and parasitic plants. Two or more factors may coincide, for instance a mechanical injury may provide entrance to a fungus, or an insect may be the vector (transmitter) of a virus. Synergism occurs when harmful causes reinforce each other, as with Panama disease, the burrowing nematode and certain bacteria of the banana (Wardlaw, 1961).

Pathogens

The biotic causes of injury are called pathogens. The total number of plant pathogens has been estimated at higher than 100,000. We shall review them briefly.

The fungi do not posses chlorophyll and hence must live as parasites, saprophytes or symbionts. A great number of very varied diseases are caused by them. According to the organ they attack, or the appearance of the symptoms, we call these diseases root rot, wilt, damping-off, leaf spot, anthracnose, fruit rot, gummosis, rust, blight, blast, mildew, etc. *Phytophthora* spp. cause rootrot, damping-off in seedlings and fruit-rot in many crops. *Glomerella cingulata*, also known as *Colletotrichum gloeosporioides* (the imperfect form), is the cause of anthracnose on leaves and fruits of various crops.

Bacteria are much smaller, their diameter is usually around 1 μm. Among the diseases they generate are moko in banana and canker in citrus.

Mycoplasmas are smaller again and contain both RNA and DNA. It is difficult, though not impossible, to grow them on an artificial medium. Typical symptoms they cause are: yellows, greening, witches' broom, stunting. Generally the Psyllidae (jumping plant lice) act as vectors.

A virus cannot be called a living organism; it is rather a molecule. It consists of either DNA or RNA, often surrounded by a thin protein coat. Reproduction takes place inside the nuclei of the host it has penetrated, but it cannot reproduce itself on an artificial medium. The symptoms are varied: mosaic, flecking, vein clearing, phloem necrosis, distortion, stunting, dieback, decline, etc. Some viruses are transmitted by vectors, usually aphids. However, the rapid spread of virus diseases in fruit crops has been promoted largely by budding and grafting. Other possible means of transmission are by natural grafts, budding knives, parasites (dodder) and seed. As transmission through seed occurs rarely, we may assume that seedling fruit trees are initially virus-free. Unfortunately it is extremely difficult to keep them so. In an experiment with presumably virus-free buds on 25 citrus seedling rootstocks it was found that most trees acquired several viruses within a a few years (Bitters, 1968). The feeding habits of insects are extremely diverse but fall into two categories: biting and sucking. Those with biting mouthparts eat leaves (locusts, beetles) or bore in stems (*Cosmopolites* in banana); those with sucking mouthparts may transmit virus diseases (aphids) or pierce holes in fruits (certain moths) which provide access to fruit flies and fungi.

Mites are small, generally less than 1 mm, but lay relatively large eggs. They reproduce very fast under favourable conditions. Certain mites feed on leaves causing them to fade and drop and sometimes causing serious injury.

Nematodes are small unsegmented worms, about 1 mm in size, which occur in large numbers in soil. Plant parasitic nematodes have a sharp stylet, which they use to penetrate their host. Some live completely inside the plant, others only half inside, on the plant surface or free in the soil. Widespread damage is done by root knot nematodes of the genus *Meloidogyne*. The burrowing nematode, *Radopholus similis*, is particularly dangerous to fruit crops; it causes spreading decline of citrus in Florida and attacks bananas everywhere. Crop rotation and mulching help to control these pests. The damage is more severe on sandy soils than on clays. In some cases resistant rootstocks can be used.

Some pathogens cannot be distinguished by their form, but act differently as pathogens. They are called races, physios, biotypes or strains. Strains of a virus differ considerably in virulence: the severity of injury caused on a susceptible host. An attack by a mild strain confers cross protection against subsequent attack by a virulent strain of the same virus. This is comparable to vaccination against smallpox in humans. Thus it is possible to protect citrus trees from a virulent tristeza virus by inoculation with a mild tristeza strain.

We now turn to pest populations. It is impossible and would cost too much to keep crops entirely free from pathogens. Control measures only become economically justified above a certain density of the pest population. The 'economic threshold' differs widely depending on pathogen, crop, market prices and ecological conditions. The latter comprise climate and soil, enemies of pathogens (predators and parasites) and competition within and between pathogen species. In other words: the whole ecosystem has to be considered before control measures can be taken intelligently.

A case in point is the relationship between ants and scales and aphids. They often live in symbiosis: to their mutual advantage. Scales and aphids exude a sweet sap which the ants relish; in return the ants protect them against enemies and transport them to fresh locations. Good control of aphids and scales is possible by denying the ants entry to the tree; this can be done by spraying diazinon on the stem. With the ants out of the way, predators and parasites keep aphid and scale populations low. A parasite develops at the cost of another organism (the host), from which it obtains food; it does not kill its host outright. A predator just uses a pathogen as food. Competition occurs when two or more pathogens utilize the same plant or plant part: a minor pest (or pathogen) may suddenly become a major one when it is released from the competing pressure of another, against which a good control method is functioning.

Climate and weather exert considerable influence on pathogen populations. Winter, in temperate regions, assures a discontinuity of pathogens which is generally lacking in the tropics. There, outbreaks may occur at any time of the year. For instance, a population of citrus rust mite can increase one hundredfold in less than two weeks of dry weather. The worst aphid attack on citrus I have ever seen took place in Surinam, in November–December 1954. It was during a fairly dry and sunny spell, following rains that had caused massive flushing of the trees. In such cases the predators and parasites cannot keep pace with the pest population and an epidemic results; spraying is urgently needed.

The relation between host and pathogen also depends on resistance and susceptibility, tolerance and sensitivity. Resistance is the sum of all factors determining whether a plant will be free of injury; susceptibility is the reverse. Tolerance is related to damage, rather than to injury. A pathogen cannot enter a resistant host, or cannot propagate there; in a tolerant host the pathogen is present, but the inflicted damage is slight. A tolerant host may be a symptomless 'carrier', from which susceptible cvs are infected. Resistance is due to vigour (the plant heals quickly), to dense hairs or a thick cuticle (the pathogen does not gain a foothold) or to toxic and repellent substances (the pathogen is killed or driven away). In hypersensitivity the reaction of the host is so severe that the affected tissue dies and the infection is not spread.

Control

Methods of control of diseases and pests comprise:

legislation; quarantine laws
sanitation; eradication, disinfection, rotation
resistance; use of resistant or tolerant cvs
mechanical means; handpicking, flaming, banding
biological means; predators, parasites
chemical means; spraying, dusting
integration; a combination of methods.

Most countries have quarantine laws which permit importation of pathogen-free plants or plant parts only. Some plants may be totally prohibited, others have to be inspected and, if necessary, fumigated. Quarantine may also operate between states of a country; e.g. California rigorously inspects all plant importations from other states. Samson (1977) suggested that the predicament facing citrus in South-East Asia, caused by the combined ravages of tristeza and greening, can be solved by isolation: designating small islands as special citrus growing areas under strict quarantine.

Sanitation aims at prevention, destruction or reduction of sources of infection. Citrus canker has been eradicated twice from Florida; millions of nursery plants and grove trees had to be destroyed to attain this end. Eradication is also used in Florida as a means to prevent the spreading of the burrowing nematode. Similarly, Israel is kept free of tristeza virus by tracing and destroying every infected tree (Raccah *et al.* 1976). Disinfection of plant materials is regularly used in banana growing, when corms are trimmed and a hot water treatment (55–60° for 10–20 minutes) is given. Suckers and corms can be treated with chemicals too. Bareroot citrus seedlings can be dipped in warm water (45°) for 25 minutes against nematodes. It may be sufficient to dry planting material in the sun; pineapple is treated in this way against base rot. Soil can be disinfected with D-D mixture, but rotation may achieve virtually the same result: nematodes die in a matter of months if no food source is available. It is unsafe to replant citrus after citrus, or papaya after papaya.

Panama disease has rendered it necessary to replace the susceptible 'Gros Michel' banana by resistant cvs. The ideal citrus rootstock should be resistant to, or tolerant for, many virus diseases, fungi and adverse soil conditions; in fact, it does not (yet) exist so one has to compromise. Besides, new pathogen strains are constantly generated; therefore, resistance is only temporary.

Mechanical means of control include handpicking of certain caterpillars on citrus; the damage is clearly visible and the hiding places are not difficult to find, at least on young trees. Nests of bees and wasps can be flamed with a torch. Glue bands around the stem are effective for a few weeks or months against caterpillars and ants. Bagging of young soursop fruits in muslin has been recommended in Surinam as a preventive control method against a caterpillar and a wasp (van Dinther, 1960). We have already mentioned that epiphytes and parasitic plants should be removed from trees in pruning.

Biological control utilizes predators and parasites, either locally present or deliberately introduced. Several spp. of the wasp genus *Aphytis* are parasites of citrus scales. The relation may be quite specific: one wasp sp. for one scale sp. In humid regions certain fungi (*Aschersonia, Fusarium*) are scale parasites. Lady beetles (adults and larvae) and larvae of Syrphidae (hover flies) can do away with large numbers of aphids. Birds help to clean up an infestation of caterpillars. Trap crops of *Crotalaria* (rattle pod) and *Tagetes* (marigold) attract and kill nematodes and are very useful in rotations; however, this method is not effective against fungi. A whole part of *Agriculture ecosystems and environment* (vol. 10, no. 2) is devoted to biological control.

Although all the methods discussed so far are useful and sometimes even adequate, they are eclipsed by chemical methods. It would be impossible to grow fruit on any scale without chemicals, which are relatively cheap and easily applied. On the other hand they do have disadvantages, among which we mention:

1. phytotoxicity, the substance is toxic to the host plant;
2. drift onto the wrong crop, or into a canal where it kills fish;
3. residue on fruits in unsafe quantities;
4. accumulation in food chains (man, birds of prey);
5. pollution of air, soil and water;
6. resistance, when ever-increasing quantities of a chemical are needed to kill a pathogen.

Chemical control is also ineffective against virus diseases. The intelligent grower should consider alternative methods before he reaches for his spraying equipment.

To describe the action of a chemical the suffix cide (killer) is used, together with the name of the group against which it is directed. Thus, a fungicide kills fungi, a herbicide weeds and an acaricide acarina (mites). The general name is pesticides. Names of pesticides are often confusing to the layman who has to distinguish between three categories of names. For instance: mangano-ethylene-bis-di-thio-carbamate (chemical name) bears the internationally agreed common name 'maneb' and has many brand names such as Agro-maneb 80, Dithane M-22, Trimangol dusting powder, etc. We shall give preference to common names in this book.

A pesticide contains one or more active ingredients (a.i.), a carrier and some additives. The carrier is generally water or oil in a spray and talcum powder in a dust; additives are spreaders (wetters), stickers, emulsifiers etc. The pesticide is either applied in dry form as a dust, or wet as a spray. Dusting is not used to a great extent. It has some advantages: no water is needed and penetration is good; however, these do not generally compensate for the disadvantage that it can only be carried out during calm weather on wet, or dew-covered plants. A pesticide can be formulated in different ways, e.g. as concentrated solution (c.s.), a wettable powder (w.p.) or an emulsifiable concentrate (e.c.).

We have to make a distinction between high- and low-volume application. In high-volume (HV) spraying the carrier is always water; amounts of 400 to 2,000 litre/ha are generally used. The droplets are forced out of a nozzle under hydraulic pressure. Their sizes vary considerably, 400 μm is an average.

In low-volume (LV) spraying the carrier is water and the spray liquid comes through a nozzle under low pressure and is broken up in droplets by a windstream; their sizes vary around 100 μm. Therefore, the weight and volume of one HV-droplet is spread over 4 $\times$ 4 $\times$ 4 = 64 LV-droplets. Each droplet, regardless of its size, has a certain sphere of action around it. Hence, the 64 LV-droplets cover a far greater leaf surface than one HV-droplet of the same volume. This means that in LV spraying far less spray liquid is required: 0.5 litre, instead of 10 litre per tree. Were the concentration to remain the same, too little a.i. would be delivered. To correct this we must increase the concentration; a tenfold increase has proved to be sufficient. Therefore, there is also a saving of chemicals in LV spraying.

When oil instead of water is used as carrier, droplets are made much finer and amounts can be reduced to 50, or even 10 litre/ha. This is the case in the control of Sigatoka disease of banana. Still finer droplets are made for ultra low-volume spraying (ULV), so that the spray amount is reduced to 2–10 litre/ha (Anon., 1972).

Machinery used for spraying is highly diverse. It ranges from small hand-operated knapsack sprayers to tractor-drawn speed sprayers. An extremely efficient unit for tropical use is the portable motorized mist blower which is carried on the back, with a 10 litre tank. Its capacity is around 2 ha/man-day (see Figs 4.5 and 4.6). Aerial application of pesticides is becoming increasingly important, especially for crops planted in large blocks such as banana and pineapple. It can be done quickly and economically by either fixed-wing aircraft or helicopters. The latter are more expensive in outlay and maintenance but have the advantage of being able to work from unprepared sites and in otherwise inaccessible terrain.

It is crucial to know just when to spray. This depends on the stages of development of host and pathogen; both are under the influence of climate and weather. It is sometimes possible to define precisely when infection will take place and a warning system can be based on this; examples are potato blight and blister blight of tea. Attempts to apply this to Sigatoka of banana have not been entirely successful, as will be shown in a later chapter. Careful counts of mites and scales can also be used as a guide to the timing of spraying.

The action of a fungicide is either preventive or curative. In the first case the crop has to be covered more or less permanently with a thin layer of the a.i.; curative treatment aims at killing the pathogen after infection has taken place. Insecticides are either contact or stomach poisons: the first kills by penetration into the insect

Fig. 4.5   Hand-operated knapsack sprayer

body, the second is taken up with the food. Some pesticides are systemic: they are transported through the plant and can act anywhere. They may be applied to the soil, or to the leaves. There are several important groups of pesticides.

Copper sulphate is phytotoxic (poisonous to plants) and will corrode metal unless it is neutralized by lime or soda. The resulting Bordeaux and Burgundy mixtures have been successful fungicides for over a century. They have been replaced nowadays by neutral compounds such as copper oxide, copper oxy-chloride and tri-basic copper; all these kill fungi effectively. Used at the wrong season they are apt to induce an outbreak of scales for two reasons: they kill fungi that parasitize scales and the residue provides a foothold for scales.

Sulphur and lime sulphur are also old remedies against rust, mildew and mites; they are both a fungicide and an acaricide. Formerly a broad spectrum of action was considered an advantage, but now the trend is towards more selective pesticides.

Dithiocarbamates are organic compounds containing a metal, either iron (ferbam), zinc (ziram and zineb) or manganese (maneb). Ferbam was found to give good control of scab and areolate spot in citrus nurseries in Surinam. Zineb works well against greasy spot of citrus; maneb is used against Sigatoka of banana.

Fig. 4.6 Motorized knapsack sprayer, vertical range 7 m

Imidazoles such as benomyl and thiabendazole (TBZ) were introduced in packing houses as a treatment against fruit rots; however,
Penicillium spp. became resistant. No resistance against imazalil has
been reported so far. In post-harvest control of decay fungistatics
(which stop fungus growth) are useful, e.g. diphenyl and SOPP
(sodium-ortho-phenyl-phenate). The first is used in fruit wrappers,
box liners and pads, SOPP is used as a wash.

Most chlorinated hydrocarbons have a broad spectrum of action,
both as stomach and contact poisons and are very persistent (they
take a long time to break down). They may accumulate in the body
fat of vertebrates, such as birds of prey, at the end of food chains.
In times of stress this body fat is used and the insecticide is released
into the blood stream with sometimes fatal consequences. This has
led to the restriction of their use in many countries. A measure for
the toxicity of pesticides to man is the LD_{50}: the lethal dose, in
mg/kg body weight, required to kill 50 per cent of the individuals
in a group of male rats. This LD_{50} ranges from 46 in dieldrin to

more than 700 in chlorobenzilate. A high LD_{50} thus means low toxicity. Lindane is the γ-isomer of BHC (HCH); it leaves no 'taint', as other isomers do.

Organophosphorus compounds are effective as contact and systemic insecticides and as acaricides. They are not persistent, but break down within a few days at high temperature. The LD_{50} is exceedingly low as in parathion (only 1), or very high as in malathion (2,800); this explains the popularity of malathion.

Insecticides derived from plants include nicotine, pyrethrins and rotenone; no resistance against their action is known. Mineral oils are phytotoxic but a 1–2 per cent solution of a refined 'white oil' is effective against scales and mites, without damaging the leaves. The range of action of some insecticides in some groups of pests is set out in Table 4.2.

Among the nematicides DD-mixture, chloropicrin, methyl bromide, and vapam are the most important. Some new nematicides are mentioned in later chapters.

A simple example of integrated control has already been described: spray-banding trees with diazinon to keep out ants, so that natural enemies get a chance to do away with aphids and scales. In Israel integrated control in citrus is accomplished as follows: purple scale is controlled by a parasitic wasp, Florida red scale by

Table 4.2   *Effect of pesticides against pests*

	Aphids	Mealy bugs	Scales	Mites	Other pests
Chlor. hydrocarbons					
Aldrin	.	x	.	.	soil insects
BHC (lindane)	x	.	.	.	ants
Chlorobenzilate	.	.	.	x	
DDT	x	x	x	.	
Dieldrin	x	x	x	.	ants
Organo-P-compounds					
Diazinon	x	.	x	x	ants
Dimethoate	x	.	x	x	psylla, black fly
Malathion	x	x	x	x	
Parathion	x	x	x	x	
TEPP	x	.	.	x	
Natural compounds					
Nicotine	x	.	.	.	
Pyrethrins	.	.	.	x	
Rotenone	x	.	.	x	
Organic oils					
Mineral oils	.	x	x	x	
Tar oils	x	x	x	.	eggs

Source: Hill (1975)   x = effective   . = not effective

another and California red scale by two other *Aphytis* spp. and a predator. Sulphur, which would kill these wasps, was replaced by chlorobenzilate, a very selective acaricide that does no harm to the wasps. The Mediterranean fruit fly is controlled by a combination of three methods: fly parasites on wild hosts, bait sprays on citrus and a sex pheromone which lures males to poisoned bait (Cohen, 1975).

All pesticides are more or less toxic and therefore dangerous. In using them, it is important to take these precautions:

(a) keep pesticides locked up in a ventilated room;
(b) do not admit children and unauthorized persons to this room;
(c) containers must be clearly labelled;
(d) they must never be used for other purposes;
(e) burn or bury empty containers after use;
(f) materials in paper bags must be stored off the ground;
(g) do not mix materials with your hands; use a stick;
(h) do not allow people with open wounds to work with pesticides;
(i) wear protective clothing when spraying;
(j) never spray against the wind;
(k) the spray must not drift to other fields;
(l) do not eat or smoke during spraying;
(m) wash hands and face thoroughly before eating or smoking;
(n) clean equipment directly after use, but not where cattle drink;
(o) take a shower after the work and put on clean clothes;
(p) observe the safety periods indicated on the label; generally two weeks before harvest is the last possible date for spraying.

If despite all precautions accidents still happen, then:

(i) call a doctor immediately and show him the container used;
(ii) take the victim into fresh air and remove wet clothes;
(iii) wash the skin where it came into contact with the poison;
(iv) if poison got into the victim's stomach, let him vomit (but only if he is conscious);
(v) do *not* give him milk to drink.

Before and after the harvest

How does a fruit grower determine the exact moment to pick his fruit? In ripening a fruit becomes bigger until it has reached its full size, the fruit flesh gets softer, starch is converted to sugar and sucrose into fructose and glucose, the acid content goes down, the green colour disappears, other colours become visible, the aroma

Table 4.3   *Range of respiration at 20 °C in mlitre O_2 kg^{-1}. h^{-1}*

	Climacteric			Non-climacteric	
	Min.	*Max.*		*Initial*	*Final*
Avocado	35	155	Lemon	10	8
Banana	20	60	Orange	13	11
Mango	22	63	Pineapple	11	17

Source: Biale, 1976

and taste develop. Briefly, there are external and internal signs of ripening. These may or may not be accompanied by a sharp rise in respiration, the so-called climacteric.

Biale (1976) places fruit crops in two classes:
  (i) non-climacteric; citrus, grape, guava, pineapple,
  (ii) climacteric; avocado, banana, fig, mango, papaya.
The difference can be great, as is shown in Table 4.3.

Figures 4.7–4.9 taken from Leopold and Kriedemann (1975), show clearly what happens during ripening. In Figure 4.7 we see that ripening proceeds rapidly in banana, moderately fast in apple and slowly in orange (which does not contain starch). Figure 4.8 shows colour changes involving loss of chlorophyll and changes in the carotenoids. In Figure 4.9 the intensities of respiration are compared.

Several plant hormones influence ripening, but ethylene is regarded as the trigger that sets it off. Ethylene is produced by the fruit, but can also be applied externally; this is done in ripening rooms of banana or in colouring rooms of citrus.

Fruit left too long on the tree becomes over-ripe and will finally fall off. However, the avocado can be 'stored on the tree' for several months because the climacteric only sets in after picking. For some citrus cultivars the harvest may be delayed for several weeks after internal maturity has been reached. But as a rule fruit is picked as soon as external and internal characteristics allow. The harvest date is generally chosen on the basis of one or more of the following criteria:
1. Number of days after fruit setting (banana);
2. Shape of the transversely cut fruit (banana);
3. Resistance to pressure (avocado);
4. Break in rind colour (citrus, mango, papaya, pineapple);
5. Ratio between sugars and acids (citrus, pineapple);
6. Minimum juice volume (citrus);
We shall return to this subject when dealing with the harvest of each kind of fruit in particular.

Fruit is picked by cutting, clipping, pulling or shaking. The last named, a labour-saving method, is usually done in combination with abscission chemicals such as cycloheximide, SADH and ethephon.

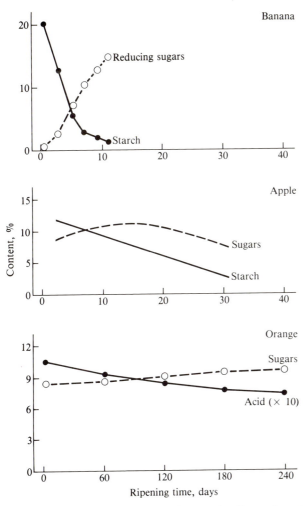

Fig. 4.7 Fruit-ripening processes in three species. *Source*: Leopold and Kriedemann (1975)

A trunk shaker operates with a short stroke at high frequency and has a capacity of 40 to 60 trees per hour. The fruit is dropped on the ground to be collected with a rake-pickup. It is limited to cannery use and has to be processed within 36 hours. Fruit caught on a padded catching frame may be suitable for the fresh market (Wilson and Coppock, 1975). In Surinam West Indian cherries were also harvested for processing by shaking into a tarpaulin (de Willigen, 1968).

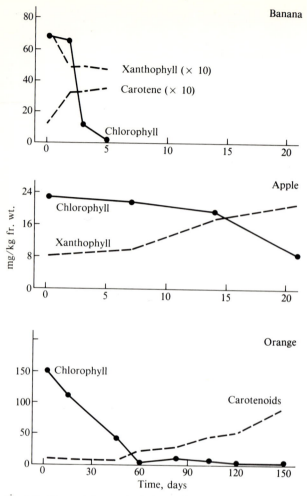

Fig. 4.8 Pigment changes in three species. *Source*: Leopold and Kriedemann (1975)

Small fruits are probably best harvested by shaking, but in the majority of cases harvesting must be done by hand. A knife is used for large fruits, such as papaya, pineapple and soursop. Clippers are used in California for citrus, but in the wet tropics they are neither necessary nor desirable for reasons we shall examine later. In 'pulling', fruit must not be pulled straight down, but twisted off with a circular upward movement of the hand in order to avoid 'plugs'. Ladders, bags, field boxes (holding 30–40 kg fruit) and field

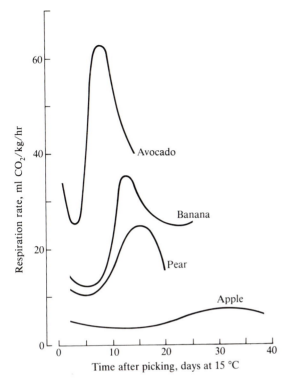

Fig. 4.9 Respiration rate during ripening of four fruits, highest peak in avocado.
Source: Leopold and Kriedemann (1975)

containers (holding up to 400 kg of fruit) are used.

In Florida bags are going out of fashion, the fruit often being dropped to the ground where it is collected by rake-pickup machines with revolving brush sweeps. It does not seem likely that this method can ever be adopted in the wet tropics, but it may be used in drier zones. In California a more efficient and less tiring 'harness and picking bag' has been proposed (Smith, 1969). The importance of efficiency in picking is shown clearly by the fact that in California picking costs 'exceed all other cultural costs combined' (Johnson, 1969) and that in Israel 'picking represents 40 per cent of the number of working days' (Alper and Sarig, 1969).

The harvested fruit is taken from the orchard in boxes or containers, either in lorries or in tractor-drawn sledges, each carrying four containers. The harvest is then delivered to a cannery, a packing house or directly to the local market. Bananas may be transported by cableway if this is cheaper than building roads.

Delivery to a market usually prevails in the tropics and generally implies that fruit is sold quickly to buyers who make no special demands on internal quality and external appearance. The fruit is also riper than when picked for export. Where demands on quality and appearance are made, as in some big cities, it is feasible to clean, size and grade the fruit in a shed or a small packing house. Let us now consider the alternatives. Citrus and banana packing will be described in more detail later. A packing-house manager should know exactly where and when picking is going on and how much fruit will be delivered to his plant; in fact he should be in control of all harvesting activities so that the work of the different field foremen can be coordinated with the work inside. Much friction will be prevented if all responsibilities and priorities are clearly indicated beforehand.

Fruit is usually washed, disinfected, dried, waxed and sorted for size and grade before it is packed. Formerly nailed wooded boxes were generally used, then wirebound 'Bruce' boxes became more common and recently cardboard boxes have become more popular for citrus. Bananas were originally shipped in whole bunches, either bare or wrapped in paper or perforated plastic bags; cardboard boxes are now in general use for detached hands. Pineapples, avocados and other tropical fruits are transported in low, padded containers.

Picked fruit is still alive; metabolism continues until a stage of over-ripeness is reached. These metabolic processes cannot be stopped, but they can be retarded by one or more of the following measures:

1. reduction of temperature;
2. packing in sealed polythene bags;
3. wax coating;
4. addition of ethylene absorbents;
5. reduced pressure;
6. controlled atmospheric (CA) storage;
7. treatment with growth regulators;
8. irradiation

We shall examine these possibilities on the basis of examples provided by Coursey *et al.* (1976). The lower the temperature, the lower respiration in storage or transport will be; but this cannot be extended too far. Not only should temperature be maintained at above freezing point, it should be above 10 °C, for most tropical fruits will suffer from chilling damage below that point. Humidity is kept high, usually between 85 and 90 per cent. However, there are considerable differences between crops and also between stages of maturity as can be seen in Table 4.4.

Somewhat different data are furnished by Wills *et al.* (1981); storage life in weeks at 5–9 °C would be: orange 6–12, mandarin

Table 4.4 *Recommended holding conditions for various fruits*

Crop	Temp. °C	Rel. hum.%	Period
Orange	-1-7	85-90	1- 6 m
Mandarin	4-7	85-90	3-12 w
Grapefruit	10-15	85-90	3-13 w
Lemon, green	11-14	85-90	1- 4 m
Lemon, yellow	0-10	85-90	3- 6 w
Lime	8-10	85-90	3- 8 w
Banana, green	11-14	90-95	10-20 d
Banana, ripe	13-16	85-90	5-10 d
Pineapple, green	10	90	2- 4 w
Pineapple, ripe	5-10	85-90	2- 6 w
Avocado	5-13	85-90	2- 4 w
Brazilnuts	0	70	8-12 m
Cashew	0- 1	85-90	4- 5 w
Guava	7-10	85-90	3- 4 w
Mango	7-10	85-90	4- 7 w
Papaya	4-10	85-90	2- 5 w
Passion fruit	5- 7	80-85	4- 5 w

Source: Hartoungh, 1978
d = days, m = months, w = weeks

4–6, ripe pineapple 4–5, avocado 3–5, mango 2–3 and passion fruit 3–5. At 10 °C it would be: grapefruit 6–12, lemon 12–20, banana 1–2 and (green) pineapple 4–5 weeks. When avocados are packed one by one in sealed polythene bags and stored at low temperature, they have a long shelf-life without chilling injury. Plantains can be kept for 25 days at a rather high temperature if an ethylene absorbent is put in the polythene bag; at 13 °C the shelf-life rises to 55 days. Skin coating with wax is used for citrus and mango and has an effect similar to that of the polythene bag: transpiration and gas exchange are reduced, even at room temperature.

Reduced pressure is applied at 100 mm mercury, say one seventh of an atmosphere; in CA the oxygen content of the air is brought down from 21 to 2 per cent and the carbon dioxide content is raised from 0.03 to 10 per cent. Combination of these methods makes it possible to prolong the storage life of some tropical fruits for many weeks.

Ripening processes can also be delayed by treatment with hormones (e.g. gibberellic acid), growth retardants (maleic hydrazide) and metabolic inhibitors (such as vitamin K). Irradiation also delays ripening, retards spoilage and minimizes insect infestation. Much research is still needed into the most effective methods, combinations and conditions for storage and transport of tropical fruits.

Another matter requiring our attention is post-harvest decay. It is responsible for major losses of fruit in storage and transportation. An infection in the field may remain latent until after picking. Fruit may be bruised or cut during and after picking so that moulds can penetrate it. Hot water or vapour treatment is used and fungicides such as SOPP (sodium-ortho-phenyl-phenate), TBZ (thiabendazole), benomyl and maneb are applied. Wrappers and liners in boxes, impregnated with diphenyl, may be successful in suppressing fungus growth. Two kinds of decay may be distinguished: that occurring on single fruits and decay spreading from a focus; we will return to that problem later. Post-harvest losses of 10 per cent are normal; they may run up to 80 per cent in developing countries (Ruskin, 1978). Much more information is provided by Hulme (1971), Pantastico (1975), Anon. (1976) and Wills *et al.* (1981).

Transportation

Transportation of tropical fruits on a large scale started about 1870, when banana ships began to ply the Caribbean Sea. The 'white fleet' of the United Fruit Company became well known. The post-war banana boat has been described by Deullin (1960). At that time the average capacity was 5,000 m³ and 180–300 kg bananas could be carried per cubic metre; this means that a ship usually carried about 1,000 tons of bananas. The maximum speed was 30 km/h. Ten years later the same author described a rather different situation. The content had increased to 10,000 or 12,000 m³ and the speed was now 37 km/h; the ships are fully automatic, polythermic (from +15 to −25 °C) and polyvalent: fit for all kinds of cargoes. The cooling speed had gone up to 0.6° an hour, so that a load could be cooled from 28 to 15 °C in just one day.

Not all fruit is transported in such sophisticated ships. Uncooled but ventilated holds are still used in many cases. This is adequate for short distances, especially in winter. But if a trip is long and the weather on the way is hot, then spoilage is bound to be high. Some freighters have refrigerated rooms that can accommodate from 100 to 500 tonnes of fruit. A fruit export business in its initial stages could be based on such ships, but then cold storage facilities must be available in the harbour to build up supplies. On the other hand, such facilities may also serve to provide fresh fruit out of season for the local market. More details are provided by Marriott and Proctor (1978).

Another possibility is the use of refrigerated containers with volumes of 22 or 45 m³. The bigger one is hard to handle, but the smaller one may soon become an important means of fruit transportation (Deullin, 1972).

Processing

Fruit processing has steadily been gaining in importance during the last decades. In Florida and Brazil more than 80 per cent of all the citrus grown now goes to processing plants. Kesterson and Braddock (1975) list about 30 products in a 'simplified chart'. Likewise, pineapple and passion fruit are generally processed. Banana, mango, avocado and papaya processing is not unknown, but it is far less important than fresh fruit marketing. Tropical fruit flavours, e.g. of durian, guava, litchi, mango, mangosteen and passion fruits, are also produced.

References

Alper, Y. and **Sarig, Y.** (1969) 'Citrus harvesting mechanization in Israel', *Proc. First Intern. Citrus Symp.*, 623–38.
Amson, F. W. van (1964) in *Jaarverslag 1963*, Med. 34, Landb. Proefst. Suriname, p. 18.
Anon. (1972) 'World crops looks at the Turbair system', *World Crops*, **24**, 252–3.
Anon. (1976) 'Handling tropical crops after harvest', *Hortsc.* **11**, 119–30.
Biale, J. B. (1976) 'Recent advances in post-harvest physiology of tropical and subtropical fruits', *First Intern. Symp. on trop. and subtropical fruits*, Lima, ISHS, 179–87.
Bitters, W. P. (1968) 'Valencia orange rootstock trial at South Coast Field Station', *Cal. Citrograph*, **53**, 172–4.
Black, J. D. F. (1976) 'Trickle irrigation – a review', *Hortic. Abstr.* **46**, 1–7 and 69–74.
Bouvries, C. des and **Rydzewski, J. R.** (1977) 'Irrigation in food crops of the lowland tropics' in *Food crops of the lowland tropics*, Oxford.
Cohen, I. (1975) 'From biological to integral control of citrus pests in 'Israel', in *Citrus*, CIBA-GEIGY Techn. Monogr., **4**, Basle, 38–41.
Coursey, D. G. *et al.* (1976) 'Recent advances in research on post-harvest handling of tropical and subtropical fruit', *First Intern. Symp. on trop. and subtrop. fruits*, Lima, ISHS, 135–43.
Davidson, M. ben David (1975) 'Mineral nutrition and irrigation in citrus crops', in *Citrus*, CIBA-GEIGY, Basle, 14–20.
Deullin, R. (1960) 'Le navire bananier', *Fruits* **15**, 99–116.
Deullin, R. (1970–1) 'Evolution du transport maritime de la banane de 1945 à 1970, *Fruits* **25**, 865–76 and **26**, 83–102.
Deullin, R. (1972) 'Transport maritime des fruits tropicaux dans des conteneurs frigorifiques', *Fruits* **27**, 383–92.
Dinther, J. B. M. van (1960) *Insect pests of cultivated plants in Surinam*, Bull. **76**, Landb. Proefst. Suriname.
Gaillard, J. P. and **Haury, A.** (1974) 'Toxicité et action dépressive de quelques herbicides à l'égard de l'ananas; étude de l'action du bromacil', *Fruits* **29**, 745–55.
Hartoungh, J. C. C. (1978) *Bewaring van vruchten en groenten in de tropen*, Lecture notes, Wageningen.
Hill, D. S. (1975) *Agricultural insect pests of the tropics and their control*, Cambridge.
Hill, D. S. and **Waller, J. M.** (1982) Pests and diseases of tropical crops, **vol. 1** *Principles and methods of control*, Longman.
Holm, L. G. (1977) *The world's worst weeds*, Univ. Press Hawaii.
Hulme, A. C. (ed., 1971) *The biochemistry of fruits and their products*, 2 vols, London.
Johnson, M. (1969) 'Systems engineering for citrus production, harvest, packing and distribution', in *First Intern. Citrus Symp.*, 619–21.

Kasasian, L. (1971) *Weed control in the tropics*, London.
Kesterson, J. W. and **Braddock, R. J.** (1975) 'Citrus fruit processing', in *Citrus*, CIBA-GEIGY Techn. Monogr. **4**, Basle, 75–80.
Koenraadt, J. (1962) 'Wieden bij boomcultures', *Surin. Landb.* **10**, 30–2.
Lange, A. H. *et al.* (1975) 'Weed control in citrus', *Citrus*, CIBA-GEIGY, 55–60.
Lee, A. N. and **Orr, L. C.** (1980) 'What is 2,4,5-T?', *Queensl. Agric. J.* **106**, 279–85.
Leopold, A. C. and **Kriedemann, P. E.** (1975) *Plant growth and development*, McGraw-Hill.
Mabett, T. H. and **Phelps, R. H.** (1983) 'The behaviour of copper deposits from reduced-volume sprays and their control of citrus greasy spot', *Trop. Pest Management* **29**, 137–44.
Marriott J. and **Proctor, F. J.** (1978) 'Transportation and conservation of tropical fruits', *Outlook on Agriculture* **9** (5), 233–9.
Pantastico, E. B. (ed., 1975) *Postharvest physiology, handling and utilization of tropical and subtropical fruits and vegetables*, AVI, Westport (Conn.).
Platt, R. G. (1973) 'Planning and planting the orchard', in *The citrus industry*, vol. III, ed. W. Reuther, Univ. Calif.
Raccah, B. *et al.* (1976) 'Transmission of tristeza by aphids prevalent on citrus and operation of the tristeza suppression programme in Israel', *Proc. 7th Intern. Org. Citrus Virol.*, 47–9.
Ruskin, F. R. (ed., 1978) *Postharvest losses in developing countries*, Nat. Acad. of Sciences, Washington DC.
Samson, J. A. (1959) 'Easier forest clearing', *World Crops* **11**, 273–4.
Samson, J. A. (1977) 'Problems of citrus cultivation in the tropics', *Span* **20**, 127–9.
Smith, R. J. (1969) 'Reducing the obstacles to efficient picking in citrus', *Proc. First Intern. Citrus Symp.*, 609–17.
Simmonds, N. W. (1966) *Bananas*, Longman.
Sizaret, A. (1983) 'Plantations fruitières sur buttes, ou le mille et une positions du collet', *Fruits* **38**, 397–415.
Wang, H. T. *et al.* (1977) 'Effects of bench terrace and agronomic practices on slopeland banana plantation', *J. Agric. Ass. of China* (Taiwan), New Series XCVII, 88–100.
Wardlaw, C. W. (1961) *Banana diseases*, Longman.
Webster, C. C. and **Wilson, P. N.** (1966) *Agriculture in the tropics*, Longman, 2nd ed. 1980.
Willigen, P. de (1968) 'Kostprijsonderzoek vruchtenproefbedrijf Boma', *Surin. Landb.* **16**, 99–109.
Wills, R. H. H. *et al.* (1981) *Postharvest*, Granada Publ. Ltd.
Wilson, W. C. and **Coppock, G. E.** (1975) 'Citrus harvesting' in *Citrus*, CIBA-GEIGY Techn. Monogr. **4**, Basle, 67–71.
Zadoks, J. C. (1975) *Gewasbescherming* (lecture notes, Agr. Univ., Wageningen, Netherlands).

Chapter 5

Citrus

Taxonomy and morphology

It is desirable to distinguish the plant genus *Citrus* (note the capital letter and italics) from the crop citrus. The latter comprises not only the edible species of *Citrus* with their cultivars, hybrids and rootstocks, but also those of *Poncirus, Fortunella* and other related genera. The distinction is quite clear in French where the word 'agrumes' is used for the crop; likewise in Spanish, Italian and Portuguese. In Indonesia, and Malaysia citrus trees and their fruits are called 'jeruk' and 'limau'.

Citrus belongs to a large family, the Rutaceae (130 genera), in which the leaves usually possess transparent oil glands and the flowers contain an annular disc. Within the Rutaceae seven subfamilies are recognized but one only, the Aurantioideae (33 genera), is of interest to us. Further divisions take us to the tribe Citreae (28 genera) and the subtribe Citrinae (13 genera), which can be split in three groups: a 'primitive', a 'near' and a 'true' Citrus group. Trials have been made with *Atalantia, Citropsis, Feronia* and *Swinglea* as rootstocks; they and other relatives may act as hosts to various citrus enemies such as insects, nematodes and viruses.

The 'true' group consists of six genera: *Eremocitrus, Poncirus* and *Clymenia* (all with one species), *Fortunella* (4), *Microcitrus* (6) and *Citrus* (16 spp.). All have orange or lemon-like fruits with stalked, spindle-shaped, inward-growing juice sacs. The flowers possess about four times as many stamens as petals. The leaves are persistent and simple except in *Poncirus* where they are deciduous and trifoliate. The six genera may be grafted on one another and can be crossed, producing hybrids.

Poncirus, which occurs naturally in Central and North China, sheds its leaves in winter and is resistant to cold. *Fortunella* (from southern China) is evergreen but owes its hardiness to a prolonged winter dormancy. *Eremocitrus* (from Australia) is very drought-resistant.

The classification of *Citrus* is difficult because of the large number of cultivars and hybrids; polyploids, mutations and polyembryony add to the confusion. Two widely differing systems of classification (Swingle's and Tanaka's) and some variants of these exist. Tanaka has given the rank of species to many evident cultivars and hybrids and recognizes more than 160 spp. Swingle's system, which admits only 16 spp., is far more convenient (Swingle and Reece, 1967). An intermediate course was taken by Hodgson (1961) who gave specific rank to rough lemon (*C. jambhiri*), sweet lime (*C. limettioides*), Rangpur lime (*C. limonia*), Cleopatra mandarin (*C. reshni*) and some others, thus increasing the number of spp. from 16 to 31. Other authors also confer species rank to *C. amblycarpa*, *C. taiwanica* and *C. volkameriana*. What follows is based on Swingle, with some exceptions.

The genus *Citrus* consists of two subgenera: *Papeda* and *Eucitrus*. In *Papeda* the fruits are inedible because of numerous droplets of acrid oil in the juice sacs; the petioles are long and broadly winged. *C. hystrix* is popular on the island of Réunion (Aubert and Lichou, 1974) where a sauce ('rougaille') is made from the fruits; in Indonesia the fruits are used for a hair wash. The 'Alemow' (Hodgson's *C. macrophylla*) is a promising rootstock for lemons.

Table 5.1 *Botanical and local names of* Citrus *and relatives*

Botanical name	English	French	Spanish
C. sinensis	sweet orange	orange	naranja dulce
C. aurantium	sour orange	bigaradier	naranja agria
C. reticulata	mandarin	mandarine	mandarina
C. paradisi	grapefruit	pomélo	pomelo
C. grandis	shaddock	pamplemousse	toronja
C. limon	lemon	citron	limón
C. medica	citron	cédrat	cidra
C. aurantifolia	lime	lime	lima
P. trifoliata	trifoliate orange	poncirus	poncil
F. margarita	kumquat	kumquat	—

There are ten spp. in *Eucitrus*, eight of which are cultivated. Their names, botanical and popular, are set out in Table 5.1. Alternative names are: Seville or bitter orange for sour orange, tangerine for mandarin, pummelo for shaddock and pomelo for grapefruit. As the names pummelo and pomelo are easily confused I prefer not to use them. To add to the confusion, Israel now produces a hybrid also called pomelo. There are many hybrids between the *Citrus* spp., for instance:

tangor   = mandarin × sweet orange
tangelo   = mandarin × grapefruit
lemonime = lemon   × lime

In these compound names 'tang' stands for *tang*erine and 'elo' for pom*elo* (grapefruit); the other parts are easily understood. Apart from these *intra*generic hybrids (within the genus), there are also *inter*generic hybrids (between genera), such as:

citrange = sweet orange × *Poncirus*
citrumelo = grapefruit   × *Poncirus*
limequat = lime   × kumquat (*Fortunella*).

There are even trigeneric hybrids, between three genera, for instance the citrangequat. It is to be noted that 'citr' in these names is derived from *Ci*trus *tri*foliata as *Poncirus* was formerly known.

Briefly, the genus *Citrus* can be described as follows: small trees, young twigs angled, later cylindrical, with single spines in leaf axils, older branches often spineless; leaves unifoliolate, petioles winged and articulated with the leaf blade (except in *C. medica*); flowers single in leaf axils or in short racemes; ovary usually in 10–14 parts; fruit a special kind of berry (hesperidium) filled with stalked spindle-shaped pulp-vesicles (juice sacs) covered by a white spongy tissue (albedo) and a peel with numerous oil glands, turning yellow or orange at full maturity; seeds contain one or many embryos.

As pointed out in Chapter 3, spines are a sign of juvenility, occurring especially on upright branches and near the centre of the tree. Farther from the centre and higher up they become smaller and tend to disappear. Budwood is therefore always taken from outside branches. Budded trees have far less spines than seedling trees. On most cultivars and some seedlings ('Shamouti', 'Cleopatra') they are practically absent.

The *Citrus* leaf is regarded as unifoliolate: a compound leaf with only one leaflet, which is attached to the petiole by articulation. Intrageneric hybrids may show one or two small side leaflets and intergeneric hybrids with *Poncirus* have a trifoliate leaf, because that characteristic is dominant over the unifoliolate form. When crosses are made, *Poncirus* must be the pollinizer (in plants always the second-named parent). In this way, hybrids with trifoliate leaves can easily be recognized, as nucellar seedlings of *Citrus* are unifoliolate.

The colour and shape of leaf blade and petiole wings are important for diagnosis. Lemon, for example, has pale green leaves that are more or less transparent. In grapefruit and shaddock the petiole wings touch or partly overlap the leaf blade, which is not the case in sour orange. Generally, the leaf is ovate (egg-shaped), but it is rather elliptic in citron and oblong in mandarin. Usually the leaf margin is entire, but it is serrated in lemon or citron and crenulated in lime. Petiole wings are missing in citron, narrow in most spp., broad in sour orange, grapefruit, shaddock and very broad in *Papeda* (Fig. 5.1).

Fig. 5.1   Young plant of *Citrus hystrix*; note the very large petiole wings

On a healthy, full-grown citrus tree there are between 150,000 and 200,000 leaves, with a total area of about 200 m^2; the LAI (see Ch. 3) is four to five. The leaves remain on the tree for two or three years. In general there are 400 to 500 stomata per mm^2 on the lower side of the leaf, but practically none on the upper side. The stomata are about 7 × 3 μm in size. They open and close mainly under influence of light; usually they are open from 10 a.m. till 4 p.m.

The flowers are white in all spp. except lemon and citron where they are purplish on the outside. The most striking differences are in the fruits, shaddock has the largest and lime the smallest. The surface may be rough as in sour orange, or smooth as in sweet orange. The peel is loose and easily detached from the segments in mandarin, but adheres in other spp. There may be a central cavity, as in sour orange. A beautiful illustration of the colours, shapes and sizes of citrus fruits can be seen on the back cover of the periodical *California Agriculture* of September 1977. The seeds also have particular shapes and textures, so that nursery people identify them easily. Seeds are polyembryonic, but not in *C. grandis* and *C. medica*; in other spp. monoembryonic cvs occur.

With the previous remarks in mind we can now construct a simple key to the eight edible species of *Citrus*:

1. petiole not winged .... *C. medica*
   petiole winged and articulated 2
2. stamens more than 4 times as many as petals .... *C. limon*
   stamens 4 times as many as petals, or less 3
3. fruit with loose peel .... *C. reticulata*
   fruit peel not loose 4
4. fruit very large, seed monoembryonic .... *C. grandis*
   fruit not very large 5
5. fruit fairly large .... *C. paradisi*
   fruit medium small 6
6. fruit small, egg-shaped, very acid .... *C. aurantifolia*
   fruit bigger, sweet, with smooth skin .... *C. sinensis*
   fruit acid and bitter, with rough skin .... *C. aurantium*

For more detailed descriptions, see Purseglove (1968) and Reuther *et al.* (1967).

Uses and composition

Citrus fruits may be eaten fresh as are the sweet orange, mandarin and grapefruit, or their segments may be canned. Juice is extracted and consumed, or it may be concentrated four or more times. Juice of lemon and lime is diluted in lemonades and other soft drinks. Segments of shaddock are used in fruit salads because the pulp-vesicles fall apart easily. Citrus pulp and molasses are used as cattle feed. Pectin and essential oils are made from the peel. One tonne of lemon fruit yields 5–9 kg of essential oil (Kesterson and Braddock, 1975).

The most expensive essential oils are derived from flowers of sour orange cvs (neroli) and from fruit of the bergamot sour orange. Citric acid is made from lemons and limes. Citron peel is candied, while kumquats are candied as whole fruits. The juice of a ripe orange or mandarin contains about 12 per cent sugars and other soluble solids, 1 per cent citric acid and 50 mg vitamin C (ascorbic acid) per 100 grams. One glass of juice is, therefore, sufficient to provide the daily requirement of vitamin C. Grapefruit and sour orange contain relatively less sugar and more citric acid; between 1.5 and 3 per cent of the latter. In lemon and lime the acid content may go up to 6 and 8 per cent respectively.

The great variety of distinctive smells in citrus leaves, flowers and fruits is caused by essential oils containing several flavonoids. One of these, hesperidin, is found in most citrus fruits; other, more specific flavonoids are neohesperidin, naringin, aurantamarin, tangeretin and limonin. The colour of citrus fruits is due to yellow, orange and red pigments called carotenoids; they are located in the chloroplasts. Some of them do not develop unless the temperature has remained below 13 °C for several hours. This is why tropical oranges are still green when fully mature. In the so-called blood oranges the red colour is caused by anthocyanin which is dissolved in the cell sap. Here too, colour does not develop under sustained high temperature.

Origin, distribution and production

The natural environment of the six genera in the 'true' citrus group is a region stretching from India and China in the north-west to Australia and New Caledonia in the south-east. *Citrus* is still found in the wild state in that area. Cultivation of citrus must have started in China; the records date back to 2200 BC. The first book on citrus appeared in China in the year AD 1178: many varieties were described and cultural methods were discussed. But long before that citrus cultivation must have spread throughout South-East Asia. After Alexander's conquest (about 300 BC) citrus seeds were taken to the Mediterranean area. First came the citron, then sour orange, lemon and lime. Sweet orange and mandarin arrived much later, probably during the Crusades. When the great voyages of discovery started, all these species were already well known in Italy, Spain, Portugal and North Africa.

Columbus introduced citrus to the western hemisphere on his second voyage (1493). Seeds of orange, lemon and citron were collected on Gomera (Canary Islands) and taken to Hispaniola (Haiti), where they were sown. From here the crop has spread to other Caribbean islands and to the continent. In the sixteenth and seventeenth centuries introductions were made along the coasts of South America and Africa and the circle was closed when citrus was (re)introduced to Australia.

All but one of the eight edible spp. of *Citrus* arose in Asia, but *C. paradisi* came into being on the island of Barbados in the West Indies, around the year 1750. It was unknown in Asia until its recent introduction and must have originated from the shaddock as a mutation or a hybrid. According to Meijer (1954) travellers arriving in 1658 in Pomeroon (Guyana) found '*Limoins, Laranjas como meloins piquenos, Bananas e outras frutas*' growing there; this probably means lemon and shaddock (an orange as big as a small melon).

Travellers and missionaries have greatly assisted the spread of citrus, especially since it became known that fresh fruits were the best remedy against scurvy. The British navy even made it compul-

Table 5.2   *The major producers of oranges (10³ tonnes)*

Country	1975	1980	1981	1982	1983	1984
USA	9,294	10,734	9,514	6,931	8,631	6,566
Brazil	6,333	8,877	9,312	9,587	9,515	13,372
Spain	2,016	1,711	1,469	1,693	1,895	1,310
Italy	1,582	1,540	1,751	1,500	1,945	1,700
Mexico	2,322	1,950	1,600	1,690	1,480	1,600
India	940	1,160	1,180	1,234	1,200	1,223
Egypt	856	921	895	916	1,205	1,300
Israel	1,052	985	785	1,050	857	921
China	818	807	896	900	1,203	1,495
Argentina	729	704	663	600	619	580
South Africa	600	588	624	630	483	495
Morocco	477	757	685	695	691	746
Greece	538	507	704	653	550	600
Turkey	551	691	687	746	691	744
World	32,702	38,463	37,751	35,952	38,171	39,679

Source: FAO Production Yearbooks
NB: indications for unofficial figure and FAO estimate have been omitted in this and following tables

sory for their sailors to drink lime juice every day; a sensible measure which, however, earned them the nickname 'limeys'.

Nowadays, citrus is cultivated throughout the subtropics and tropics; roughly between 40° North and South latitude. The total area planted with citrus now amounts to about two million hectares. Among the fruits of the world citrus takes second place; only grape, most of which is used for wine, has higher production figures. Table 5.2 gives an idea of the most important producers of oranges during the last 22 years (FAO, 1983). It is clear from Table 5.2 that oranges are predominantly grown in the sub-tropics. Even in India, Brazil and Mexico a great part of the production comes from regions that are not fully tropical. Tropical countries which produce more than 500,000 tonnes are Ecuador and the Philippines; Venezuela produces more than 400,000 tonnes and Cuba more than 300,000 tonnes.

Mandarins too (see Table 5.3), are a mainly subtropical crop: Japan supplies more than half the world production. Yet a considerable quantity is probably produced and consumed in tropical regions and does not appear in the statistics. Lemons and limes are lumped together in FAO statistics, but differ greatly in their ecological requirements; limes are almost exclusively tropical, while lemons thrive in cooler climates. Mexico produces mainly limes, the other countries in Table 5.4 chiefly lemons. FAO statistics do not distinguish between grapefruit and '*po*melo' (for which one should of course read *pum*melo, or shaddock). I assume that China produces principally shaddocks, the other countries in Table 5.5 chiefly

Table 5.3   *The major producers of mandarins (× 1,000 tonnes)*

Country	1975	1980	1981	1982	1983	1984
Japan	3,823	2,892	2,819	2,847	3,169	2,187
Spain	652	906	723	837	1,117	868
Brazil	310	327	570	572	580	530
USA	619	756	564	532	532	446
Italy	350	305	379	320	400	370
China	228	256	267	260	269	248
Morocco	106	267	280	294	245	243
Argentina	230	214	207	202	239	225
Pakistan	177	200	200	205	205	205
World	7,209	7,267	7,108	7,218	8,047	6,620

— No data available
Source: FAO Production Yearbooks

Table 5.4   *The major producers of lemon and lime (× 1,000 tonnes)*

Country	1975	1980	1981	1982	1983	1984
USA	1,053	756	1,122	902	921	787
Italy	880	747	845	683	770	690
Mexico	623	473	500	560	580	600
India	450	485	490	512	500	500
Spain	450	336	433	428	522	283
Argentina	339	396	405	390	450	320
Turkey	290	283	301	314	330	300
World	5,093	4,873	5,324	5,234	5,477	4,907

Source: FAO Production Yearbooks

Table 5.5   *The major producers of grapefruit and shaddock (× 1,000 tonnes)*

	1975	1980	1981	1982	1983	1984
USA	2,271	2,709	2,503	2,625	2,220	1,945
Israel	417	509	494	500	450	409
Argentina	185	164	115	135	146	140
China	119	110	127	153	157	161
South Africa	87	110	114	120	108	81
Mexico	—	73	163	115	110	105
World	3,682	4,470	4,356	4,484	4,088	3,775

Source: FAO Production Yearbooks

grapefruit. From these tables it appears that the explosive growth is over, although some expansion still continues. However, seasonal fluctuations in fruit production are great and a trend cannot be predicted on the basis of only a few years. It seems best, therefore, to accept the predictions made by Cadillat (see Ch. 1).

Finally, let us consider the citrus production per head in four tropical regions (Africa, West Indies, South America and South-East Asia), compared with some sub-tropical countries (Table 5.6). In

Table 5.6   *Citrus production (1975) in 10^3 tonnes and kg/head in selected countries*

Country	or.	mand.	l + l	g + s	nsp	tot.	pop.	k/h
Ghana	165	.	31	.	.	196	9	22
Madagascar	91	.	2	1	.	94	8	12
S. Leone		.	.	.	110	110	3	37
Zaire	102	.	5	8	.	115	24	5
Belize	33	.	.	20	.	53	0.14	379
Cuba	119	10	16	26	1	172	9	19
Jamaica	23	4	19	31	.	77	2	38
Trinidad	12	.	1	19	.	32	1	32
Ecuador	240	20	28	.	90	378	7	54
Paraguay	123	31	17	23	.	194	3	65
Peru	236	20	88	.	.	324	15	22
Venezuela	241	.	.	.	.	241	12	20
India	924	.	450	20	28	1,422	613	2
Pakistan	330	115	27	2	.	474	71	7
Phillipines	11	14	19	27	.	71	44	2
Thailand	49	.	1	12	.	62	42	2
Australia	404	34	39	18	.	495	14	35
USA	9,294	616	1,053	2,264	.	13,227	214	62
Spain	1,946	628	295	7	.	2,876	35	82
Israel	1,116	.	35	435	.	1,586	3	528

Key: or. = orange; mand. = mandarin; l + l = lemon and lime; g + s = grapefruit and shaddock; nsp = not specified citrus; tot. = total citrus; pop. = population (in millions); k/h = kg/head.
Source: FAO Production Yearbook (1976)

Chapter 1 we found that a minimum daily intake of 100 g fruit was desirable; this amounts to 36 kg/year. If half of this quantity is provided by citrus, then some 18 kg/head, preferably well spaced over the year, should be available. Asian countries are remarkably poorly supplied. Indonesia was formerly not mentioned in FAO statistics, but is known to have a chronic shortage of citrus fruits. It seems that countries at the centre of origin are more in need of sound fruit development programmes than other countries; Central Africa probably comes next.

Growth and development

Citrus seeds have no dormant period and are injured by drying, they should be sown fresh. The optimal temperature is about 32 °C. Within a few days after sowing the seed visibly swells from the absorbed water. The cotyledons remain underground but the primary root elongates, emerges from the seed and bends downwards to form a taproot. The shoot becomes visible above the ground and forms its first leaves in the third or fourth week. Meanwhile the taproot begins to branch and puts out secondary roots.

Most *Citrus* spp. and cvs are polyembryonic: from one seed two, three or more plants may emerge. One of these may be a 'sexual' embryo as a result of crossing. All the others, and frequently all embryos, have been formed in the nucellus (see Ch. 3) and are therefore called 'nucellar' embryos; genetically they are identical with the mother plant. This has several important consequences:

(a) seedling rootstocks are largely uniform in the nursery;
(b) clones are 'rejuvenated' by sowing (virus is rarely passed on by seed);
(c) breeding work becomes more difficult.

Exceptions to the rule are *C. medica, C. grandis* and cvs such as 'Temple' and 'Clementine' mandarins and 'Meyer' lemon: they are monoembryonic and their seedlings are very heterogeneous; that is why they are not used as rootstocks. Yet virus-free clones can be produced by tissue culture techniques.

The root system of a citrus tree varies with the rootstock, the sp. or cv. grafted on it and the nature of the soil. Seedlings generally retain their taproot, until they meet a hardpan or water-soaked soil layer. In well-aerated soils, citrus roots may extend over 4 m deep. Layered trees, however, have shallow root systems. This is not altogether a disadvantage, for it may enable a tree to survive on soil with a high water table which is sometimes found in the humid tropics.

There are two types of lateral roots: big branching and small fibrous ones. On the latter few root hairs are found. Their function has been taken over by mycorrhiza: fungal threads (*Glomus* spp.) growing inside the root; these supply the citrus plant with nutrients, especially P, taken up from the soil. Nemec (1978) reports that top growth of six rootstocks was increased 6 to 21–fold in absence of P and 1.8–fold in its presence; sour orange and 'Cleopatra' depended the most on mycorrhiza. Kriedemann and Barrs (1981) found an abundance of abnormally short root hairs.

Roots and shoots show definite, alternate cycles of growth; when one is growing, the other is at rest. In the subtropics there is a completely dormant period caused by winter conditions; growth stops when the temperature drops below 13 °C. In the tropics dormancy is induced by a dry spell. A growth cycle lasts from four to six weeks and there are up to five of them in a year. At each flush bloom may occur and this usually happens in the lemon. However, the citrus tree first passes through a stage of juvenility during which no flowers can be formed. In seedling trees this stage lasts six to eight years, whereas a budded tree will generally start flowering in the second or third year after planting. Massive blooming usually occurs about three weeks after an inducing period of cold or drought has ended. The flowers sit alone or in groups (racemes). Several types of flowering shoots exist; some bear only flowers, others may also bear one or many leaves. For instance, 'Valencia' inflorescences carry only flowers, while in 'Washington Navel' a flowering shoot bears a few flowers and many leaves. The top flower is the biggest and the first to open, followed by the bottom flower and then upwards from there. In 'Washington Navel' and 'Satsuma', the flowers have no viable pollen, they are male sterile. Orchards of these cvs produce seedless fruits formed by parthenocarpy. Along the margin of such an orchard cross pollination can take place, causing these fruits to contain seeds. Female sterility occurs in 'Marsh' grapefruit, 'Eureka' and 'Lisbon' lemons; their fruit is practically seedless, regardless of the pollen.

Ecology and physiology

It is customary to distinguish three zones of citrus cultivation:
1. Subtropical: California and the Mediterranean coastal areas, with a cool, rainy winter and a hot dry summer; they generally lie between 30° and 40° latitude.
2. Semitropical: Florida, Brazil and Transvaal (South Africa) with a cool dry winter and a rainy hot summer; located between latitudes of 20° and 28°.
3. Tropical: alternating rainy and dry seasons but no cold period; located within 20° of the equator.

In all three zones aridity and altitude modify the climate considerably. The 15 °C isotherm for the coldest month forms the geographical limit of citrus cultivation; this more or less coincides with latitudes 35 ° to 40 ° North and South. Near the equator, citrus can be grown from sea level to an altitude of about 2,000 m above the sea. At that altitude the average temperature is 10 °C lower than at sea level, about 17 °C. However, the tropical mountain climate, equable as it is, differs greatly from the subtropical climate with its seasonal fluctuations. Thus, average yearly temperature gives us insufficient information about the climatic requirements of a crop.

The optimal temperature for growth of citrus is between 25 °C and 30 °C. Above and below this range growth diminishes and stops completely at the *maximum* temperature of 38 °C and at the *minimum* temperature of 13 °C. Yet, citrus trees can withstand more extreme temperatures fairly well; that is to say up to about 50 °C and down to the freezing point or slightly below it. High temperatures, certainly at night, are harmful for two reasons:
1. Respiration and transpiration continue at a high rate, while photosynthesis decreases sharply.

2. Pigmentation of fruit does not take place; on the contrary, coloured fruit may regreen.

Fruit exposed to very high temperatures will have a scorched appearance. A remarkable effect of high temperature has been described by Samadi and Cochran (1975) in the south of Iran where bloom and fruit set occurred only in the centre of the tree, under the protection of the leaf canopy.

It is generally assumed that crops can only be produced if a certain quantity of energy (heat) is available. That quantity has been calculated for citrus by Webber (1946); he deducted the minimum temperature for growth (13°C) from the average daily temperature and added up those differences over the year, or rather the growing season. This 'heat index' is shown in Table 5.7 (figures adapted to centigrade):

Table 5.7   *18-year average of heat indices available for citrus in six locations of California*

Place	Section	March–Nov.	Dorm. per.	Best crop
Oakland	Coastal North	659	−290	All poorly
Santa Ana	Coastal South	1,516	−50	Lemon
Oroville	Interior North	1,749	−426	Navel
Porterville	Interior Central	1,978	−359	Navel, Valencia
Riverside	Interior South	1,783	−121	Navel, Valencia
Imperial	Desert South	3,377	+6	Grapefruit

Source: Webber (1946)

It must be pointed out that average and maximum temperatures in California increase along two gradients: from north to south and from the coast via the interior valley to the desert region. We learn from this table that lemon cvs ('Eureka' and 'Lisbon') can be grown under a low heat index (around 1,500) while grapefruit needs a high index (over 3,000); 'Navel' and 'Valencia' oranges occupy intermediate positions. Reuther (1973) points out that the concept of 'heat summation' has serious limitations. Air temperature is but one of many factors such as radiation, air velocity and relative humidity which, when integrated, determine temperatures of specific plant cells, tissues, or organs in relation to time and position.

Yet this concept is still used to some extent as an indication of whether citrus growth is a possibility. Mendel (1969) has remarked that the rate of development of nursery plants is quite closely correlated with the heat index. In cooler areas it took a seedling from 12–15 months to reach buddable size (12 mm at 10 cm above ground), whereas in warmer areas 9 months were sufficient.

Cassin *et al.* (1969) have calculated the heat index (or total effective temperatures) between mid-bloom and maturity for early and late cvs in West Africa, the West Indies and some other places. Less time was required for growth and maturation of the fruit in the tropics. The distinction between early and late cvs also tended to disappear; maturation time was reduced from eight to six months for early cvs and from eleven to seven months for late cvs. Praloran (1968) also compared the yearly heat indexes of citrus regions all over the world. A sampling of his figures gives us the following data (Table 5.8):

Table 5.8   *Heat indices for six locations*

Place	Lat.	Alt. (m)	Heat index (year)
Arivonimamu, Madag.	19°	1,460	1,817
Sao Paulo, Brazil	23°	795	2,090
Amoy, China	24°	5	3,313
St. Leo, Florida	28°	54	3,468
Dibrugarh, Assam	27°	106	3,928
Patna, Ganges valley	25°	53	5,019

Source: Praloran (1968)

Reuther and Rios-Castaño (1969) compared Colombia with California; they found the upper limit for commercial navel orange production in Colombia (3° N) at an elevation of 2,100 m. For grapefruit the limit was 1,200 m and for 'Valencia' 1,600 m.

Summing up, we find that heat index is not an important criterion for citrus growth in the tropics, except at higher altitudes. Rainfall is far more important. How much is needed and what is the best pattern of distribution? Purseglove (1968) states that an average annual rainfall of at least 35 in. (875 mm) is required if citrus is to be grown without irrigation. Mendel (1967) puts the minimum at 1,200 mm, while Garcia Benavides (1971) suggests between 1,000 and 1,400 mm.

It seems best to use potential evapotranspiration (E_p) as a yardstick. This has been estimated in California as 760 mm for the coastal and 1,200 mm for the desert climate. In Trinidad 1,425 mm was reported and in Upper Volta 1,650 mm. In the very arid citrus region of south Iran E_p climbs to 2,000 mm/year.

Thus, total rainfall per year cannot be regarded as decisive. As we saw in Chapter 2, only part of surplus rainfall can be stored: the water storage capacity of the soil seldom exceeds 100 mm. The pattern of rainfall distribution should be the main consideration. In most months the rainfall should be equal to or slightly higher than E_p, although a few dry months are beneficial as they concentrate blooming. Where rainfall is much lower than E_p during more than three months, irrigation is required.

Cassin (1958) found that the dry season in Kindia, Guinea (10° N, 400 m above the sea) had a mean duration of 128 days, from 8

December to 15 April. On 9 May citrus was generally in bloom. However, a dry period of two months was sufficient for flower induction. When this period was followed by irrigation, the trees would bloom within 20 to 28 days. Thus it was possible to advance the harvest by two months.

In Surinam, which has two wet and two dry seasons, Van der Weert *et al.* (1973) examined the relationship between citrus yield and available water in the root zone, for two localities (latitude 5° N and 6° N, sea level). They distinguished three periods:
1. A dry period to initiate flower buds.
2. A rainy season with high soil moisture during six to eight weeks for bloom and fruit set.
3. A period during which the soil may dry out to the wilting point without causing a significant reduction in yield.

Once the trees are in bloom, or young fruits hang on the tree, a regular supply of water must be assured. Irrigation is necessary if the rains temporarily recede. However, Van der Weert *et al.* found that irrigation would seldom be needed on the deep, light soils of the interior in Surinam.

Light intensity and the number of hours of sunshine are generally much lower in humid than in arid regions, yet they do not appear to limit citrus growth anywhere. Relative humidity is important: Baker (1940) based his prediction of scab attack in Trinidad on this and on temperature. These are the main factors influencing growth of fungi and insects, and the incidence of diseases and pests. Strong wind severely damages citrus crops, so windbreaks should be planted.

Citrus grows well on a wide variety of soils, from coarse sands (as in Florida) to heavy clays (as in Surinam). On coarse sand irrigation may, within minutes, leave hardly a trace on the topsoil but on heavy clays a light rain may make the soil slippery for hours. The main requirement is that no waterlogging should occur. But there are exceptions to this rule. In coastal areas of Thailand and Indonesia, for instance, air-layered mandarins and shaddocks are grown on very wet soils. The shallow root system is not harmed by the high water table. And in Surinam citrus is grown on cambered beds in polders with an intricate system of drainage and irrigation canals. When Burke (1956) saw this 'swamp citrus culture' in operation he exclaimed: 'I see them growing, but I still don't believe it.'

Good drainage is essential for sustained high yields, but do not assume that good drainage is never possible on heavy clay. Freshly impoldered heavy clays in Surinam generally have a good drainage capacity, provided the overlying organic matter (pegasse) has not been burned. Burning and management practices such as ploughing have damaged clay soils which initially had good physical properties (van Amson, 1966).

'It is generally agreed, however, that the best soil for citrus is a medium-textured soil of recent alluvial origin, uniform, reasonably deep and fertile, having good internal drainage, and free from injurious salts' (Platt, 1973). This was written in and for California, but it is good advice for the tropics too. It should be remembered that citrus can be grown in conditions which are far from ideal. This is especially true of chemical fertility, which can be bought in the fertilizer bag; physical properties, on the other hand, cannot be changed overnight and should be the first consideration.

The best pH for citrus soils ranges from 5 to 6. In soil which is too acid, citrus roots do not grow well and nutrients are leached out, or may even become toxic, e.g. copper. At a pH above 6, on the other hand, fixation of nutrients (especially zinc and iron) will take place and trees develop deficiency symptoms. A low pH can be corrected by adding lime or dolomite to the soil. Dolomite works more slowly than lime, but has the advantage of containing magnesium too. However, liming may not be practicable on heavy soils. In Surinam, for instance, a single application of 2,000 kg/ha lime on sandy soil raised the pH from 4.3 to 5.9 within months, after which it gradually sank to 5.0. On clay soil yearly applications had no effect until the third year (Table 5.9).

Table 5.9 *Effect of lime on pH of clay soil in Surinam*

	1958	*1959*	*1960*
Check	4.2	4.1	4.2
10 kg CaO per tree	4.2	4.2	5.5
20 kg CaO per tree	4.2	4.3	6.2

Source: van Suchtelen (1961)

If rainfall is insufficient, then irrigation water must be applied. It should be of good quality, which means that the salt content must be low. It should contain little or no boron, anything over 0.5 ppm is considered dangerous. More than 150 ppm of chloride is also undesirable.

Cultivars

There are literally thousands of citrus cultivars. Thanks to polyembryony and vegetative propagation the great majority are well-established clones. Mutations appear regularly and are usually undesirable. Budwood should therefore only be taken from carefully selected mother trees. On the other hand, one must have an eye open for those mutations that might be an improvement. In all citrus growing countries a large number of citrus cvs have been imported, sometimes as seed, usually as budwood. Only a few of these were adapted to the local environment. Useful mutations must have appeared, but they were seldom noticed and propagated.

Citrus cultivars are classified according to their ripening season into early, mid-season and late cvs or according to shape, colour, taste and other properties. Ripening season is relative, it depends on the locality. Terms like early and late relate to the plant's requirement for a low or high heat index. Thus a cv. which belongs to the early group in a warm country may shift to the middle group in a cooler environment.

This shift may also happen with the other criterion. For instance the 'Kwata' orange, a member of the common group in Surinam, proved to be a blood orange in California (Bitters, personal communication); under the hot conditions of Surinam the 'blood' colour could not develop.

Some cvs are 'stored on the tree', which means that they are left hanging, although fully mature, and may therefore be classified in a later group. Of course, this practice is only possible where the crop is not attacked by fruit-piercing moths, fruit flies and such. Furthermore, this on-tree storage depends on the reaction of the cv. to the climate. Thus 'Washington Navel' fruit holds well on the tree in California, but rapidly senesces in Florida and in Surinam. In a humid climate oranges generally have smoother and thinner skins and a higher juice content than under arid conditions.

The description of a cv. is almost wholly concerned with its fruit. In order to distinguish one cv. from another we must know something about the size, weight and shape of the fruit. A useful indication is the D/H index (diameter divided by height), although its range may be quite large. An outline for fruit description is given by Hodgson (1967), and by Cassin and Lossois (1978).

The reader should note that the brief descriptions of cvs given here have mostly been taken from Hodgson (1967) who worked in California and from Camp et al. (1951) and Hume (1957) who had a Florida background. Their accounts cannot be applied to other places without modification. But, we shall have to be content with them, since good cv. descriptions from the tropics are not available. Furthermore we must not overlook the influence of rootstock and soil type on fruit properties such as acidity, sugar content and so on. We should also remember that some apparent differences in form or behaviour may be ascribed to virus diseases.

Sweet orange

Three groups of cvs are recognized by Hodgson (1967); the common, blood and navel oranges (a fourth group, acidless oranges, is disregarded here). Hume (1957), on the other hand, speaks of

Spanish, Mediterranean, Blood and Navel oranges; we shall follow Hodgson. In blood oranges the flesh, juice and rind show pink or red colour, caused by anthocyanin; the colour does not develop in the tropics. A navel is a small secondary fruit, pushed towards the top of the primary fruit. All other cvs belong to the common group. Alternative names are 'beladi' (North Africa, Near East), 'caipira' (Brazil), 'criolla' (rest of South America) and 'corriente' (Mexico). In France, Italy and Spain the common group is known as 'blonde', 'biondo' and 'blanca' respectively.

The most important common cv. by far is 'Valencia'. It is very adaptable in spite of its high heat requirement. In the tropics it matures at sea level in 7–9 months; this takes progressively longer with altitudes up to 1,600 m. In the subtropics 'Valencia' may ripen in a year, as in Florida, or it may take longer as in California. In California, in fact, two crops hang on the tree simultaneously and 'Valencia' thus becomes a summer-ripening fruit. The fruit is medium-large, oblong to spherical, with few seeds, or none, with abundant juice and a good flavour, but slightly acid. It is the latest maturing commercial cv. 'Lue Gim Gong' and 'Pope' are practically indistinguishable from 'Valencia'. Furthermore, nucellar budlines of 'Valencia' exist: 'Cutter', 'Frost' and 'Olinda'. There are also nucellar lines of several of the cvs described below.

Next in importance is 'Shamouti', but this cv. has a narrow range of adaptation to climate and soil and only grows well in the coastal region of Israel. It is therefore highly unlikely that 'Shamouti' can grow successfully anywhere in the tropics. This cv. is sold under the commercial name *Jaffa* but 'Jaffa' is the name of a mid-season cv. quite distinct from 'Shamouti'; they should not be confused. 'Hamlin' is the principal early cv. of Florida and thrives in Brazil and many other locations; it requires high temperature and humidity. Its main disadvantage is its rather small fruit. However, its earliness, high productivity and few seeds are in its favour. 'Mosambi' is the main orange cv. of India. It is moderately seedy, early maturing and has low acidity, which makes it unsuitable for export and processing. 'Parson Brown' is very early and moderately seedy. Although its popularity is declining in Florida, it may still be considered a good cv. for tropical countries.

'Pera' is the main cv. of Brazil, it has very few seeds, is late maturing, holds well on the tree and is very productive. It is probably identical to 'Lamb Summer', a cv. that went out of favour in Florida but is still seen in tropical countries. 'Pineapple' is moderately seedy, highly productive and excellent for processing. It does well in the humid tropics. 'Sanford's Mediterranean', a mid-season cv. with few seeds, was once popular in Florida. 'St. Michael' (also called 'Paperrind') is small and moderately seedy. These cvs are still found in tropical areas, e.g. Ghana and Sierra Leone.

Navel oranges are seedless 'resulting from the fact that functional pollen is lacking and viable ovules are rare. Other distinctive characteristics of most navel oranges include a crispness of flesh texture, ease of peeling and separation of the segments, and richness of flavour. They are less vigorous than other oranges and poorly adapted to the humid tropics, semi-tropics, or intense desert heat. They are used primarily in the fresh form. Juice develops bitterness in storing and is not fit for processing.' (Hodgson, 1967). However, the bitter principle (limonin) can now be changed into a non-bitter substance by means of a bacterial enzyme (Anon. 1981). The only cv. we need to consider is 'Washington', also known as 'Bahia' (it stems from Brazil). It has large fruit, spherical to ellipsoid, sometimes with a protruding navel, is seedless, has a thick rind and is early maturing. In the tropics it can only be grown at an elevation of 1,000 to 2,000 m.

Among the blood oranges 'Ruby' can be found here and there in the tropics, but its fruit does not differ in colour from that of the common oranges; it is a mid-season cv. with few seeds.

In several tropical countries local selections have been made. 'Kwata' may serve as an example. In 1920 fruits of seedling trees were sent from Surinam to the Royal Institute for the Tropics in Amsterdam. 'Kwatta 202' was judged to be the best, while 'Kwatta 71' and some others were also received favourably. The first named was chosen for propagation on a large scale. It was described thus: 'it is thin-skinned, not completely seedless, rich in juice, with a high sugar and a rather low acid content and has thick membranes between the segments' (Oppenheim, 1939). When Samson and Byron (1958) made a survey of 10,000 'Kwatta 202' trees in an experimental orchard, they found many different forms: some with few and some with many seeds, flattened or spherical fruits, thick or thin peel, hard or fairly soft membranes (Table 5.10). Seven mother trees were selected and one (no. 7) was later designated as 'Kwata' (with one t, as local spelling rules had changed). Kwata's

Table 5.10   *Properties of seven mother trees Kwatta 202*

Tree	Av. D/H	Skin		Av. seeds	Membranes	Juice colour
		Texture	Thickness			
1	1.01	Rough	4 mm	6	Tough	Orange
2	1.03	Medium	4	5	Medium	Orange
3	1.03	Rough	5	4	Medium	Orange
4	1.01	Medium	5	3	Soft	Yellow
5	1.03	Rough	4.5	6	Tough	Yellow
6	1.02	Rough	6	3	Tough	Orange
7	1.02	Smooth	4	4	Soft	Orange

Source: Samson and Byron (1958)

Table 5.11 *Properties of sweet orange cvs in Surinam (in sequence of maturity)*

Cultivar	Group	Season	Form	D/H	Seeds	Remarks
Washington Navel	Navel	Early	Elliptical	0.96	5	Big fruit
Kwatta 202	Norm.	Early	Flat	1.09	10	
Parson Brown	Norm.	Early	Spherical	1.00	14	
Pineapple	Norm.	Mid	Spherical	1.01	26	Smooth skin
Ruby	Blood	Mid	Spherical	1.02	13	
Jaffa	Norm.	Mid	Elliptical	0.96	4	
Sanford	Norm.	Mid	Spherical	0.98	16	
Kwatta 71	Norm.	Late	Spherical	1.03	10	
Lamb Summer	Norm.	Late	Elliptical	0.88	7	Small fruit
Valencia	Norm.	Late	Elliptical	0.93	6	
Lue Gim Gong	Norm.	Late	Spherical	0.99	7	

Source: Samson (1966)

main bloom is usually in December and January, sometimes in May. Ripening is gradual, so that harvest takes place from June to October; the fruit holds well on the tree. Some years earlier 'Kwatta 202' had been compared to other cvs present in Surinam (see Table 5.11). The 'Kwata' cv. has been exported over a period of about 40 years to Holland in quantities seldom exceeding 100,000 boxes a year. In spite of its green, or at best yellow colour, it has enjoyed a modest degree of popularity because it is very juicy.

In contrast to 'Kwata' the Naranja Criolla of Peru is not a cultivar, but a mixture of seedlings, or buddings from seedlings. The fruit contains many seeds and is very juicy (Wolfe *et al.*, 1969). Evidently the need for selection was less urgent in mountainous Peru than in lowland Surinam, where not one of the imported cvs was really satisfactory. There must be many countries where this need for selection arises, but so far Surinam seems to be one of the few that has done anything about it.

Ghana has some named local selections, but as far as I know they have not been propagated to a large extent. Table 5.12 gives some details (Opoku, 1971). The best selections of Nigeria are 'Agege' and 'Umudike' (Adigun, 1978).

Table 5.12 *Yields in fruits/tree of four Ghana selections (7 years old) of sweet orange on three rootstocks*

Cultivar	Rootstock		
	Rough lemon	Rangpur lime	Walters gf.
Asuansi	661	381	112
Anomabo	630	317	80
Jumapo	190	170	—
Shama	622	55	12

Source: Opoku (1971)

Mandarins

The edible mandarins can be divided into four groups: Satsuma, King, Willowleaf (Mediterranean) and Common. According to Hodgson (1967) these are botanical spp., while Swingle puts them in one sp. There certainly are several hybrids among them. They generally have a short on-tree life so that the mandarin season – around Christmas – does not last long. Almost all mandarins are eaten fresh, there is little processing.

The fruit is mostly oblate (flat), with a D/H index around 1.40. Cotyledons are greenish, with some exceptions. Alternate bearing frequently occurs in all four groups. The critical temperature for colour lies higher than in oranges, perhaps near 18 °C; 'King' develops some colour even in the humid tropics.

The Satsuma group is derived from Japan and is the most cold-tolerant. However, it is possible to grow them successfully in the tropics (Rey, 1982). The fruit is seedless, of medium size and often has a navel; the tree has a spreading habit. Well known cultivars are 'Owari' and 'Wase'.

The King group originated in Indo-China and is well suited to tropical climates. The main cv. 'King' (of Siam) is a tangor (a hybrid of mandarin and orange) with large fruits and many seeds with white cotyledons. The fruit keeps well on the tree and withstands storing. As regards taste, it is one of the best citrus fruits. The tree has an open growth habit and a high heat requirement; it produces much better quality fruit on sour orange stock than on rough lemon.

The Willowleaf group is not important for the tropics, but the Common group is all the more so. Its most prominent member is 'Clementine', a small seedless tangerine of outstanding quality. It has a low heat requirement and grows best in the coastal region of West Morocco. As it is self-incompatible, it needs a pollinizer for satisfactory bearing. This cv. should be tried in tropical mountain areas with moderate humidity. The 'Monreal Clementine' is self-compatible and averages 12 seeds per fruit.

'Dancy' is the most important mandarin cv. of the United States. It grows well in Florida and needs humidity. It matures in mid-season, is of excellent quality, productive, with a tendency to alternate bearing. This cv. seems suitable for the tropics and should be tried there.

'Ponkan', the foremost tropical mandarin, is also known as 'Nagpur' santra (santra being the Indian name for mandarin). It has large fruit, sometimes with a navel. There are few seeds, but these are polyembryonic, so that this cv. can be grown as unbudded seedling trees, e.g. in the Coorg district of Mysore State, India. It matures in mid-season, is highly productive, but strongly alternate in bearing.

'Ortanique' originated on Jamaica. Its name is a composite of *Or*ange-*tan*gerine-un*ique*; it is a tangor. The fruit is large, sometimes with a navel, has about ten seeds (with white cotyledons), matures in mid-season and holds well on the tree; it has a rich and distinct flavour. As it is self-incompatible it has to be interplanted with a pollinizer, e.g. 'Valencia' (Lange, 1974). Other well known common cvs are: Beauty, Ellendale, Kara, Kinnow, Lee and Wilking.

Ochse (1931) mentions three types of mandarin on Java: djeroek djepoen (which is the 'King'), a djeroek keprok with tight, and another with loose skin. It is strange that Molesworth Allen (1967) does not mention any mandarins for Malaya at all. However, Santiago (1962) has named seven cvs under the Malayan name limau kupas. In Surinam the 'Curacao orange' is a tight-skinned mandarin, whereas 'Pompon' and 'Surino' are loose-skinned types.

Tangors are hybrids of mandarin and orange. The best known are 'Temple' and 'Umatilla'. Among the tangelos (mandarin × grapefruit hybrids) 'Minneola', 'Orlando' and 'Ugli' (from Jamaica) have to be mentioned. Hybrids (of 'Clementine' × tangelo) released by the US Department of Agriculture bear the names Fairchild, Lee, Nova, Osceola, Page and Robinson. They have not been tried in the tropics yet.

The small-fruited mandarins are a fifth group. 'Cleopatra' is a good rootstock and the tree has ornamental value, but the fruit is barely edible. 'Calamondin', a hybrid of kumquat, is popular in the Philippines and is frequently grown as a house plant.

Grapefruit

Two groups of grapefruit are known: the white and the pigmented cultivars. In both groups seedy and seedless cvs occur. In general seedy cvs mature earlier than seedless ones; this takes about eight months in favourable conditions, but over a year if there is insufficient heat.

The seedy cvs, such as 'Duncan' still occur in tropical countries, but have gone out of fashion in Florida, where they were almost totally replaced by 'Marsh'. This latter cv. has medium-sized fruit, flattened to spherical in shape, with few or no seeds. It is very juicy, has a good flavour, holds unusually well on the tree and ships and stores well. It is very productive and has a high heat requirement. It grows well in Florida and in the desert region of California, but also in Trinidad and Surinam. Because of its high quality, high productivity and good colour it is one of the best citrus cvs in the humid tropics. A disadvantage is that the local population has not yet learned to appreciate this breakfast fruit. In Surinam this was overcome by processing; the juice proved to be very popular.

Fig. 5.2   Branch of the 'Marsh' grapefruit

A pink-fleshed mutation of 'Marsh' was propagated under the name 'Thompson'. Soon afterwards a sport of 'Thompson' was found with much deeper colour of flesh and rind and this 'Redblush' is now a very popular grapefruit cv. in Florida and Texas. Unlike 'Marsh' it is unsuitable for processing because the pigment is not retained. It matures earlier than 'Marsh'.

'Chironja' is a natural hybrid between *Chi*na (orange) and to*ronja* (grapefruit), found in a mountainous region of Puerto Rico. It has large, pear-shaped fruits with few seeds (polyembryonic), peels easily, is very juicy, not bitter, matures in mid-season and holds well on the tree. It deserves a trial in other tropical areas.

Lemon

The lemon is very sensitive to heat and cold and thrives only in coastal locations with a mediterranean-type climate. In Surinam a round-fruited lemon cv., obtained from Trinidad in 1934 under the name 'Woglum', produced good crops for some years. However, the fruit was so unlike subtropical lemons such as 'Eureka' and 'Lisbon' that it was hard to sell and the trees were rebudded. 'Eureka' and 'Lisbon' can be grown in the tropics for processing; however they could not compete as fresh fruit against subtropical lemons.

Lime

In the tropics the lemon is replaced by the lime. There are two natural groups: acid (sour) limes, and acidless (sweet) limes. The 'West Indian' lime, also called Mexican and Key lime is round, small-fruited, moderately seedy and highly polyembryonic; it has a thin, smooth rind, greenish flesh and a citric acid content ranging from 7 to 8 per cent. It is usually grown as a seedling, as no satisfactory rootstock is known, but in Ghana it is grown on Rough lemon stock. Fennah (1940) found wild grapefruit to be the only serviceable rootstock for lime on three Caribbean islands. The tree is highly susceptible to anthracnose (wither tip) and tristeza, but resistant to scab. 'Tahiti' or Persian lime is seedless, large-fruited and well adapted to the climate of South Florida. The sweet limes are only used as rootstocks.

Citron

Citron is very sensitive to heat as well as frost and its culture is limited to islands like Corsica, Corfu and Crete. It cannot be grown successfully in tropical lowlands, but is now cultivated in Puerto Rico.

Sour orange

This is mainly used as a rootstock; as such it is highly susceptible to tristeza. However, as a seedling it does not suffer from tristeza because of hypersensitivity (see Ch. 4). The fruit is used for marmalade, drinks and essential oil. Neroli, from which the most expensive perfumes are prepared, is made of the flowers of certain sour orange cvs. The bergamot is regarded as a subspecies by Swingle and Reece (1967); its fruit yields oil too. It is chiefly grown in Calabria, South Italy, but the culture was developing successfully in Ivory Coast and Guinea (Praloran, 1971).

Fig. 5.3 Branch of the 'Eureka' lemon; note the wingless petiole

Shaddock

Shaddock (or pummelo) is a very varied group, having seedy and seedless, common and pigmented cvs. Shaddock is mono-embryonic and must be propagated vegetatively or inferior seedlings will result. The shaddocks of South-East Asia are widely grown in reclaimed coastal marsh areas as air-layered plants. The best known cvs are 'Kao Pan' and 'Kao Phuang', both white, from Thailand and the pigmented 'Pandan Bener' and 'Pandan Wangi' of Java. The tree is highly tolerant to salt and very susceptible to citrus canker (Nat. Acad., 1978).

Rootstocks

Thanks to polyembryony most citrus cvs can be propagated from seed and this method was followed for many centuries, even though grafting was already known to the Chinese around 1000 BC (Hartmann and Kester, 1975). Seedling trees are still found in every tropical country, but new plantings of citrus, except on a small scale, generally use budded plants. This has several advantages, but also one great disadvantage (see Table 5.13).

Table 5.13 *Comparison of budded and seedling citrus trees*

	Budlings	*Seedlings*
Trueness to type	Great	Moderate
Thorns	Few, small	Many, large
Bearing	From 3rd year	From 7th year
Habit	Spreading	Upright
Picking cost	Low	High
Reaction to foot rot	Resistant on suitable stock	Susceptible in most spp.
Virus diseases	Dangerous	Initially none

Cuttings of lemon and citron are easily made, but they suffer from foot rot. Shaddock and mandarin are air-layered for shallow, heavy soils with a high water table. Grafting is possible but expensive and budding is the general method of vegetative propagation in citrus.

Every citrus cv. can be budded on practically any member of *Citrus* and related genera; however, some rootstocks are greatly preferred. What are the most important requirements for a citrus rootstock? Next to good seed production, one expects:
1. high degree of polyembryony;
2. good union with the main cvs;
3. ability to grow on various soils;
4. tolerance to virus diseases;
5. tolerance to fungus diseases;
6. tolerance to nematodes;
7. good nursery performance;
8. tolerance to drought and wind.
The rootstock should also induce good yields of high quality fruit in the graft. No single rootstock is satisfactory in all respects. 'Finding the best possible rootstock for each fruit variety in every

location is so difficult as almost to defy accomplishment' (Batchelor and Rounds, 1948). Thirty years later this is still true.

Two species, *C. grandis* and *C. medica*, can be left out of the discussion as they are monoembryonic. The same may be said for 'Clementine' and 'Temple' mandarins and for 'Meyer' and 'Ponderosa' lemons. And, of course, seedless cvs, or those producing few seeds, are also unsuitable as rootstocks.

In a good union, stock and scion must have more or less the same growth rate. But, *Poncirus* and grapefruit stocks grow faster than the sweet orange scion; sour orange stock, on the other hand, grows somewhat more slowly than sweet orange. A difference in growth rate has a dwarfing effect which may lead to close planting, hence higher production and lower picking cost.

Many instances are known of the influence of stock on scion and we shall encounter several in this section. Far rarer is an example of the influence of scion on stock: Klotz (1967) has shown that certain stocks were, if budded with lemon, more resistant to foot rot than the same stocks when budded with sweet orange.

Let us now take a look at *Citrus* and relatives in order to discover how far they meet these requirements.

Sour orange

This has been a popular rootstock in southern Europe and Florida for centuries. At the beginning of this century it was used almost universally, except in South Africa where it suffered from a destructive disease (later to be called tristeza). It is still widely used around the Mediterranean. Sour orange grows well on heavy soils and can withstand flooding better than any other rootstock; on the other hand it is fairly drought-resistant, thanks to its deep and spreading root system with several tap roots. Fruit from trees budded on this stock is juicy, thin-skinned and of excellent quality, but productivity is not high.

Sour orange is resistant to foot rot and to several virus diseases (exocortis, cachexia, psorosis), but subject to tristeza virus. In regions where this virus is present together with the aphid *Toxoptera citricidus*, this rootstock becomes untenable. It is also susceptible to scab, a fungus disease, which greatly hinders growth. In the nursery plants may differ in leaf shape, width of the petiole wing, etc., but a sowing is fairly uniform, as a rule. Sour orange seedlings in Surinam grow perceptibly more slowly than those of 'Rough' lemon or 'Rangpur' mandarin, but definitely faster than those of 'Cleopatra' mandarin.

Apart from the ordinary sour orange, the fruit of which contains about 3 per cent citric acid, there is also a *bittersweet* form with fruit containing only 1.5 per cent citric acid. The Nanshôdaidai (*C.*

Fig. 5.4 Plants of four citrus rootstocks, four months old. From left to right: 'Rangpur' mandarin, *Poncirus trifoliata*, 'Rough' lemon, sour orange

taiwanica) is probably a hybrid of sour orange; it may have some merit as a rootstock.

Sweet orange

This would be an almost ideal rootstock for sweet orange buds, but for foot rot. Most sweet orange seedlings are susceptible to the disease, rendering them unsuitable as rootstock; however some are fairly resistant, e.g. 'Indian River' (Fawcett, 1936). Sweet orange is tolerant to tristeza and exocortis but not to psorosis and cachexia. It grows best on loamy soils with good drainage. Fruits produced on this stock are of excellent quality. The best rootstock cvs are probably 'Bessie', 'Caipira' (from Brazil), 'Koethen' and 'Madam Vinous' (an Indian River seedling). From 70 to 90 per cent of the plants in a seedbed are nucellar.

Mandarin

The mandarins usually exhibit a high degree of nucellar embryony, from 80 to 100 per cent. Most cvs are resistant to scab and tolerant to tristeza. I have noted in Surinam that they are more prone to manganese deficiency than other rootstocks. The best known mandarin rootstock is 'Cleopatra'. It forms a round-topped tree, is practically thornless and has small fruit with many polyembryonic seeds. It grows well on clay soils, but is not resistant to foot rot. Trees on this stock remain small and can therefore be planted closer than on other stocks. They start to bear fruit later than others but usually make up for it after about ten years.

'Sunki' is widely used as a rootstock in China and Taiwan. In Surinam it grew much faster in the nursery than 'Cleopatra'. 'King' and 'Surino' were good rootstocks for 'Kwata 71' on sandy soil in Surinam (Samson, 1951).

'Rangpur' mandarin is probably a hybrid of lemon and mandarin and certainly not a lime. It is called 'Rangpur' lime in the United States, 'Cravo' lemon in Brazil and Japansche citroen in Dutch. It is necessary to remark here that the Dutch word *'citroen'* means lemon and not citron as some authors think. Although it does resemble the 'Rough' lemon in some respects, it seems better to call it a mandarin. This most drought-resistant of all rootstocks stimulates early fruiting and gives good yields of good-quality fruit. Its sensitivity to exocortis, however, is a great handicap. This stock can only be used when exocortis-free budwood is available and even then great care has to be taken to prevent infection.

'Volkameriana' is also a lemon × mandarin hybrid, it is one of the most promising rootstocks. Two small-fruited mandarins (*C. amblycarpa* and *C. depressa*) seem to have some promise as rootstocks, but little is known of their performance in the tropics.

Lemon

Lemons generally make few nucellar seedlings, e.g. 'Eureka' 34 per cent, 'Lisbon' 20 per cent. They are susceptible to foot rot and shell bark, and intolerant to tristeza. For all these reasons they are unsatisfactory rootstocks. An exception has to be made for the rough lemon, which is placed in a different species (*C. jambhiri*) by Tanaka, Hodgson and many others. It has medium-sized rough fruits, with a hollow axis and many small highly polyembryonic seeds. From 90–100 per cent of the seedlings are nucellar, they grow very rapidly, straight up and are easy to handle in the nursery. The root system is very extensive, usually without a tap root. This rootstock is tolerant to tristeza, but subject to foot rot. Fruit grown on rough lemon stock is often coarse and has a rather low sugar and acid content; it tends to dry out and exhibit granulation when held on the tree too long. Nevertheless, this is the most important rootstock for light soils in Florida, India, Brazil and other countries.

Since the burrowing nematode has started its devastating attacks in Florida, it is being replaced by 'Milam', a nematode-resistant rough lemon selection; 'Schaub' rough lemon is tolerant to *Phytophthora* (Carpenter *et al.*, 1982) and to greening (Cheema *et al.* 1982). Rough lemon is generally regarded as a good rootstock for limes.

Grapefruit

This has never been used much as a rootstock, although it is highly polyembryonic, has a good root system and produces fruit of good quality. The reason is that it is shy bearing, unless the trees receive a lot of fertilizer. In the West Indies 'Wild grapefruit' is sometimes used as a stock, especially for lime. In Surinam a red-tinged, polyembryonic cv. named 'Alamun', probably belonging to *C. paradisi*, showed promise as a rootstock, but was not widely used. The names 'Alamun' and 'Alemow' should not be confused.

Lime

Acid limes have hardly figured as rootstocks, no doubt because of their great susceptibility to cold injury and tristeza. Sweet lime once was a popular stock on light soils in Israel, but has not been used much in the tropics. Its good points (early bearing, heavy fruiting, excellent quality) are probably overshadowed by the bad ones (susceptibility to foot rot, tristeza and cachexia).

Citron

Cuttings of citron were once used extensively as a rootstock in India and Egypt but there were problems such as overgrowth of the scion, dwarfing, short life, many diseases and inferior fruit; their use has been discontinued.

Poncirus

Poncirus trifoliata is deciduous, very cold-resistant and a good stock for marginal citrus regions. The seedlings grow slowly and are very thorny, but are for the greater part (70% or more) nucellar. Seed from plants with large flowers is preferred. This stock is very resistant to foot rot and tolerant to tristeza, but susceptible to canker, scab and exocortis. The last named disease causes serious dwarfing, but even with virus-free buds the trees remain small. Fruit produced on this stock is rather small, but of excellent quality. Nevertheless, this stock cannot be recommended for the tropics.

Citranges

These hybrids of sweet orange and *Poncirus* have trifoliate leaves, the central leaflet being much bigger than in *Poncirus*. They are evergreens with highly polyembryonic seeds, the seedlings coming true to type. They are resistant to foot rot, tolerant to tristeza and subject to exocortis, but less so than *Poncirus*. Cvs Carrizo and Troyer look very much alike, although Bitters (1968) reports that their field performances differ. They originated from the same zygotic seedling, but a mutation may have caused genetic differences. Both are incompatible with 'Eureka' lemon. 'Carrizo' gave slightly higher yields in solids per tonne of fruit and was somewhat more resistant to *Phytophthora* and *Radopholus* than 'Troyer' (McCarty *et al.* 1974). 'Troyer' is now the main rootstock of California, while in Texas and Florida 'Carrizo' has become an important rootstock. Another citrange, 'Rusk', has performed well in Florida but has few seeds.

'Swingle' citrumelo, a grapefruit × *Poncirus* hybrid, is a very promising rootstock in Florida and Texas. It is highly polyembryonic, resistant to footrot, citrus nematode and drought, and tolerant to tristeza. It is not suitable for heavy or lime-rich soils. Fruit grown on this stock is of good quality (Wutscher, 1981).

'Alemow', a hybrid between *Papeda* and *Eucitrus* (also called *C. macrophylla*) is one of the best rootstocks for lemon in California. In an experiment by Bitters (1968) it produced the highest of 25 rootstocks for 'Valencia', but the fruit quality was not good. It is resistant to *Phytophthora* and tolerant to exocortis. Other Papeda rootstocks of promise are *C. macroptera* and 'Yuzu', a hybrid of *C. ichangensis*.

Among the more remote *Citrus* relatives *Microcitrus*, *Citropsis*, *Swinglea* and *Severinia* have given interesting results (Bitters, 1969), but nothing is known of their performance in the tropics. In Table 5.14 properties of the main rootstocks are set out. Only tristeza and exocortis out of 30 virus diseases appear in this table as one may start with virus-free budwood, but aphids will soon spread tristeza again; it is also difficult to prevent the spread of exocortis on knives, shears, etc. This rapid spread, by other means than budding, does not apply to most other virus diseases.

We draw the following conclusions from this table:

Table 5.14   *An evaluation of properties of the main rootstocks*

Rootstock	Foot-rot	Tris-teza	Exo-cortis	Nema-todes	Drought & salt	Produc-tion	Quality
Sour orange	4	0	4	2	2	2	4
Sweet orange	1	4	3	2	2	3	4
Rough lemon	0	3	3	1	2	4	1
Milam lemon	0	4	3	3	2	3	2
Cleopatra mand.	2	4	3	2	3	2	3
Rangpur mand.	2	4	0	2	4	4	3
Sweet lime	3	2	1	1	2	2	2
Troyer citrange	3	4	1	3	1	3	4
P. trifoliata	4	4	0	4	0	2	4

0 = very bad, 1 = bad, 2 = moderate, 3 = good, 4 = very good

1. Where foot rot is prevalent, lemons and sweet orange should be avoided.
2. If a severe strain of tristeza is present, sour orange cannot be used.
3. With 'Rangpur', *Poncirus*, sweet lime and the citranges, exocortis-free budwood must be used and all tools must be disinfected regularly.
4. Where quality carries more weight than quantity, lemon should not be used as rootstock.

'Cleopatra' comes late into bearing. No rootstock is resistant to greening, but *C. grandis* and some mandarins are fairly tolerant. For an other evaluation of rootstocks, see Hearn *et al.* (1974).

The observations made here are the outcome of practical experience and rootstock experiments in many places. But these results are often unreliable, as the budwood was generally suffering from virus diseases. Thus, one could not be sure whether a certain effect was due to a rootstock, or a virus. This is particularly true for all experiments started before 1946, the year in which the virus nature of tristeza was proved. Other objections are that: the stocks were not always nucellar, the buds might, or might not have come from a certified mother tree, and the ecological conditions were sometimes not clearly stated. In many cases, therefore, these experiments have only local significance.

How to start a rootstock trial? Briefly, one collects seed of the most likely rootstocks. The seedlings are checked for nucellar embryony, and aberrant types are removed. Some seedlings are kept to provide seed in the future, the others are grafted with buds originating from one (preferably certified) tree of each cv. Everything must be clearly labelled. Seed of the cvs is also sown so that 'new lines' can be made, if necessary. Budwood must be indexed for the main virus diseases. On uniform soil a 'plot' of three to four trees is generally large enough and each treatment (i.e. a combination of rootstock and cv.) should be repeated several times. The 'design' of the experiment, in other words how treatments and replications are to be distributed over the plots, is outside the scope of this book; we refer the interested reader to Pearce (1976). He should also consult Sizaret (1974) on rapid propagation.

The most important characteristics to be considered in a rootstock experiment are, according to Batchelor and Rounds (1948): (1) yield per acre, (2) quality, grade and size of fruit and (3) productive life of the tree. Diameter of stock and scion, height and spread of tree, time of bloom and maturity and incidence of diseases and pests must also be recorded. The canopy volume (C) can be computed from Turrell's formula $C = 0.5238 \times \text{Height} \times \text{Diameter}$.

Probably the most comprehensive rootstock experiments ever made, were performed in California from 1922 onwards; a detailed summary is presented by Batchelor and Rounds (1948). The experiments were continued, mainly by Bitters, but have not yet been reported in Reuther's *Citrus Industry*. One report referred to a coastal station (Bitters, 1968), and was not very interesting from our tropical point of view.

In the Netherlands East Indies (now Indonesia) rootstock research was done by Magielse and Ochse (1930), Terra (1932), De Jong (1933) and Toxopeus (1937), but in all cases tristeza, unknown to them, decisively influenced the outcome. Sour orange was 'incompatible' and the choice was between rough lemon and 'Rangpur' (the so-called Japansche citroen); the latter is now used almost universally on Java.

The first rootstock trials in Surinam date back to 1924. 'Kwata 71', 'Kwata 202' and other cvs were budded onto 11 stocks and set out with replications. Unfortunately the Agricultural Experiment Station ran out of funds, the trials were neglected and no reports were published after 1932. In 1949 the trial on 'Kwata 71' was restored and two mandarin stocks ('King' and 'Surino') were shown to be substantially better than the others; trees on grapefruit were unproductive and those on citron had died or were in very bad condition (Samson, 1951, 1954).

In 1953 new experiments were started in Surinam; 'Kwata 202' was budded on sour orange, 'Cleopatra' and 'Rangpur' mandarins on three soil types (sandy young coastal, sandy old coastal and clay). On the first soil type 'Marsh' was included. The results were disappointing because nearly all trees on 'Rangpur' suffered from exocortis; moreover, an 'old' citrus soil had been used (young coastal) or leaves contained excessive amounts of aluminium (old coastal). The only treatment to give normal yields was sour orange stock on clay (Table 5.15 and Fig. 5.5). The trials led to the supposition (afterwards confirmed) that several virus diseases were present in Surinam: in succession exocortis, tristeza and psorosis were found, later followed by others (Childs, 1964). This made it necessary to set up a new series of rootstock trials, which was started in 1961.

Now these rootstocks were included: sour orange, 'Caipira' sweet orange, 'Dominica' wild grapefruit, rough and 'Milam' lemons,

Table 5.15   *Average yields (fruits per tree) in Surinam, 1963*

Soil type/Rootstock	Rangpur	Cleopatra	Sour
Clay soil	183	261	546
Sand, old coastal plain	70	125	86
Sand, young coastal plain	42	68	196

Source: Samson (1966)

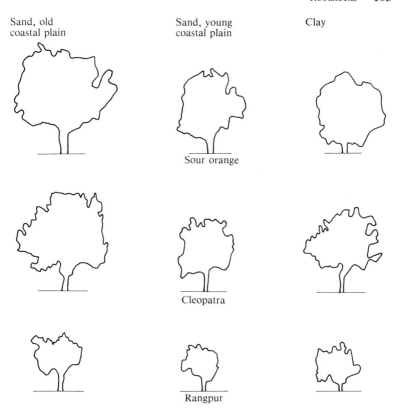

Sand, old
coastal plain

Sand, young
coastal plain

Clay

Sour orange

Cleopatra

Rangpur

Fig. 5.5 Outline of ten-year-old 'Kwata' orange trees on three rootstocks on three soil types of Surinam; trees on 'Rangpur' dwarfed due to exocortis, scale 1:100

'Cleopatra', 'Surino', 'King', 'Sunki' and 'Rangpur' mandarins, 'Troyer' citrange and *Poncirus trifoliata*. Again several soil types were chosen and the trials were partly done with virus-free budwood. Fung Kon Sang and Nanden-Amattaram (1975) report about three experiments on clay soil: 'When budded with a nucellar top, Rangpur lime induced the highest yield followed by Caipira, Sunki, Rough lemon and King. With an old line as top, Caipira, Sunki and Rough lemon induced stunted growth and much reduced production. King, Cleopatra and Surino behaved uniformly either with a nucellar or an old line top and might serve as rootstocks in case no virus-free budwood is available.' As far as I know, Surinam is the only country in which such good results with 'King' have been reported. Later Fung-Kon-Sang (1978) rated cvs King, Troyer and

Alemow as 'poorly adapted to the light textured soils of the interior, where they suffer from a sharp dry spell'.

According to Moreira and Salibe (1969) a rapid evolution towards tristeza-tolerant stocks is occurring in Peru, where tristeza was recently introduced. In the dry coastal area rough lemon and 'Cleopatra' mandarin are used but in the humid Amazonic region 'Rangpur' mandarin and 'Troyer' citrange are preferred.

The biggest organization doing rootstock research (among other things) in the tropics is the French institute IRFA (formerly IFAC). According to its Index (1970–71) there are seven of these trials in Cameroon, two in the Ivory Coast, five in Dahomey, three in Madagascar, seven in Mali and two in Niger, started 1968–70. Results are published in the monthly periodical *Fruits*, e.g. Fouqué *et al.* (1977) reported a lot of foot rot in the Ivory Coast; *C. volkameriana* and 'Troyer' were promising, but fruit quality was mediocre.

A year later the following yields (t/ha) on six- to nine-year-old trees were reported: 'Marsh' 16, 'Valencia' 15, 'Hamlin' 13, 'Pineapple' 13, 'Orlando' 9 and Common mandarin 8; all on 'Cleopatra' rootstock. The highest marks for quality were given to 'Ortanique' on Cleo stock. In Niger, with its very dry and hot climate, six-year-old 'Marsh' on Volkameriana stock produced 26 tonnes/ha grapefruit of good quality. For the Sahel region, Haury *et al.*, (1982)

Fig. 5.6   View of a citrus rootstock experiment at Azaguié, Ivory Coast

advised to use 40 per cent Volkameriana, 30 per cent Rangpur and 30 per cent sour orange rootstocks.

The 1977 Proceedings of the International Society of Citriculture (1978) reported on rootstock trials made in California, Texas, Florida, Surinam, Brazil, Corsica, Israel, South Africa and Australia. Of particular interest to tropical growers are the trials by Krezdorn on mandarins, of Hutchinson on 'Valencia' and by Pompeu on 'Piralima'. Wutscher presented an extensive review of citrus rootstock research in 1979.

If we may draw a general conclusion from all these experiments, it is that the rootstock problem is very intricate and highly dependent on climate and local circumstances. It is far from solved, especially in the tropics.

Cultivation measures

Propagation

In the majority of cases citrus is propagated by budding on seedling rootstock. Fruits from healthy trees typical for the rootstock are taken. They are cut around the 'equator' of the fruit, not deeply, in order not to damage the seeds. The fruits are squeezed out over a sieve and the seeds are washed to remove the pulp. Seeds are best sown fresh as they lose viability on drying out (however, they can be held for several weeks if kept cool, e.g. in moist charcoal at 10 °C). Disinfection by warm water at 52 °C for 10 minutes, is beneficial to the seed. The minimum temperature for germination is 15 °C and the optimal temperature is around 32 °C.

In the subtropics seeds are sown in spring. Germination in the seed bed takes about a month. Five to six months later the young plants are transplanted to the nursery bed. In the autumn of the second year the seedling rootstock is ready for budding. After union has taken place the plant becomes dormant. In the next spring the bud is forced by cutting the rootstock back. The shoot then develops and the plant can be delivered by the spring of next year, three years after sowing. A fourth year may be needed, depending on the size required.

In the tropics there is no dormant season, if irrigation is provided, and the whole process can be completed in two years or less. Seeds are sown at the start of the rainy season, transplanted three months later and budded about nine months after that. Another nine to twelve months of growth, and the plant is ready for delivery.

Seedbeds are preferably located on virgin soil, or soil that has not been used for citrus before; they are 1.80 m wide, so that the middle of the bed can be reached easily. If necessary, the soil is ploughed and harrowed to a depth of 10 cm. A path 70 cm wide is left between two beds. On each bed six rows are sown 30 cm apart with a distance of 5 cm in the row. Thus, a piece of land of

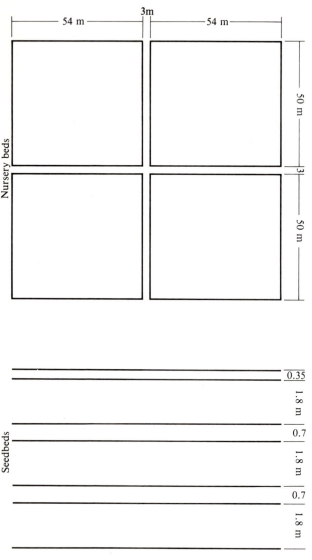

Fig. 5.7 Layout of a citrus nursery

100 m × 10 m (0.1 ha) can accommodate four beds, each with 12,000 seeds; 48,000 seeds in all. Not all will germinate and losses of 10–20 per cent will occur at each stage. After three months the plants are lifted with a digging fork. They are carefully inspected and all plants with crooked or otherwise defective roots are discarded. The rest are transplanted to nursery beds at 30 cm × 100 cm. Let us assume 36,000 plants are left; they need 36,000 × 0.3 m² = 10,800 m² of space (not counting the paths). Two paths of 3 metres width are made (see Fig. 5.7) so that total space required is 111 m × 103 m = 11,433 m².

At both stages weed control is essential. Chemical weed control is expensive and will damage the plants if not carefully done. However, where labour is scarce or expensive, it is used more and more. The recommended treatment in Florida was 4 to 6 kg/ha dichlobenil, followed by irrigation (Ryan and Davis, 1967). Vullin (1978) used trifluralin, 1.7 kg/ha a.i. at weekly intervals with no phytotoxic effects on 'Carrizo' seedlings.

Fertilization on seedbed and nursery bed is also necessary. Florida practice is to apply NPK 4-7-5, 22 kg/1,000 seedlings once a month (Platt and Opitz, 1973).

Budwood is taken from healthy, if possible virus-free, mother trees; it should be round and brown, not angular and green. The inverted T-method is commonly used, although it is not clear that this is better than the T-method. In order to prevent the spread of exocortis, knives and shears must be dipped in a 10 per cent solution of household bleach. To prevent corrosion, rinse with vinegar, water and emulsifiable oil (Klotz *et al.* 1972). All buds are inserted on the same side; this facilitates further work. In Surinam, where the wind is mostly from the east, the buds are put on the west side to protect them from the rain.

The inserted bud is wrapped so that no water can penetrate, but the eye of the bud is left free. After two weeks the plant is inspected and if the bud is still green, union has probably been effected. The tape is now partially unwrapped and the bud is forced into growth. In the subtropics this is generally done by topping, which effectively destroys apical dominance (see Ch. 3). It is better not to top in the tropics, but to *lop* the rootstock, i.e. cut halfway through it, just above the inserted bud. This breaks apical dominance selectively, on the upper side only, meanwhile ensuring continued transport of assimilates and auxin along the lower side to the roots (Samson and Bink, 1976), see Fig. 5.9. The inserted bud on a topped plant often grows sideways, while on a lopped plant it generally grows straight up. Topping causes many buds of the rootstock to grow out, these have to be rubbed off; after lopping far fewer buds of the stock do this. Bending and half-ringing have similar effects as lopping, but are more difficult to perform. In 'breaking over' a so-called snag,

Fig. 5.8*a* and 5.8*b*   Citrus nurseries in Surinam

Fig. 5.9 Bud shooting on a lopped rootstock

10 cm of rootstock, is left above the bud; the shoot can be tied to it.

A bud that does not 'take', turns black and will be cast off. In that case the wrap is wholly removed as a sign to the budder that he may use the rootstock again.

How high should the bud be inserted? High budding, at 50 cm above the ground, is favoured in places where foot rot is rampant, because the bud is generally susceptible to the disease: the higher the resistant rootstock, the better the protection it gives. On the other hand, high budding delays production while low budding induces early bearing. As a compromise, budding is generally done at 20 to 30 cm above the ground. The rootstock is considered buddable when it has reached pencil size, or a diameter of 8 mm, at the desired height. Stakes are used to support growing shoots until they harden and the danger of being blown off has passed.

A modification of the system described above was introduced in Surinam, in order to overcome the fierce competition of weeds and the handicap of waterlogged soils. Seeds are germinated for two to three weeks between wet jute sacks and sown one by one in baskets or plastic bags, when the rootlet is showing. The bags, filled with soil mixed with fertilizer, are set closely together in double rows. As a protection against showers the surface is covered with a grass mulch. Weeding and disease control become much easier this way

Fig. 5.10 Sowing pregerminated sour orange seed in Surinam

and no transplanting is necessary: budding is done *in situ*, about eight months after sowing (Samson, 1966). Thus budding has been advanced by four months. A good potting mixture is made as follows: mix equal parts of fine sand and peat, add per m^3: superphosphate 1.5 kg, dolomite 2.25 kg, lime 0.45 kg, copper sulphate 90 g, zinc sulphate 30 g, manganese sulphate 30 g, iron sulphate 50 g, boric acid 1 g and ammonium molybdate $\frac{1}{2}$ g; disinfect the mixture (Nauer *et al.* 1968). N and K are added once a week.

Nursery plants may be pruned back just before delivery, to branch out in the field. However, it is better to prune sooner and deliver plants with a fully developed framework. This generally means 30 cm of unbranched rootstock, 20 cm of unbranched scion and three to four well placed branches. Especially in mandarins these branches have a tendency to sit too close together. A good formation pruning is essential to prevent difficulties later. It is likely that the final number of plants delivered will be half the number of seeds sown. Therefore 24,000 plants may finally be expected from a nursery area of 12,000 m^2: 2 plants/m^2.

For delivery the plants are carefully lifted with a digging fork and bagged, if necessary. In wet regions transporting them bare rooted is easier and cheaper, provided the roots are kept covered with wet sacking and the plants stay out of the sun. The delay between digging in the nursery and planting in the field should be as short as possible. Bare root plants may be treated against *Phytophthora* by immersion in hot water (43.2 °C) for ten minutes (Reuther *et al.* 1978).

Planting

The field where planting out is to take place must have been cleared, weeded and drained; indeed, drainage is the first consideration. Plants suffering from wet feet will never grow well and may contract foot rot and other diseases.

The planting density usually varies from 200 to 400 trees/ha but much closer planting is sometimes seen. This induces high yields for a number of years, but when the trees become crowded they start to decline. Thinning to a normal density may assist in restoring the orchard to its former vigour. An example is provided by Samson (1966), see Fig. 5.11. In 1940 six plots of 0.5 ha had been planted in three densities: 7 × 3.5, 5.5 × 4 and 4 × 4 m. In 1953 the number of trees was halved in three plots. Their production did not decrease; on the contrary it rose not inconsiderably above that of the plots retaining their original density.

A system of close planting for the early years, followed by thinning out, has certain advantages; however, growers usually wait too

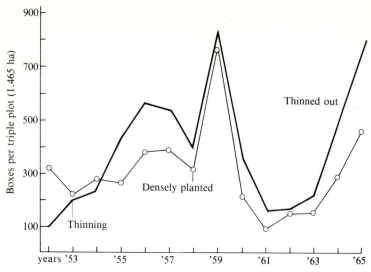

Fig. 5.11   Comparison of fruit production during 13 years in densely planted versus thinned-out plots

long to eliminate the unwanted trees. Maximum yield was achieved in Florida at densities of 225–250 trees/ha up to the 24th year; optimal density gradually decreased to 150/ha for trees 40 years old. In the tropics citrus trees seldom reach that age. Passos and Boswell reviewed this problem in 1979. Hutton and Cullis (1982) report a yield of 135 tonnes/ha from dwarf trees spaced at 1,250/ha; at a spacing of 3,000 trees/ha the yield was 260 tonnes/ha but the management of this spacing was extremely difficult.

The planting system can be square, rectangular or triangular; it matters little as the trees quickly fill the available room. A spacing of 5 × 6 m is frequently used for orange and mandarin, and 7 × 7 m for grapefruit. However, 5 × 7 m and 6 × 8 m leave more space between the rows. Hedgerows, maintained by mechanical pruning and topping, are now common practice in subtropical citrus-growing areas.

The field is lined out and stakes are placed to locate the planting holes. The size of the holes is 30 × 30 × 30 cm if the drainage is good, otherwise it should be 50 × 50 × 50 cm. Top soil is kept apart and later returned to the top. In the bottom of the hole 1 kg of rock phosphate or basic slag is applied.

Citrus trees should be planted on a mound, 10 cm high. After the soil has settled they must be a little higher than they were in the nursery; otherwise foot rot might develop. The best time for planting is the beginning of a rainy season. The soil around the plant must be thoroughly wetted and irrigation should be used if some dry days follow. As a rule of thumb: 10 litres of water/plant is needed, once a week. The direction of the rows (assuming rectangular spacing) is also important. In the subtropics a north–south direction ensures maximal sunlight. In the tropics rows are set perpendicular to the prevailing wind, which facilitates spraying, but on slopes they follow the contours.

Soil management

This is a rather difficult matter in the tropics, because heavy showers cause run-off and leaching. The subtropical practice of clean cultivation must be ruled out, except in dry regions. In wet areas it is better to grow a permanent cover crop of *Pueraria* and other leguminous creepers.

The cover crop keeps the soil cool, adds organic matter, crowds out weeds, improves structure and fixes nitrogen; so much, in fact, that no nitrogen needs to be included in the fertilizer. However, there are three disadvantages. The first, a tendency to climb the tree, is easily corrected by monthly inspections and a few strokes of the machete. The second, competition for water during the dry season, is counteracted by mowing the cover crop at the start of the dry period. The third is a fire hazard during drought. I would therefore recommend a cover crop where rainfall is sufficient (at least 1,500 mm/year) and the dry season is not severe.

In drier regions and on flat land weeds can be controlled by spraying with oil, or herbicides such as dalapon, diuron, simazin and paraquat. But on slopes weeds are best left to protect the soil against erosion. They should be cut only periodically to act as a mulch. Additional mulching, e.g. with elephant grass, is another possibility.

Irrigation

Citrus trees can withstand a drought of four months if grown on deep soil with good water-holding capacity; especially if 'Rangpur' is the rootstock. However, irrigation will be necessary if the dry season lasts longer than three months. The amount of water to be supplied depends on rainfall, evapotranspiration and soil type. As a rule the maximum amount is 100 mm at intervals of three weeks. Too much water is just as harmful as too little; therefore, checks with tensiometers must be made regularly. The soil should be kept moist, but not wet, to a depth of at least 1 metre. It is also important to ensure that the water is evenly distributed over the tree area. Irrigation is also useful for inducing bloom, provided at least six weeks of drought have preceded.

Crop fertilization

This subject is most important in citrus growing. No general rules can be given as so much depends on the soil type. Let us therefore start with Florida, where citrus is grown on deep, chemically poor, sandy soils; everything must be provided in the fertilizer. The recommended formula (de Geus, 1973) for non-bearing sweet orange trees is

$N:P_2O_5:K_2O:MgO:MnO:CuO:B_2O_3 = 8:2:8:2:0.5:0.25:0.1$.

During the first year 150 g of this formula is applied five times. In the second and third year four applications are made (450 g and 900 g). From then on three applications/year are made: 1,800 g in the fourth year, increased by 225 g for each further year, until 3,150 g is reached in the tenth year. By then the tree is receiving 756 g N/year and the orchard gets 166 kg N/ha.

The formula for bearing trees is 180 g N/field box (orange). As a box weighs 40 kg, we may translate this figure into 4.5 kg N/harvested tonne. Assuming a yield of 1,000 boxes or 40 tonnes/ha/year, we apply 180 kg N/ha/year. The other elements are given in the proportion set out above, although phosphate is usually reduced to half the amount for bearing trees. Grapefruit needs less N, namely 136 g/box (3.4 kg/tonne), but more K than the orange.

Zinc has to be sprayed on the tree, often together with copper and manganese. Iron deficiency is cured with iron-chelate. Sometimes even molybdenum is applied. Calcium and sulphur are also needed, but are already present in the other fertilizers. Thus, twelve elements in all are administered in the fertilizer or as a spray in Florida: N, P, K, Mg, Ca, S, Zn, Mn, Cu, Fe, B and Mo. In other regions several of these elements can be omitted. If soil is rich in potassium, there is no need to include that in the fertilizer. As a matter of fact, it could be harmful, as it might induce a deficiency of magnesium or calcium. The tree reacts very quickly to a zinc spray, but slowly to magnesium fertilizers: it takes about two years before the symptoms have disappeared.

How does one determine what fertilizers are needed in a particular case? To begin with, one must keep a constant eye on the orchard and watch for deficiencies. With some training most are easily recognized. Zinc and magnesium deficiencies are especially harmful and render a tree unproductive. Labanauskas (1979) found a harvest of 24 tonnes of 'Valencia' to remove the following minerals in kg/ha: N 44, P 4, K 43, Ca 19 and Mg 4; and Zn 30, Mn 16, Cu 5, B 176, Fe 62 in g/ha.

Table 5.16 *Results of leaf analysis in citrus from Mokanji, Sierra Leone, compared to a standard*

Cultivar	Stage	%N	P	K	Ca	Mg	S	Fe ppm	Mn
St Michael	Non-Fr.	2.79	0.19	2.28	1.63	0.45	0.032	190	12
St Michael	Fruiting	2.26	0.13	1.45	3.07	0.50	0.038	130	16
Lue Gim Gong	Fruiting	2.55	0.17	1.75	2.20	0.48	0.048	200	14
Marsh gf.	Fruiting	2.08	0.11	1.58	2.35	0.42	0.054	150	3
Standard	Fruiting	2.40	0.12	1.00	5.00	0.40	0.300	60	35

Source: A. van Diest, personal communication

Furthermore, leaf samples must be analysed at fixed intervals. They should be of uniform age and location, e.g. from a five-months-old non-fruiting, terminal branch. Other factors such as climate, soil, cultivar and rootstock influence the outcome of the analysis. Therefore, each area must set its own standard of minimum, optimum and maximum amounts of elements in dry matter of leaves. For California optimum amounts (per cent) are about 2.50 N, 0.15 P, 1 K, 4 Ca, 0.50 Mg and 0.25 S. Note that these figures relate to elements, not their oxides. Samples from Sierra Leone were compared against these standards (Table 5.16).

The conclusion was to discontinue potash in the fertilizer, to apply lime instead of dolomite, to apply N as ammonium sulphate (to correct the low S-content) and to spray more often with manganese. Fertilizer experiments will supply additional information on the requirements of the trees on a particular soil type. As an example some treatments of a fertilizer experiment in Surinam (Samson, 1966) are compared (Table 5.17): minor elements alone tripled the yield, the addition of N-K-Mg boosted it another 50 per cent and P had no effect.

Pruning

This should start in the nursery and continue in the field during the early years. In order to prune well, one must have an image of the ideal tree in mind. My ideal tree looks as follows. The rootstock is unbranched and 30 cm high, the scion has an unbranched portion

Table 5.17 *Yields in fruits per tree (Kwata orange), Surinam*

Fertilizer	Spray	1951	1952	1953	1954	1955	1956	Average %
None	None	103	92	147	117	77	138	112	100
None	Zn–Cu–Mn	187	635	198	491	250	191	350	312
N–K–Mg	Zn–Cu–Mn	190	870	340	626	493	604	520	464
N–P–K–Mg	Zn–Cu–Mn	189	770	345	742	514	470	505	451

Source: Samson (1966)

of 20 cm and then splits into three or four main branches that share
the surrounding space equally and go up at an angle of 40°. The
main branches split into two or three outgoing branches and these
split again. The final result resembles a half globe, its flat underside
about 50 cm above the ground, having a diameter of about 5 m.

However, every tree is an individual and poses its own problems.
The art of pruning consists of approaching the ideal, while doing
the least possible damage to the tree. Not all suckers must be
removed, some are best left on the tree; if growing outward they
will later acquire a horizontal habit and become fruitful. Inward
growing branches, dead wood and half-parasites of the Loranthaceae family are removed twice a year. Nests of ants, termites and
wasps must also be eliminated. Unlike apples and pears, citrus trees
have small reserves of carbohydrates. Heavy pruning should therefore be avoided; it delays production. Only declining trees are
severely pruned back in order to rejuvenate them.

A few words must said about hedge pruning. In this dense
planting system, e.g. 2 × 5 m, it is customary to prune mechanically
once a year on one side: east, west and top successively. The bottom
is kept wider than the top, knives are set at an angle of 10° with
the vertical up to 4 m; then a steeper slope goes to the top as 5 m.
Thus two sides of the hedge are in production, while the third is
recovering from the pruning. Although citrus trees survive such
shock treatment, it is likely that individual care will give better
results in the long run.

Growth regulation

There are many ways to use growth regulators in citrus cultivation,
but few can be recommended to the tropical grower. Experimental
results from one area cannot simply be applied in another; temperature, humidity, cultivar, rootstock and other factors affect the
outcome. We therefore refrain from giving details on doses, time
and duration of application and just mention some important
practices.

2,4-D at low concentrations prevents pre-harvest drop of mature
fruit. Alternate bearing in mandarins is countered by spraying NAA
in on-years. An abscission-inducing chemical, e.g. cycloheximide,
is sprayed on trees just before harvest and the fruit is shaken down
and picked up by machines; such fruit is only fit for cannery use and
cannot be sold on the fresh-fruit market. Ethylene and its precursors
are used for degreening of fruit.

The reader who desires more exact information is referred to
Moss (1975) and Monselise (1979).

Yields

It is far from easy to get exact information on citrus yields. They are frequently given in fruits per tree; average fruit weight and number of trees per hectare are not always stated. Or, they may be given in boxes (sometimes of unknown weight content) per tree or acre. Furthermore, there is much variation between years. On the whole climate, soil and management practices are decisive but weather, social and economic factors also affect the issue.

In Florida 400 boxes of 90 lb per acre is regarded as a good yield for orange; this comes to 40 tonnes/ha. The average for the United States was 31 tonnes/ha and for Florida 34 tonnes/ha in 1975/76. For grapefruit these figures were 40 and 44 tonnes/ha respectively (Powell and Huang, 1978).

Purseglove (1968) mentions two or three boxes per tree as annual average for Trinidad; for grapefruit four to five boxes per tree. If we put the net content at 27 kg and the density at 200 trees/ha, these figures come to 14 tonnes/ha for orange and 24 tonnes/ha for grapefruit. Similar yields were achieved by plantations in Surinam during 1959, but small-holders only averaged 7 and 14 tonnes/ha respectively (Samson, 1966). Even lower was the yield in Indonesia: 5 tonnes/ha in 1980 (Internal report).

Citrus yields in the tropics are well below those of the subtropics. However, with good planning and management they should reach a level close to that prevailing in subtropical areas.

Fig. 5.12 Foot rot on a citrus tree, Ivory Coast

Diseases and pests

Great numbers of fungi, bacteria, mycoplasmas, viruses, insects, mites and nematodes attack citrus. We cannot deal with all of them, so we shall select a few from each group. For the detailed descriptions of diseases, etc., the interested reader should consult Klotz (1973), Knorr (1973), Hill (1975), Kranz *et al.* (1977) and Reuther *et al.* (1978).

Fungi

One of the most dangerous citrus enemies is the genus *Phytophthora*. Its members cause foot rot and gummosis on stems, blight of seedlings and brown rot of fruit. The main spp., in the order of their adaptation to higher temperatures, are: *P. citrophthora*, *P. parasitica* and *P. palmivora*; their pathogenic effect, however, is similar.

Foot rot (Fig. 5.12) begins as a discolouration on the stem of susceptible cvs. The spots become bigger and exude gum (hence the name gummosis). As Table 5.14 shows, sour orange and *Poncirus* are resistant; the citranges, 'Cleopatra' and 'Rangpur' are tolerant, while sweet orange and lemon are susceptible. Control consists of cutting out diseased parts and some surrounding healthy-looking tissue; the wound is treated with a disinfectant or Socony fat 2,295 A. Prevention is better. A resistant stock should be used and soil particles, which may contain spores, must not spatter onto the scion. Bare soil furthers this and the practice of forking around the stem is therefore harmful, but a cover crop and mulch are beneficial.

As a last resort, preventive treatments with copper compounds on stems may be necessary. Hough *et al.* (1979) proposed to use a blow torch, heating the bark to 60 °C, as a remedy. Weathers *et al.* (1982) achieved good control of *Phytophthora* with systemics such as metalaxyl and fosetyl-aluminium. Laville and Chalandon (1982) report long-term control (up to three months) from a 2000 ppm foliar spray.

Seedling blight proceeds downwards from the leaves (which turn brown), in contrast to damping-off, caused by *Rhizoctonia*, which rings the young plant just above soil level. Spraying with copper compounds or dithiocarbamates during the wet season is indicated in both cases.

Brown rot causes fruit to drop and turns them light brown; the smell is not unpleasantly acid. Cultural practices like drainage, pruning, removal of shade trees and windbreaks are generally sufficient to prevent it; otherwise a copper spray is needed. But the biggest danger lies in the presence of these fruits in the packing house. The decay spreads rapidly on contact and can cause enormous losses in packed boxes. The only method of control after harvest is a warm water treatment (four minutes at 48 °C), but some fruits are injured by this. It may be necessary to temporarily exclude orchards with brown rot from use of the packing house. Two other forms of fruit decay are important, stem-end-rot and mould; both occur in two forms. The fungi causing stem-end-rot are always present in the orchard, but do not affect the fruit until it is picked. Then fungus threads start to grow inside, from just below the calyx. *Diplodia natalensis* is better adapted to hot conditions than *Diaporthe citri* (Phomopsis). The first is prevalent in Surinam, both do damage in Florida, but in California they are of minor importance. The disease is recognized by a brown discolouration of the fruit axis, the skin then becoming slippery while the fruit smells unpleasantly acid. This form of decay does not spread on contact. Good control is achieved by debuttoning (removing the calyx). Table 5.18 shows the decay percentages reported by Samson (1966) in Surinam.

Table 5.18 *Percentage decay of citrus fruits after four weeks, held at ambient temperatures in Surinam*

Exp. No.	Cultivar	None	Debuttoned	Borax	Dowicide A
			Treatment		
6	Kwata orange	12.0	2.0	—	0
7	Kwata orange	38.9	11.1	43.0	13.9
10	Kwata orange	56.0	—	61.2	29.6
13	Kwata orange	29.2	—	—	6.7
19	Kwata orange	38.0	—	—	2.5
20	Woglum lemon	75.6	35.0	—	—
21	Woglum lemon	42.0	6.0	—	—
22	Woglum lemon	48.7	2.7	—	—

Source: Samson (1966)

Debuttoning is not a practical measure, but in pulling many buttons are removed. Clipping is therefore not recommended for the wet tropics. Fruit can also be disinfected in the packing house with Dowicide A (SOPP 2 per cent, hexamine 1 per cent, NaOH 0.4 per cent), but for maximum effect this must be done within one day after picking. Wrappers of diphenyl are also effective, but washing with soda or borax has little effect.

A new group of fungicides, the benzimidazoles, was introduced in the 1960s. Thiabendazole, or TBZ, was soon widely used as a dip or spray in citrus packing houses. Benomyl was regarded as promising but had not been registered for use on citrus in many countries (Dupuis, 1975). Unfortunately, the moulds and other fungi have developed resistance to both compounds. Recently, imazalil came forward as a fungicide against which no resistance has (so far) been found. Meanwhile, SOPP has retained its position as one of the most reliable treatments for post-harvest decay in citrus fruits.

Moulds (*Penicillium* spp.) are present everywhere, but they are far more dangerous in dry than in wet regions. Picking with clippers and careful handling are required. Green mould (*P. glaucum*) does not spread on contact, but blue mould (*P. italicum*) does. After thirty-five holding experiments, involving 105,000 fruits checked four times at weekly intervals, Samson (1955) concluded that *Diplodia* was responsible for 90 per cent of the decay in Surinam, *Penicillium glaucum* for most of the rest.

Three fungus diseases cause serious blemishes on leaves and fruits: melanose (*Diaporthe citri*), greasy spot (*Mycosphaerella citri*) and scab (*Elsinoe fawcetti*). Temperature is again a major factor in determining the severity of injury. Melanose, which gives a sand-papery feel to leaves and fruits, is very serious in Florida and does perceptible damage in Trinidad, but was seen only once in Surinam. Scab attacks sour orange everywhere and also grapefruit in Trinidad, but not in Surinam; this, however, could be due to different strains. Greasy spot causes leaf fall and lowers the quality of fruit. In all three cases control can be achieved by spraying with copper compounds or dithiocarbamates. Mabett and Phelps (1983) report that low and ultra-low-volume sprays deposited most fungicide on the lower surface of 'Marsh' grapefruit leaves in Trinidad; after five months a substantial amount of fungicide was still present. This provided excellent control of greasy spot, which penetrates via the stomata on the under-side of new leaves.

Citrus blight (also known as 'young tree decline') is a very destructive disease in Florida; the cause is unknown (Childs, 1979). It seems likely that it also occurs in South America and South Africa (Wutscher, 1980).

Bacteria

Citrus canker is a destructive bacterial disease in many countries. The only effective control method is eradication; this was done twice in Florida. Where this is impossible, copper sprays may keep it in check.

Virus and mycoplasma

More than twenty virus and virus-like diseases of citrus have been described (see Klotz, 1972). All these have been spread over the tropical and subtropical world with infected budwood. Some of them, particularly tristeza and greening which are the most dangerous diseases of citrus, can also be spread by insects. Exocortis is spread by budding knives, pruning shears and other tools, and possibly even by the hands of workers. For prevention, tools must be dipped in a solution of one part household bleach in nine parts water.

Tristeza is transmitted by several aphids, most of which are not very efficient vectors: they spread the disease only slowly. However, the black citrus aphid (*Toxoptera citricidus*) is a very efficient vector; where that insect exists in the presence of a source of tristeza virus, the disease will sooner or later destroy all susceptible stock-scion combinations of citrus. Either the virus source can be destroyed (as was done in Israel) or sour orange can be replaced by tolerant rootstocks (see Table 5.14) as was done in several South American countries. As there are mild and severe strains of tristeza, it is possible to inoculate young trees, grown in isolated nurseries, with a mild strain. This will confer protection against the more virulent strains. Biological control of *T. citricidus* by a fungus (*Verticillium lecanii*) and some predators has been reported from Venezuela (Rondon *et al.* 1980).

Tristeza is probably indigenous to Asia, but was first found in South Africa, at the beginning of this century. Around 1920 it was reported from Indonesia and Australia. These countries had to abandon the sour orange rootstock very early. In the 1930s infected budwood was imported into Argentina and then the virus began to spread like wildfire through South America. Within a few years millions of trees had been killed in Argentina, Brazil and Uruguay.

However, the spread may also be gradual, as in California and Spain, or in several tropical countries. The reason for the slow spread can be the absence of the black citrus aphid, or high temperatures which mitigate tristeza symptoms. In Africa the disease was, until 1960, limited to (former) English colonies. These countries had imported infected budwood from South Africa, while the (former) French colonies got clean budwood from the Mediterranean region. But a virus has no regard for language or political barriers and so the disease started to work its way into French-speaking Africa too. From the Ghanaian border it was half-way into Ivory Coast (A. Fouqué, personal communication, 1977) and from Surinam, where it was first seen in 1963, it has reached French Guiana (J. Bové, personal communication, 1975).

A plant infested by tristeza cannot recover, except by heat treatment. This is only possible with small plants in heat chambers. In the field, tristeza can be countered by inarching a tolerant rootstock or by cutting back to the rootstock and rebudding with lemon. The sour orange-lemon combination survives because both members are hypersensitive to tristeza. In most cases it is cheaper to start anew with virus-free material on tristeza-tolerant stocks. Aphids may later infect these trees with tristeza, but the absence of other viruses (not transmitted by insects) allows a wide choice of root-stocks. It is therefore wise to start a nucellar programme even if

Fig. 5.13   CVPD (greening) of citrus in Central Java

tristeza is not a direct threat; this was done in Trinidad (Hosein, 1968).

Greening is dangerous, as no resistant rootstocks or cultivars are known. The name refers to the fruit which remains greenish; other symptoms are discoloration of young leaves, drying-up of branches and a low yield. The disease was first recognized in South Africa but occurs all over tropical Asia as 'citrus decline' (India), 'leaf mottling' (Philippines), 'likubin' (Taiwan) and citrus vein phloem discoloration (CVPD, Indonesia), see Fig. 5.13.

The pathogen is a small bacterium that is spread by two psyllid vectors: *Trioza erytreae* in Africa and *Diaphorina citri* in Asia. *Trioza* does not thrive under hot and arid conditions, while *Diaphorina* prefers high temperatures and an annual rainfall of about 1,500 mm. There are two distinct strains of the pathogen; no symptoms appear with the African strain at temperatures above 25 °C by day and 20 °C at night. This makes it unlikely that greening will become as destructive in tropical Africa as in Asia.

Both vectors exist in Réunion (21° S); *Trioza* above 300–500 m, *Diaphorina* below that altitude. Wasp parasites, genus *Tetrastichus* were imported and within a few years they had nearly eliminated the two vectors. Such a favourable biological equilibrium can only be maintained if the parasites are not killed by spray materials directed against other diseases and pests.

Diseased trees can be cured by the injection of penicillin or tetracycline into the stem (Schwarz and van Vuuren, 1971); however, the use of antibiotics in agriculture is not permitted. Another control method is the establishment of isolated nurseries under quarantine. For recent reviews the reader may consult Moll and van Vuuren (1978) and Aubert *et al.* (1980).

The countries of South–East Asia are particularly hard hit, because here tristeza and greening occur side by side with their most efficient vectors. Millions of trees were rooted-up in East and Central Java (Soelaeman, personal communication, 1981), but this is only a partial solution. A far better approach, in my opinion, would be to designate suitable small islands as special citrus growing areas (Samson, 1977); such an island should be cleared of all citrus and other Rutaceae for a few years and kept under strict quarantine; then 'clean' rootstock seed and virus-free buds can be introduced to set up a large modern citrus plantation. There are many islands off the coast of Java and Luzon that suit the purpose. One 10,000 ha well-run citrus orchard could provided 50–60 kg of fruit per year to every inhabitant of Jakarta or Manila.

Exocortis causes bark scaling, stunting and certain leaf and twig symptoms. Its presence precludes the use of 'Rangpur' and *Poncirus*; the citranges are also susceptible. To prevent exocortis one has to start with virus-free budwood, but this is not enough: the rootstock must come from certified seed and all the tools must be sterilized regularly with a 10 per cent solution of household bleach (Klotz, 1972). Of course, whenever possible exocortis-resistant rootstocks should be preferred. In 1966 the use of a mild exocortis strain was proposed by Mendel for control of tree size. Growers in New South Wales, Australia are doing this now (Cohen, 1981).

Psorosis is a whole complex of virus diseases: psorosis A, blind pocket, concave gum, etc. One member of the complex gives no protection against the other members. Symptoms are bark scaling, holes on the trunk, gumming, leaf patterns etc. Sweet orange and mandarins are very susceptible. Scraping of the bark will give temporary relief from psorosis A, but the use of virus-free budwood is the only practical measure. The crinkly leaf and infectious variegation complex caused a 40 per cent decline in yield on Corsica (Vogel and Bové, 1977).

A citrus tree may carry several viruses or mycoplasmas without showing signs of disease; it is then said to be tolerant to these diseases. A sweet orange infected with exocortis will not exhibit symptoms if budded on sour orange. However, if budwood is taken from this tree and used on *Poncirus* or 'Rangpur' stock, the symptoms will soon appear. It is therefore necessary to inspect prospective mother trees by indexing them on very sensitive cvs, e.g. West

Indian lime seedlings for tristeza, 'Hamlin' seedlings for psorosis and certain citron clones for exocortis. For more information see Childs (1968) and Klotz *et al.* (1972).

Insects

Among insects the scales do most damage to citrus. We may distinguish armoured (Fig. 5.14) and soft scales, round, comma-shaped and globular ones etc. Chemical control is possible with light mineral oils, parathion and other products. However, it is far cheaper and less harmful to the environment to use biological and integrated control methods. *Aphytis* wasps are specific parasites of scales. In humid regions two fungi: a red *Aschersonia* and a *Fusarium*, are parasitic on scales and keep them at least partly in check. Care should be taken not to harm these beneficial fungi. Spraying with copper during a dry period would kill them and is bound to be followed by a massive attack of scales. Natural enemies of scales and aphids,

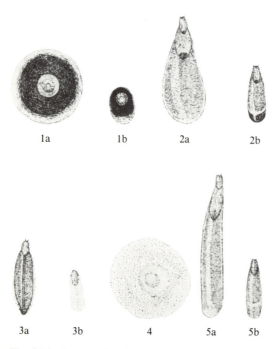

Fig. 5.14   Armoured scales prevalent on citrus trees in Surinam. 1. *Chrysomphalus ficus* 2. *Lepidosaphes beckii* 3. *Fiorinia* sp. 4. *Selenaspidus articulatus*, semi-transparent scale, vaguely showing adult female 5. *Lepidosaphes gloverii*. (a = scale of adult female, b = scale of male). *Source*: van Dinther (1960)

such as lady beetles and hover fly larvae, are hampered in their action by ants. It is therefore imperative that permanent ant control is maintained by spraying stems, but not leaves, at least three times a year with diazinon. The citrus blackfly, *Aleurocanthus woglumi* provides another example of effective biological control by predators (*Amitus* and *Prospaltella*).

Aphids transmit tristeza and cause injury to young leaves, shoots and flowers; good ant control usually keeps them in check. However, they multiply exceedingly fast, so they must be watched at each flush and be controlled if necessary. This is done by spraying malathion or lindane (BHC) along the skirts of the tree (on the flush only).

Widespread damage is brought about by fruit-piercing moths. Operating at night they bore holes forming an entrance for fungi, fruit flies and such; this leads to fruit fall and rot. The moths live in forest trees and cannot be controlled, but early picking prevents damage. Interplanting with mango and cashew nut was advised by Taylor (1969) in Sierra Leone; the moths seem to prefer these fruits to citrus.

Fig. 5.15 Citrus trees in Surinam, defoliated by leaf-cutting ants (*Atta sexdens*), note the nest openings in the soil; *A. cephalotes* makes fewer but far bigger nests

Leaf-cutting ants have been known to defoliate a tree in a single night (Fig. 5.15). Their nests must be found and destroyed. Caterpillars are generally only troublesome on individual trees and can be picked off by hand without trouble.

It is necessary to know scientific names of pests in order to find more information on them. Therefore, a list of major pests, adapted from Hill (1975) and van Dinther (1960), is given in Table 5.19.

Table 5.19   *Names of major citrus pests and their distribution*

Scientific name	English name	Family	Continents
Phyllocoptruta oleivora	Rust mite	Eriophyidae	All
Lepidosaphes beckii	Purple scale	Diaspididae	All
Lepidosaphes gloverii	Glover's scale	Diaspidae	All
Aonidiella aurantii	California red scale	Diaspididae	All
Chrysomphalus ficus	Florida red scale	Diaspididae	All
Saissetia oleae	Black scale	Coccidae	All
Coccus viridis	Soft green scale	Coccidae	All
Toxoptera aurantii	Black aphid	Aphididae	All
Toxoptera citricidus	Black aphid	Aphididae	All
Pseudococcus adonidum	Mealybug	Pseudococcidae	All
Aleurocanthus woglumi	Blackfly	Aleryodidae	All
Dialeurodes citri	Whitefly	Aleyrodidae	Most
Trioza erytreae	Psyllid	Psyllidae	Africa
Diaphorina citri	Psyllid	Psyllidae	Asia
Papilio demodocus	Orange dog	Papilionidae	Africa, Asia
Papilio anchisiades	Orange dog	Papilionidae	America
Ophideres fullonia	Fruit piercing moth	Noctuidae	Africa, India
Othreis fullonica	Fruit piercing moth	Noctuidae	America
Othreis procus	Fruit piercing moth	Noctuidae	America
Atta cephalotes	Leaf-cutting ant	Formicidae	America
Atta sexdens	Leaf-cutting ant	Formicidae	America
Solenopsis geminata	Fire ant	Formicidae	All
Camponotus sp.	Tree ant	Formicidae	America
Anastrepha ludens	Mexican fruitfly	Tephritidae	America
Ceratitis capitata	Mediterranean ff	Tephritidae	All (Asia?)
Nasutitermes surinamensis	Termite	Termitidae	America

Source: Hill (1975) and van Dinther (1960)   ff = fruit fly

Mites

The rust mite (*Phyllocoptruta oleivora*) is ever present in tropical citrus orchards, but is hard to find in the rainy season. It cannot be seen with the naked eye as it is very small, about 0.1 mm long. During dry periods it multiplies fast as a new generation develops in seven days; forty generations may arise in one year. Within three or four weeks of favourable weather, high temperature and high humidity but little rain, a massive infestation takes place. The mites destroy epidermal cells of leaves and fruits and thus cause 'russetting' (dark spots, in lemon silvergray). Lesions formed on young fruits expand and render the fruit unsaleable. They also provide entry for the greasy spot fungus.

Another fungus, *Hirsutella thompsonii*, is a parasite of rust mite. However, it takes weeks before this fungus can catch up with the mite and suppress the population sufficiently. It is therefore of vital importance to apply a miticide as soon as a dry period has started. Formerly sulphur was used, but this broad-spectrum pesticide kills useful fungi; it was first replaced by zineb, then by chlorobenzilate at 0.2 per cent a.i.

Van Brussel (1975) had good results in Surinam from spraying with fragmented *Hirsutella* mycelium; the mite population declined within five days. A *Hirsutella* product called Mycar is commercially available, it is compatible with oil (Anon., 1983), but copper and other metals should be withheld. In Brazil, control of rust mite was obtained by applying granulated aldicarb (10 g per tree) to the soil (Brunelli *et al.* 1978).

Nematodes

Two nematodes are important in citrus. *Radopholus similis* causes a serious disease called spreading decline in Florida that has not yet been reported elsewhere. The notorious banana nematode belongs to the same sp., but to a different race. Use of resistant stocks, such as 'Milam' lemon, and eradication of diseased trees are the measures adopted in Florida. *Tylenchulus semi-penetrans* causes slow decline. *Poncirus* is very resistant and the citranges are fairly resistant. Hot water treatment of nursery plants and applying lime have also been recommended. And of course, it is strongly recommended not to plant citrus after citrus.

Disease control

The following materials are generally used:
1. Copper compounds, e.g. cuprous oxide, copper-oxy-chloride and tri-basic-copper against fungi, algae and lichens (these materials must be used with great care during dry weather);
2. Chlorobenzilate against rust mite;
3. Malathion against scales and aphids (tree skirts only);
4. Diazinon against ants (on stems only)

To all sprays, except the last, the sulphates of zinc, manganese and copper may be added to correct a deficiency. As a rule, hydraulic spraying is done in concentrations of 1–2 per cent, while mist blowing uses 5–10 per cent solutions. Recipes are listed in Table 5.20.

For general use this spray programme may be followed:

Table 5.20   *Quantities required per 100 litre solvent (water)*

Chemical	Formulation	Spraying High volume		Low
1 Zinc sulphate	Crystal	0.5 kg		5 kg
2 Manganese sulphate	Crystal	0.25 kg		2.5 kg
3 Hydrated lime	Powder	0.25 kg		2.5 kg
4 Copper oxy-chloride	Powder	0.25 kg		2.5 kg
5 Malathion	Miscible oil 20%	1 litre		10 litres

1, 2 and 3 are dissolved and 4 may be added; 5 is applied separately.
Source: Samson (1966)

1. During bloom or flush – malathion (tree skirts only).
2. After bloom – zinc sulphate* and chlorobenzilate.
3. In dry period – chlorobenzilate.
4. Before harvest – diazinon on stems.
5. In dry season – diazinon on stems.
* Manganese and copper should be added when necessary.

   This programme must be adapted to local conditions as more becomes known about the relation of pests to climate.

   Of the main citrus producers, Florida comes closest to tropical conditions. A dozen pests are considered to be of major importance especially rust mite, red and purple scales, mealy bug, aphids and whitefly. Formerly ever more spraying had to be done to control these pests but nowadays red and purple scales are kept in check by *Aphytis* parasites. This has given other scales, such as glover's, snow and chaff scales, the chance to become prominent. An integrated pest management system was adopted, in which two main sprays are generally sufficient:
1. in spring, against rust mite and disease;
2. in summer, against rust mite, disease and scales.
The timing depends on the stage of growth. Frequent inspections are necessary, possibly resulting in a third application of acaricide. Chemical control of aphids is usually warranted only for young trees, or where tristeza might be involved (Bullock and Brooks, 1975).

   In conclusion we mention physiological disorders (not caused by pathogens) e.g. sunburn, granulation, creasing and chilling.

From harvest to consumption

A citrus fruit stays on the tree from six to twelve months, in the subtropics even longer. How can the correct moment for harvest be determined? Colour could be an indicator but this is not a good criterion in the tropics. Orange colour does not develop there, as cool nights are required for this to happen. Yet, there is a colour break, from hard green to light green, usually coinciding with maturity. Later the fruit will become yellow. Taste is another possibility, but this too is hazardous. As taste and colour cannot be relied on it is desirable to establish objective standards. These have been found in the percentage of total soluble solids (the so-called Brix) and the percentage water-free citric acid of the juice. Sometimes the percentage juice is also used (Harding *et al.* 1940). In Florida samples of 20 oranges are taken, but I have found 10 to be sufficient (unless they are small). They are weighed and squeezed and the percentage juice is calculated; it should be about 50 per cent. A Brix hydrometer calibrated at 20 °C is used and a correction factor has to be applied for other temperatures. A refractometer with a Brix scale is useful for checking individual fruits (personal communication by James Soule).

Twenty-five ml juice are titrated with sodium hydroxide. At a titre of 0.4063 each ml hydroxide would be equivalent to 0.1 per cent citric acid. The ratio of Brix to acid (B/A) is generally accepted as a measure of maturity. In Florida and California eight is regarded as the minimum ratio, but in the tropics this point may quite safely be set at ten. Nearly everybody accepts an orange with such a B/A ratio as good to eat, although some may prefer somewhat sweeter fruit.

Citrus fruit has a lower acid content in the tropics than in a Mediterranean climate. This effect is enhanced by rain during the two months before the fruit reaches maturity. Altitude also influences maturity as was shown by observations in Réunion on 'Orlando' tangelo: a B/A ratio of 7 was reached on 17 March at 125 m and three weeks later at 325 m; at 600 m maturity came another month later (Études, 1978).

What happens while an orange is ripening? As long as the fruit hangs on the tree the Brix continues to rise, at first quickly, then gradually slower; it may go up to 13°, or even higher. Meanwhile the acid content is steadily coming down from 2.5–1 per cent and lower. As long as the B/A ratio lies between 10 and 16 a large majority of eaters will be satisfied. But if fruit remains too long on the tree, it becomes overripe and unpleasantly sweet with ratios of 20, or higher. Such fruit is not acceptable for processing or export.

We must remember that citrus fruits contain no starch; unlike bananas they cannot be picked green for after-ripening. Once an orange is picked, it is as good as it will ever be. Yet the taste becomes slightly sweeter in holding, because acid is broken down faster than sugar.

Most growers know when their fruit is ripe, but the criteria given above may help to make decisions at the beginning and end of the harvesting season. A very useful summary of these tests is given by Soule *et al.* (1967).

Picking is either carried out using snub-nosed clippers, or is done by pulling. Mandarins are easily damaged and must always be clipped to avoid plugs. Oranges and grapefruit are clipped in dry regions where green mould (*Penicillium*) causes rot. In humid regions wounds heal quickly and it is better to pull, as many buttons are thus removed; this reduces stem-end-rot. However, the fruit should not be pulled straight down (which may cause plugs), but up, with a rotating movement of the wrist (the thumb moves from below to above the fruit).

Picking teams must have good equipment: light, but sturdy ladders, picking bags with open bottoms that can be folded up and hooked and trucks with low-loading platforms. Smith (1969) described an efficient picking bag with harness and girdle, based on the results of a time and motion study. The best pickers in California are able to seek out, clip, put into a bag and carry to a container something like 100 boxes a day; that is 20,000 fruits. Pulling is much faster. In the tropics the picking capacity is likely to be 30 to 50 boxes (900–1500 kg) a day.

Shaking and mechanical harvesting have become standard practice in Florida, Australia and other places with high wages, but this fruit is only fit for cannery use. Before harvest, ethephon and other chemicals that promote abscission are applied. This causes not only ripe fruit, but also immature fruit and leaves to fall. The damage is slight on early and mid-season cvs. On 'Valencia', where two crops may hang simultaneously on subtropical trees, it may lead to serious problems (Grierson and Wilson, 1983).

Fig. 5.16 Grapefruit packed in Bruce (wire-bound) boxes for shipping

To prevent mould, picking must not be done during or shortly after rain; one has to wait until the fruit is dry. Field boxes should be protected from sun and rain by tarpaulins. Fruit must be handled carefully during and after picking. An experiment by van Suchtelen with oranges, held at room temperature (23–31 °C) during three weeks in Surinam (Samson, 1966) showed that with careful handling decay was only 7 per cent, while it rose to 12 per cent if fruit was dropped on concrete, 21 per cent if fruit was picked wet, 29 per cent if fruit was overripe, and 38 per cent if fruit had been in the sun one day. Smoot (1976) in Florida dropped oranges, grapefruit and tangerines and inoculated them with green mould; decay rose tenfold in the worst case.

Where possible fruit should be pregraded in the orchard; this will save much in transportation costs. All rotten and damaged, too small and too big, hard green or overripe and badly spotted fruit are removed at this stage. Then the boxes are stacked four high on a truck and taken to the market, or a packing house and (or) processing plant. In the packing house the fruit undergoes several treatments, some of which may be omitted: degreening, washing, brushing with soap, disinfection, drying, 'color-adding', waxing, grading, sizing and packing.

Borax and soda have been standard disinfection for almost 50 years, and SOPP has been used extensively for the last 20–30 years. Recently a great step forward was made with the introduction of thiabendazole and benomyl in sprays, suspensions, waxes and foams. However, it has been shown that some fungi may develop resistant strains. Imazalil also shows promise, but has not everywhere been approved for use yet. A very good review of the subject is presented by Eckert (1978).

Formerly, wooden boxes of two cubic feet capacity, holding about 30 kg fruit were used in many countries. Then the wire-bound ('Bruce') box (same capacity) was adopted almost universally. Nowadays most citrus fruit is packed in cardboard boxes holding 18–20 kg fruit.

Packed fruit can be stored for several months at temperatures of 3–8 °C, but grapefruit must be held at 10–15 °C, see Table 4.4. The relative humidity should be kept at 85–90 per cent. Ships carrying citrus fruit must be cooled quickly to 15 °C, below which most decay organisms stop growing. Good ventilation is essential.

Citrus fruit can be processed in several ways. Single strength juice has to be sterilized and acquires a cooked taste. The same is true of 'hot' concentrates. A far superior product is made under the name 'frozen concentrate': nine parts of juice are concentrated under vacuum at room temperature to a Brix of 56° (the starting point is around 12°); the tenth part is added fresh and the mixture is quickly frozen at −20 °C. It is not sterile, but will not spoil at

Table 5.21   *Labour requirement for bearing citrus, Surinam (man-days/ha)*

Month	J	F	M	A	M	J	J	A	S	O	N	D	Total
Cover crop	1	1	1	1	1	1	1	1	1	1	1	1	12
Pruning	1	1	1	2	1	2	.	.	.	.	1	1	10
Ditches, roads	2	3	3	3	1	1	.	.	.	2	3	2	20
Fertilizing	.	.	.	.	2	.	.	.	.	.	.	2	4
LV* spraying	.	.	1	.	.	1	.	1	.	.	1	.	4
Harvesting	2	1	.	.	.	.	4	5	6	4	.	.	22
Total	6	6	6	6	5	5	5	7	7	7	6	6	72

* LV = low volume
Source: Samson (1966)

that low temperature. This way a concentrate of 44° B is manufactured that retains much of the taste and aroma of fresh juice when diluted with three parts of water. About 80 per cent of the orange production in Florida and Brazil is converted to frozen concentrate.

How much labour is required for the activities discussed in this chapter? This depends, of course, on the degree of mechanization. Data are provided by Samson (1966), see Table 5.21, for a small farm in Surinam. These figures do not include the labour requirement for clearing, planting and transport of fruits. We may therefore conclude that, once the orchard is planted, one man with a motorized portable sprayer (see Fig. 4.6) can take care of 4 ha of citrus; the work is spread evenly over the year.

References

Adigun, O. O. (1978) 'Evaluation of five local sweet orange cultivars in Nigeria', *1977 Proc. Internat. Soc. Citriculture* **2**, 640–1.

Amson, F. W. van (1966) *Some aspects of clay soils in the Demerara formation in Surinam*, Bull. 84, Agr. Exp. Station, Surinam.

Anon. (1981) 'Breakthrough: a method for debittering Navel juice', *Citrograph* **67**, 35.

Anon. (1983) 'New biological miticide for citrus mite', *Citrograph* **68**, 59–60.

Aubert, B. and **Lichou, J.** (1974) *Données générales concernant l'agrumiculture aux Mascareignes; inventaire des potentialités agrumicoles de la Réunion*, IFAC, Paris.

Aubert, B. *et al.* (1980) 'La lutte contre la maladie du greening des agrumes à l'isle de Réunion', *Fruits* **35**, 605–24.

Baker, R. E. D. (1940) 'The influence of climatic factors on Citrus scab disease', *Tropical Agr.* (Trinidad) **17**, 83–6.

Batchelor, L. D. and **Rounds, M. B.** (1948) *Choice of rootstocks* in: The Citrus Industry II, Un. Cal., 169–222.

Bitters, W. P. (1968) 'Valencia orange rootstock trial at South Coast Field Station', *Cal. Citrogr.* **53**, 172–4.

Bitters, W. P. *et al.* (1969) 'The Citrus relatives as Citrus rootstocks', *Proc. First Intern. Citrus Symp.* 411–15.

Brussel, E. W. van (1975) *Interrelations between citrus rust mite, Hirsutella thompsonii and greasy spot in Surinam*, Wageningen.

Brunelli, J. *et al.* (1978) 'Granulated systemic pesticides applied to the soil for the

control of *Phyllocoptruta oleivora* in mature orange trees', *Solo* **70**, 15–19.
Bullock, R. C. and **Brooks, F. R.** (1975) Citrus pest control in the USA, pp. 35–7 in *Citrus*, CIBA-GEIGY Tech. Monogr. 4, Basle.
Burke, J. H. (1956) *Citrus industry of Surinam*, USDA Foreign Agr. Rpt. 89.
Camp, A. F. *et al.* (1951) *Citrus industry of Florida*, Tallahassee.
Campbell, C. W. (1977) *Tahiti lime production in Florida*, Univ. Florida Gainesville, Bull. 187.
Carpenter, J. B. *et al.* (1982) 'Performance of rootstocks inoculated with virus', *Citrograph* **67**, 101–16.
Cassin, J. (1958) 'Influence du climat sur la floraison des citrus en Guinée', *Fruits* **13**, 286–92.
Cassin, J. *et al.* (1969) 'The influence of climate upon the blooming of Citrus in tropical areas', *Proc. First Intern. Citrus Symp.*, 315–23.
Cassin, J. and **Lossois, P.** (1978) 'Method of nucellar selection used in Corsica', *1977 Proc. Intern. Soc. Citriculture* **2**, 536–40.
Cheema, S. S. *et al.* (1982) 'Evaluation of rough lemon strains and other rootstocks against greening-disease of citrus', *Scientia Hortic.* **18**, 71–5.
Childs, J. F. L. (1964) 'Observations on citrus culture and problems in Surinam (Dutch Guiana)', *Surin. Landb.* **12**, 57–61.
Childs, J. F. L. (1968) *Indexing procedures for 15 virus diseases of citrus trees*, USDA Handbook 333, Wash. DC.
Childs, J. F. L. (1979) 'Florida citrus blight', *Plant Dis. Rprt.* **63**, 560–9.
Cohen, M. (1981) 'Beneficial effects of viruses for horticultural plants', *Hortic. Reviews* **3**, 394–411.
Dinther, J. B. M. van (1960) *Insect pests of cultivated plants in Surinam*, Landb. Proefst. Suriname, Bull. 76.
Dupuis, G. (1975) 'Pesticide residues in citrus', pp. 81–8 in: *Citrus*, CIBA-GEIGY Techn. Monogr. 4, Basle.
Ebeling, W. (1959) *Subtropical fruit pests*, Un. Cal.
Eckert, J. W. (1978) 'Post-harvest diseases of citrus fruits', *Outlook on Agriculture* **9**(5), 225–32.
Études sur agrumes (1978) *Fruits* **33**, 671–800.
Fawcett, H. S. (1936) *Citrus diseases and their control*, McGraw Hill.
FAO (1978) *Production Yearbook 1977*, Rome and other yearbooks.
Fennah, R. G. (1940) 'Observations on behaviour of root stocks in St. Lucia, Dominica and Montserrat', *Trop. Agric.* (Trin.) **17**, 72–6.
Fouqué, A. *et al.* (1977) 'Resultats préliminaires des essais de port-greffe d'agrumes en Côte d'Ivoire', *Fruits* **32**, 335–49.
Fouqué, A. and **Combres, J. C.** (1978) 'Études sur l'irrigation en Côte d'Ivoire', *Fruits* **33**, 845–7.
Fung-Kon-Sang, W. E. (1978) 'Promising citrus rootstocks and scion cultivars in Suriname', *1977 Proc. Int. Soc. Citricult.* **2**, 648–50.
Fung Kon Sang, W. E. and **Nanden-Amattaram, T.** (1975) 'Citrus rootstock performance with old and nucellar tops on heavy clay soils of the coastal plain in Surinam', *Surin. Landb.* **23**, 109–18.
Garcia Benavides, J. (1971) 'Clima agricola de Citrus sinensis', *Agronomis tropical* *XXI*, 77–89.
Geus, J. G. de (1973) *Fertilizer guide for the tropics and subtropics*, Zürich.
Godfrey-Sam-Aggrey, W. and **Haque, I.** (1976) *Citrus production manual*, Njala, Sierra Leone.
Grierson, W. and **Wilson, W. C.** (1983) 'Influence of mechanical harvesting on citrus quality: cannery vs. fresh fruit crops, *Hortsci.* **18**, 407–9.
Harding, P. L. *et al.* (1940) *Seasonal changes in Florida oranges*, USDA Techn. Bull. 735.
Hartmann, H. T. and **Kester, D. E.** (1975) *Plant propagation*, 3rd edition, Englewood Cliffs.

Haury, A. *et al.* (1982) 'Resultats des essais de porte-greffe des agrumes entrepris au Niger', *Fruits* **37**, 509–22.

Hearn, C. J. *et al.* (1974) 'Breeding citrus rootstocks', *Hortsc.* **9**, 357–8.

Hill, D. S. (1975) *Agricultural insect pests of the tropics and their control*, Cambridge.

Hodgson, R. W. (1961) 'Taxonomy and nomenclature of citrus' in: *Proc. 2nd IOCV conf.*, 1–7.

Hodgson, R. W. (1967) 'Horticultural varieties of citrus' in: *The Citrus Industry I*, Un. Cal. 431–591.

Hosein, I. (1968) 'Citrus budwood improvement by nucellar seedlings', *Citrus Research*, Un. W. Indies Bull. 10.

Hough, A. *et al.* (1979) 'Heat treatment for the control of *Phytophthora* gummosis in citrus', *Plant Dis. Rprt.* **63**, 40–3.

Hume, H. H. (1957) *Citrus fruits*, New York.

Hutton, R. J. and **Cullis, B. R.** (1982) 'Tree spacing effects on productivity of high density dwarf orange trees', *Proc. Internat. Soc. Citriculture 1981*, 186–90.

IFAC (1970–1) *Index des activités*, Paris.

Jong, W. H. de (1933) 'Rootstocks for mango and Citrus', *Landbouw* (Buitenzorg) **5**, 266–7.

Kesterson, J. W. and **Braddock, R. J.** (1975) 'Effets du cultivar, du porte-greffe, de l'irrigation et de la fertilisation sur le rendement et la qualité des huiles essentielles d'agrumes', *Fruits* **30**, 103–107.

Klotz, L. J. (1973) *Color handbook of citrus diseases*, 4th ed. Riverside.

Klotz, L. J. *et al.* (1967) 'Orchard tests of citrus rootstocks for resistance to Phytophthora', *Cal. Citrogr.* **53**, 38 and 55.

Klotz, L. J. *et al.* (1972) *Virus and viruslike diseases of Citrus*, Cal. Agr. Exp. Station, circular 559.

Knorr, L. C. (1973) *Citrus diseases and disorders; an alphabetized compendium with particular reference to Florida*, Gainesville.

Kranz, J. *et al.* (1977) *Diseases, pests and weeds in tropical crops*, Berlin.

Kriedemann, P. E. and **Barrs, H. D.** (1981) 'Citrus orchards' in: *Woody plant communities*, vol. **6** of *Water deficits and plant growth* (T. T. Kozlowski ed.), Academic Press.

Labanauskas, C. K. (1979) 'Nutrient removal by Valencia orange fruit from orchards', *Citrograph* **64**, 251–2.

Lange, J. H. (1974) 'The influence of cross-pollination on fruit set and seed content of Citrus cv. Ortanique', *Sc. Hortic.* **2**, 285–92.

Laville, E. and **Chalandon, A. J.** (1982) 'Control of *Phytophthora* gummosis in citrus with foliar sprays of fosetyl-Al, a new systemic fungicide', *Proc. Intern. Soc. Citricult. 1981*, vol. **I**, 346–9.

Mabett, T. H. and **Phelps, R. H.** (1983) 'The behaviour of copper deposits from reduced-volume sprays and their control of citrus greasy spot', *Trop. Pest Management 29*, 137–44.

McCarty, C. D. *et al.* (1974) 'Comparison between Troyer and Carrizo citrange', *Citrograph* **59**, 294, 310.

Magielse, M. M. and **Ochse, J. J.** (1930) 'Results of budding and grafting of fruit trees in the Ragoenan Experiment Garden', *Landbouw* (Buitenzorg) **6**, 944–71.

Meijer, J. (1954) *Pioneers of Pauroma*, Paramaribo.

Mendel, K. (1967) 'Zitrus' in: *Handbuch der Landwirtschaft und Ernährung in den Entwicklungsländern II*, 415–31, Stuttgart.

Mendel, K. (1969) 'The influence of temperature and light on the vegetative development of citrus trees', *Proc. First Intern. Citrus Symp.* 259–65.

Molesworth Allen, B. (1967) *Malayan fruits*, Singapore.

Moll, J. N. and **van Vuuren, S. P.** (1978) 'Greening disease in Africa', *1977 Proc. Internat. Soc. Citriculture* **3**, 903–12.

Monselise, S. P. (1979) 'The use of growth regulators in citriculture', *Scientia Hortic.* **11**, 151–62.

Moreira, A. and **Salibe, A. A.** (1969) 'The contribution of research for the progressive changes in citrus rootstocks for South America', *Proc. First Intern. Citrus Symp.* 351–7.

Moss, G. I. (1975) 'The use of growth regulators in citrus culture', in: *Citrus*, CIBA-GEIGY Techn. Monogr. 4.

Nagy, S. *et al.* (1977) *Citrus science and technology*, 2 vols, AVI, Westport (Conn.).

National Academy of Sciences (1978) *Underdeveloped tropical plants with promising economic value*, Wash. DC.

Nauer, E. M. *et al.* (1968) 'Growing citrus in modified UC potting mixtures', *Citrograph* **53**, 456–61.

Nemec, S. (1978) 'Response of six citrus rootstocks to three species of *Glomus*, a mycorrhizal fungus', *Proc. Fla. State Hortic. Soc.* **91**, 10–14.

Ochse, J. J. and **Bakhuizen van den Brink, R. C.** (1931) *Vruchten en vruchtenteelt in Nederlandsch-Oost-Indië*, Batavia.

Opoku, A. A. (1971) *Annual report*, Agr. Research Station Kade, Ghana.

Oppenheim, J. D. (1939) *Ontwikkelingsmogelijkheden van een citruscultuur in Suriname*, Dept. LEZ Med. 2.

Passos, O. S. and **Boswell, S. B.** (1979) 'A review of citrus spacing', *Citrograph* **64**, 211–18.

Pearce, S. C (1976) *Field experimentation with fruit trees and other perennial plants*, East Malling.

Platt, R. G. (1973) 'Planning and planting the orchard' in: *The Citrus Industry III*, Un. Cal. 48–81.

Platt, R. G. and **Opitz, K. W.** (1973) 'The propagation of citrus', *ibidem* 1–47.

Powell, J. V. and **Huang, B. W.** (1978) 'Citrus production in the United States: 1977 and 1985', *1977 Proc. Intern. Soc. Citricult.* **2**, 333–6.

Praloran, J. C. (1968) 'Les besoins en chaleur et en lumière des agrumes', *Fruits* **23**, 107–17.

Praloran, J. C. (1971) *Les agrumes*, Paris.

Proceedings of the International Society of Citriculture (1978) Lake Alfred, Florida.

Purseglove, J. W. (1968) *Tropical crops, Dicotyledons*, Longman.

Reitz, H. J. (1984) 'The world citrus crop', *Outlook on Agric.* **13**, 140–46.

Reuther, W. (1973) 'Climate and citrus behaviour' in: *The Citrus Industry III*, 280–337.

Reuther, W. and **Rios-Castaño, D.** (1969) 'Comparison of growth, maturation and composition of citrus fruits in subtropical California and tropical Colombia', *Proc. First Intern. Citrus. Symp.* 277–300.

Reuther, W. *et al.* (1967) *The citrus industry*, vol. I, *History, Botany and Varieties*, Univ. Cal.

Reuther W. *et al.* (1973) *The citrus industry*, vol. III, *Production technology*, Univ. Calif.

Reuther, W. *et al.* (1978) *The citrus industry*, vol. IV, *Crop protection*, Univ. Calif.

Rey, J. Y. (1982) 'Notes sur la collection standard d'agrumes de Nyombé au Cameroun', *Fruits* **37**, 167–79 and 217–20.

Rondon-G., A. *et al.* (1980) 'Comportamiento del *Verticillium lecanii*, patógeno del afido *Toxoptera citricidus* en fincas citricolas de Venezuela', *Agronomia tropical* **30**, 201–12.

Ryan, G. F. and **Davis, R. M.** (1967) *Guide for use of herbicides in Florida citrus*, Agr. Ext. Serv. circ. 303A, Gainesville.

Samadi, M. and **Cochran, L. C.** (1975) 'An unusual flowering and fruiting habit of grapefruit in the Khuzestan area of Iran', *Hortsc.* **10**, 593.

Samson, J. A. (1951) 'Citrus, onderstammen-onderzoek' in: *Jaarverslag Dept. LVV Suriname*, 57–8.

Samson, J. A. (1954) 'Citrus, onderstammen-onderzoek' in: *Jaarverslag Dept. LVV Suriname*, 48.

Samson, J. A. (1955) 'De invloed van verschillende factoren op het bederf van citrusvruchten', *Surin. Landb.* **3**, 306–12.

Samson, J. A. (1966) *Handleiding voor de citruscultuur in Suriname*, 3rd ed., Landb. Proefst. Med. 39.

Samson, J. A. (1977) 'Problems of citrus cultivation in the tropics', *Span* **20**, 127–9.

Samson, J. A. and **Bink, J. P. M.** (1975) 'Citrus budding in the tropics: towards an explanation of the favorable results of lopping', *Proc. 7th IOCV conf.* 213–16.

Samson, J. A. and **Byron, D. G.** (1958) 'Veredeling van sinaasappels in Suriname', *Surin. Landb.* **6**, 98–102.

Santiago, A. (1962) *An illustrated guide to Malayan citrus species and varieties*, Bull. 111, Div. Agr. Fed. Malaya s.1.

Schwarz, R. E. and **van Vuuren, S. P.** (1971) 'Decrease in fruit greening of sweet orange by trunk injection of tetracyclines', *Plant Dis. Reptr.* **55**, 747–50.

Sizaret, A. (1974) 'Suggestions pour la multiplication rapide de trois espèces fruitières (agrumes, avocatiers, manguiers), lors des premières introductions de matériel végétal', *Fruits* **29**, 767–70 (abstract 75013309, Royal Tropical Institute, Amsterdam).

Smith, R. J. (1969) 'Reducing the obstacles to efficient picking in citrus' in: *Proc. First Intern. Citrus Symp.* 609–17.

Smoot, J. J. (1976) 'Citrus: drop the subject'; *Agr. Research USA* **24**(11), 3–5.

Soule, J. *et al.* (1967) *Quality tests for citrus fruit (what every grower should know)*, Circ. 315 Agr. Ext. Serv., Gainesville.

Soule, J. (1979) *Rootstock-scion relationships*, Gainesville (Fla).

Suchtelen, N. J. van (1961) 'Citrus, bemesting' in: *Jaarverslag 1960 Landb. Proefst. Suriname*, 42–3.

Swingle, W. T. and **Reece, P. C.** (1967) 'The botany of Citrus and its wild relatives of the orange subfamily' in: *The Citrus Industry I*, Un. Cal., 190–430.

Taylor, W. E. (1969) *Agricultural pest handbook*, Njala Univ. Coll., Sierra Leone.

Terra, G. J. A. (1932) 'Growth and variation of shaddocks on different stocks', *Landbouw* (Buitenzorg) **8**, 31–4.

Toxopeus, H. J. (1937) 'Stock-scion incompatibility in citrus and its cause', *Jnl. Pomology and hort. sci.* XIV, 360–4.

Vogel, R. and **Bové, J. M.** (1977) 'Nouvelles données sur la frisolée-panachure-infectieuse des agrumes en Corse', *Fruits* **32**, 93–103.

Vullin, G. (1978) 'Multiplication des agrumes sous climat méditerranéen', *Fruits* **33**, 790–800.

Weathers, L. G. *et al.* (1982) '75 years of citrus research', *Calif. Agric.* Nov.–Dec. (CRI-16).

Webber, H. J. (1946) 'Plant characteristics and climatology' in: *The Citrus Industry I*, Un. Cal. 41–69.

Weert, R. van der *et al.* (1973) *Opbrengst van citrus in relatie tot bodemvocht*, Landb. Proefst. Suriname Bull. 90.

Weir, C. C. (1967) 'A general citrus fertilizer programme', *Citrus Research*, Un. W. Indies Bull. 8.

Weir, C. C. (1974) *Citrus growing in Jamaica*, Kingston.

Wolfe, H. S. *et al.* (1969) *El cultivo de los citricos en el Perú*, Min. Agric. y Pesq. Boletin Tecnico 72.

Worku, Z. *et al.* (1982) 'Comparative tolerance of three citrus rootstocks to soil aluminum and manganese', *Haw. Univ. Res. series* 017.

Wutscher, H. K. (1979) 'Citrus rootstocks', *Hortic. Reviews* **1**, 237–69.

Wutscher, H. K. (1980) 'Blightlike citrus decline in South America and South Africa', *Hortsci.* **15**. 588–90.

Wutscher, H. K. (1981) ''Swingle' in Texas and Florida', *Citrograph* **66**, 197.

Chapter 6

Banana and plantain

Taxonomy and morphology

In this chapter *Musa* stands for the genus and musa for the crop. The plantain, which is a vegetable rather than a fruit (see page 1), is included for two reasons: firstly, it has much in common with the banana and secondly, it is seldom mentioned in books devoted to vegetables.

Banana and plantain belong to the family Musaceae. They are large herbs with pseudostems built up from leaf sheaths; the leaves stand in a spiral and new leaves arise from an underground true stem or rhizome. There are two genera in the family, but we are concerned here with only one: *Musa*. The following description is based on Purseglove (1972), Simmonds (1966) and Champion (1963).

Musa has pseudostems composed of tightly clasping leaf sheaths, slightly swollen at the base. Suckers are freely produced by the true stem. The genus can be divided into five sections, in four of which the inflorescence stands straight up. In the fifth section *Eumusa*, the inflorescence hangs down.

All edible bananas and plantains, with the exception of the unimportant Fe'i banana, descend from a wild ancestor: *Musa acuminata*. The plantains also carry genes of another wild ancestor: *Musa balbisiana*. Let us call their genomes respectively A and B. We then find most edible bananas to be triploids that can be described as AAA; in other words they carry three sets of chromosomes derived from *M. acuminata*. The plantains are generally triploids too: AAB and ABB. Other combinations exist, namely AA, AB, AAAA and ABBB, but they occur less frequently.

In older literature one usually finds the names given by Linnaeus: *Musa sapientum* for the banana and *M. paradisiaca* for the plantain. Linnaeus did not (and could not) know that both his descriptions were based on AAB hybrids; his names may therefore not be used as general names for either banana or plantain. Later names (*M.*

cavendishii, M. nana, M. sinensis and *M. corniculata*) are also dubious and should be avoided. Simmonds proposes to abandon formal Latin names altogether and to use a cv. name, preceded by the genus name and a group indication, e.g. *Musa* (AAA group, Cavendish subgroup) 'Robusta'.

We shall now briefly describe the structure of the banana. It is a tree-like herb, perennial but monocarpic, from two to six metre high, with milky juice in all parts. Adventitious roots spread to all sides and down, forming a dense surface mat. The underground stem (corm, rhizome) has short internodes. The corm's terminal growing point produces leaves in a spiral succession; in the axil of each leaf a bud is present which may grow out a shoot. Normal leaves, of which about thirty are formed, consist of a sheath, a

Fig. 6.1. Inflorescence of 'Robusta'; note the rolled back bract and flower parts on fruits

petiole and a blade. The sheaths are nearly circular and tightly packed into a non-woody pseudostem; they are much longer than the blades. The petiole is 30–90 cm long and U-shaped. The blade emerges from the middle of the stem as a rolled cylinder ('cigar') and unfurls slowly. Older leaves are pushed aside until they hang down and their blades shrivel. For a time the number of functioning leaves remains more or less constant. The last leaf is short and spade-like.

Musa is monocarpic: a shoot flowers only once and dies after it has borne fruit. Yet the plant is perennial, as the corm's life is perpetuated by suckers; a whole clump may thus develop. After a sucker has formed a fixed number of leaves, its meristem changes into a flowering stem that begins to push up through the pseudostem. About a month later it emerges in the leaf crown and hangs down under the influence of gravity. It is a complex spike consisting of a stout stalk with flower clusters in a spiral. Each cluster has 12–20 flowers in two rows, covered by a large reddish bract.

All flowers possess a pistil and stamens, but the pistil becomes progressively shorter. In the first five to fifteen clusters it is large enough to become a fruit; those flowers are functionally female, called female for short. The other flowers are called neuter or male, although they have small pistils and the stamens rarely contain pollen. One by one the bracts rise to expose the flowers; they usually fall after one or two days, but in some cvs they stay on. At the lower end they form a bulbous 'male bud'. The axis above that is generally bare, but in some cvs male flowers and bracts are retained.

The banana fruit is a berry. It contains many ovules, but no seeds; the fruit develops by means of parthenocarpy, i.e. without fertilization. In contrast to the bunch, which grows down, the fruits curve upwards. This has led to an ideal shape in cv. Gros Michel, so that the bunch may be transported unwrapped. In most cvs the fruits stick out in all directions and are easily damaged in transport.

A fruit cluster is generally called a 'hand' and a single fruit a 'finger'. They differ from cv. to cv. in characteristics such as shape, size, colour of skin and flavour. A good bunch consists of eight hands of 15 fingers, average weight 150 g; the fruit thus weighs 18 kg, the entire bunch about 20 kg. However, bunch weights of 30 kg and more are not at all rare.

Uses and composition

After citrus, musa is the most important fruit in world trade. It is grown in home gardens for home consumption or the local market,

Fig. 6.2   Banana as an intercrop in rubber, Ivory Coast

but also in very big plantations for export. Between these extremes we find small farms delivering fruit for export on a regular or more casual basis. The crop may be grown in monoculture or in a mixed cropping system with maize and beans, or as a shade or nurse crop for coffee, cacao, rubber (Fig. 6.2) and citrus.

Bananas are harvested in a nearly ripe state for home use, but for export they must be shipped unripe. On arrival in the port of destination the bunches (now more frequently the boxes) are placed in ripening rooms. After a few days they are sold to retailers who sell detached hands to the public. The European and American consumer knows only one or two of the best export cultivars and has no idea of the enormous diversity existing among bananas and plantains in the tropics.

Bananas are chiefly eaten raw, as a dessert fruit; in the ripe state they are sugary and easily digestible. Plantains are much starchier and can be eaten either ripe or unripe; they are cooked, fried or roasted. In world trade plantains are insignificant, but for home use they are as important as bananas. In the equatorial belt of Africa, stretching from Kilimanjaro to the Atlantic, plantains and cooking bananas are the main staple food; daily consumption here may surpass 4 kg per head. Bananas and/or plantains are also important in parts of South-East Asia and on some Pacific islands.

In Africa large quantities of beer are brewed from musa. It has a low alcoholic but high vitamin content and is therefore of some significance in nutrition. Special cvs are used for this purpose. Other products derived from musa are 'figs' (dried slices of ripe bananas), powder (ground from ripe bananas), chips and flour (both made from unripe fruits), flakes, juice and purée. The banana may be used as a dietary food, e.g. against stomach complaints; it is said to contain an unidentified compound called, perhaps jokingly, vitamin U (against ulcers). On the other hand musa fruits contain serotonin and other compounds that increase blood pressure and might contribute to certain heart conditions if used in large quantities.

In parts of Asia male buds are eaten as a vegetable; several changes of water are required during their preparation to reduce the astringent taste. Male buds and chopped pseudostems are also fed to cattle.

Von Loesecke (1950) reports that 'during the period of plentiful supply, banana jam was prepared in considerable quantities'. His recipe was tried out in Surinam and modified by de Vos (1953) who used lime instead of lemon juice. Banana catsup is now more popular than tomato catsup in the Philippines (Valmayor, 1976).

Green leaves are often used as umbrellas and plates, or for wrapping. Fibre made from the pseudostem is not as strong as abacá or manila hemp (from *Musa textilis*), but can be used for making bags, rope and other textiles. The Banana Institute in Hamburg (Germany) reported that the biological fibre of cv. Robusta was 15–25 cm long, the technical fibre 100 cm, and the output was 1.5 per cent (Small, 1964).

Banana sap causes brown indelible stains on clothes and this explains the Surinam saying '*Mama mofo na banawatra*'. Literally this means: 'mother's mouth is [like] banana water' and freely translated: 'heed your mother's words'. According to Purseglove the sap can be used as marking ink.

The pulp of the ripe banana contains about 70 per cent water and for the rest mainly carbohydrate. One hundred grams of banana produce 100 calories. There is little fat or protein, but it is a fairly good source of vitamins A, B_1, B_2 and C. Plantain has practically the same nutritional value but, like the potato, carbohydrate is mainly present as starch rather than sugar.

Origin, distribution and production

Musa's wild ancestors are still found in the forests of the Malesian area; several subspecies of seeded diploid *M. acuminata* occur there. Long ago some of these subspp. crossed; as a result there was an increase in female sterility, parthenocarpy and triploidy. Inhabitants of that region then discovered that some of the plants had edible fruits and could be propagated by suckers. Selection by man has, over millennia, profoundly altered the properties of *Musa acuminata* in the humid tropics.

Meanwhile these plants had also been taken by man to drier monsoon areas, where *Musa balbisiana* is native. The two spp. crossed and the genome groups AB, AAB and ABB came into existence. There are a few edible forms of pure *M. balbisiana*. That sp. confers hardiness and drought resistance to its hybrids: they also have starchier fruits.

Malayan sailors probably took bananas to Madagascar about the fifth century AD and they spread to the east coast and mainland of Africa from there; plantains arrived much later. Both forms were already known on the west coast of Africa when the Portuguese arrived in the fifteenth century. Later musa was introduced into the western hemisphere and into other parts of the world. Nowadays, it is found in every tropical country. One cultivar only, the Dwarf Cavendish, can be grown successfully outside the tropics, e.g. in the Canary Islands, Egypt, Israel, Natal and Southern Australia. Twelve countries produce more than a million tonnes of banana per year (Table 6.1). In some cases plantain may be included in the banana figures as 'Unfortunately, several countries make no distinc-

Table 6.1   *The major producers of banana and plantain (× 1,000 tonnes)*

Banana	1980	1981	1982	1983	1984
Brazil	6,721	6,696	7,088	6,692	6,968
India	4,830	4,500	4,724	4,500	4,606
Philippines	3,977	4,000	4,100	4,200	4,100
Ecuador	2,269	2,010	2,265	2,000	1,924
Thailand	2,014	2,021	2,028	2,035	2,045
Indonesia	1,977	2,501	1,800	1,800	2,000
Mexico	1,501	1,562	1,621	1,624	1,500
Honduras	1,330	1,330	1,338	1,250	1,250
Colombia	1,030	1,155	1,274	1,280	1,200
Costa Rica	1,092	1,144	1,150	1,021	950
Panama	1,050	1,080	1,100	1,100	1,100
Vietnam	895	900	1,000	1,100	1,200
World	40,051	40,575	41,244	40,700	41,113

Plantain	1980	1981	1982	1983	1984
Uganda	3,330	3,350	3,380	3,400	3,410
Colombia	2,348	2,400	2,490	2,500	2,300
Rwanda	2,050	2,100	2,158	2,170	2,200
Zaire	1,434	1,454	1,469	1,480	1,480
Nigeria	2,250	1,360	1,395	1,420	1,420
World	21,789	21,986	22,582	19,777	20,268

Source: FAO Production Yearbooks
NB: FAO indications for estimates have been omitted

tion in their statistics' (FAO, 1976). Other large producers in 1977 were (× 1,000 tonnes); Madagascar (449), Western Malaysia (432), Argentina (374), Canary Islands (371), Uganda (320), Dominican Republic (315) and Nicaragua (310). In 1982 Cameroon, Ghana, Peru, Tanzania, Ivory Coast, Ecuador, Dominican Republic and Venezuela produced more than a half million tonnes (FAO, 1983). Table 6.2 shows the high yielding countries. Among them we find several on the fringe of the climatic zone considered fit for bananas. This proves that care and good cultivation practices may compensate for unfavourable climatic conditions. It is also worthy of note that three countries (Senegal, Cuba and Surinam) which were far below world average in 1961–65 have moved up to over twice that average.

Table 6.2 *Countries with high banana yields (kg/ha)*

Country	1961–5	1973	1974	1975
Argentina	23,601	32,015	33,325	34,027
Costa Rica	17,450	33,138	31,053	33,170
Paraguay	23,308	31,875	32,250	32,350
Senegal	5,574	28,434	27,778	32,222
Honduras	24,091	27,200	36,757	30,000
Israel	21,223	20,484	17,711	28,947
W. Malaysia	12,554	34,767	26,349	26,875
Sudan	30,000	27,211	26,667	26,667
Cuba	9,417	25,457	25,667	26,000
Surinam	9,960	22,657	24,429	25,556
Brazil	18,754	22,547	25,077	25,287
Bolivia	23,553	23,970	25,308	25,231
Papua N. Guinea	25,326	25,294	25,143	25,210
Port. Azores	25,410	25,063	25,125	25,188
World yield	11,237	12,321	12,719	12,690

Source: FAO (1976)

Banana is also important in international shipping and trade. This began early in the 19th century in schooners from Caribbean Islands to the United States. Later, railroads were built in Central America and farmers were encouraged to grow banana to provide freight for the trains. This highly perishable fruit must reach the consumer within two weeks after the harvest; thus, a high degree of organization in planting, cutting, transportation and distribution is essential.

Several small companies were founded to grow or buy bananas and transport them to the United States. The first special banana boat was built in 1888 and the first refrigerated ship room was installed in 1901. Meanwhile a Boston company had been buying up smaller companies and in 1899 the amalgamated United Fruit Company was set up. Its 'great white fleet', also carrying passengers, soon became well-known. UFC had enormous concessions in Central America and dominated politics in the 'banana republics' for decades. Its competitor, Standard Fruit Company, played a similar role but later and on a smaller scale. Nowadays these companies are restricting their activities mainly to buying and transporting the fruit. For more details, see May and Plaza (1958) and UPEB (1980).

Research activities of UFC and SFC are continued by UPEB (Union de Paises Exportadores de Banana) which was founded by six countries: Colombia, Panama, Costa Rica, Honduras, Guatemala and República Dominicana. A coordinated research programme was set up on phytopathology, entomology, nematology, soils and fertilizers, management techniques, post-harvest physiology, processing and social-economic affairs (UPEB, 1979). Bibliographies and a monthly newsletter are published.

The banana is a labour-intensive crop, giving regular work to many people; the labour requirement is 1 person/2 ha. Investment costs in Surinam were calculated at about $5,500/ha or $11,000/labourer; lower than for a mechanized rice farm (Klasen and Mulder, 1975).

Growth and development

There are often 400 roots on a single corm and sometimes as many as 700. Even shoots without functional leaves may have 200 roots whose total length may reach 600 m. A strong correlation was observed by Lassoudière (1978) between number of roots put out by the young sucker and subsequent bunch weight. When flowering begins, root formation is terminated. The total number of leaves on a banana plant is between 60 and 70; at first scale leaves are formed, then narrow (sword) leaves (this stage is sometimes missed) and finally about 30 normal leaves. The rate of production is about one leaf per week, but it may take considerably longer if conditions are unfavourable. In a greenhouse in Wageningen (52°N) the time between two leaves in summer averaged 8.3 days for 'Dwarf' and 6.5 days for 'Gros Michel' (J. Mohede, personal communication).

Meanwhile, total leaf surface increases until a maximum is reached and then it starts to decline. Leaf size is about 1 m^2 in 'Dwarf' and 3 m^2 in 'Gros Michel'. There are 10–15 functioning leaves on a healthy plant, so that leaf surface per plant is 10–45 m^2 or 20,000 to 45,000 m^2/ha; in other words the LAI (see Ch. 3) is 2–4.5, but usually between 3 and 4. Gietema-Groenendijk (1970) reports that 80 per cent of the photosynthesis of the plant takes place in leaves number two to five. Stomata are found on both sides of the lamina, about 54/mm^2 on the upper and 200/mm^2 on the

Fig. 6.3   Root system of a young 'Dwarf' banana, seen through a glass pane

lower surface. Along the midrib lie cells that can lower the blade halves in case of drought; thus reducing transpiration. Perpendicular to the midrib are thousands of veins; although the leaves tear easily, a connection with the midrib is normally maintained.

As leaves get older, they begin to wither and chemical elements are transported to the corm and up again to young leaves. Van der Vorm and van Diest (1982) found that in cv. Gros Michel the following elements are strongly redistributed: N, S, P, K and Ca; Mg, Cl, Fe and Zn were redistributed only moderately and Cu not at all. No data were obtained on Si: the authors assumed that it was not redistributed but returns to the soil.

As with sugarcane and pineapple, the first sucker set in the field is called the 'plant crop', subsequent suckers are 'ratoons'. Under ideal soil and climate conditions, and where nematodes and diseases are not a problem, many ratoons may be harvested. Simmonds reports banana fields over 100 years old in India and of 50 and 60 years in Uganda. A 40-year old plantation in Honduras was producing 60 tonnes/ha on over 3,000 ha! (Darthenucq *et al.* 1978). On the other hand, some fields are replanted at every crop; examples are the one-crop cultivation of 'Dwarf Cavendish' in Bombay and plantain in Surinam. However, in most commercial fields four or five ratoons are taken before the land is cleared for replanting.

Because suckers generally appear on the middle or upper part of the corm, the stool tends to rise gradually; bad anchorage results, which limits its life. Suckers can be classified in several ways. In Jamaica the following terms are used (Simmonds, 1966):
1. Maiden sucker, a large but non-fruiting ratoon;
2. Sword sucker, a sucker bearing narrow leaves;
3. Water sucker, a superficial sucker with broad leaves;
4. Peeper, a very young sucker with scale leaves.
All these may be used as planting material, but they are not equally good. Bits, i.e. pieces of corm bearing at least one bud, are also used. Champion (1963) uses quite different criteria for his classification of suckers, namely origin (central or lateral) and attached part (corm or grown-out pseudostem).

We have already described the flower stalk, flowers and fruits. In fact there are many variations, but we shall describe only those that help us to recognize cvs easily or have some significance for crop production. For instance, in certain cvs male flowers and bracts are retained on the male axis, or only on the lower part. The inflorescence does not always go straight down, but sideways. In AAA cvs the petiolar canal is open, in AAB cvs the margins touch. The stages of maturity are briefly: about 80 days after the appearance of the inflorescence, the bunch may be harvested at an immature stage suitable for shipping; ten days later the bunch may still be shipped but not too far. After another ten days the bunch is 'full', unsuitable for export, but just right for home consumption. These figures refer to equatorial lowlands.

Ecology and physiology

The banana is predominantly a tropical crop, unlike citrus. We have already noted one exception, the 'Dwarf', which also grows well in some subtropical areas. There are no exceptions in the case of the

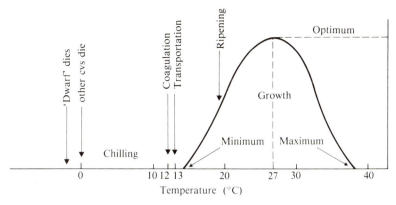

Fig. 6.4 Relation between temperature, growth and other processes in banana culture

plantain. We may therefore say that musa is a crop of the tropical lowland. Any extension of its culture into the mountains or subtropics will prolong the growth cycle considerably, i.e. one month/100 m altitude.

The relationship between temperature and some processes in banana growing is shown in Fig. 6.4. Except for the Dwarf cv., banana cannot withstand frost. Chilling injury occurs at temperatures below 12° C, when latex coagulates. Bananas are transported at temperatures between 12 °C and 13 °C, when respiration is low. Growth begins at about 18 °C, reaches an optimum at 27 °C, then declines and comes to a stop at 38 °C. Still higher temperatures cause sun-scorch. Ganry and Meyer (1975) have shown that banana fruits increase in girth up to 29 °C. Turner and Lahav (1983), working with cv. Williams in growth chambers, observed chilling injury at 17°(day)/10 °C(night) and heat injury at 37°/30 °C; the highest dry weight was found at 25°/18° C and the greatest leaf area at 33°/26 °C.

It is remarkable that the relation of banana growth to light has not been the subject of much research. Purseglove (1972) categorically states: 'Bananas require a high light intensity', but gives no source in support of this statement.

We all know that bananas grow well as shade or nurse crops in cacao plantations, but hardly anyone has asked the question: 'how much do they produce there and how long is the growth cycle?'. Simmonds mentions one experiment by Murray where shade, down to 50 per cent of full sunlight, had no effect on growth or yield; 80 per cent shade probably slowed growth but did not affect yield.

Shade has a beneficial effect on leaf spot disease and this may have confused the issue.

We also know that some major producers, especially Ecuador and Guinea, do not have sunny climates. In both countries the sky is overcast during a large part of the year. Brun found in a laboratory experiment that banana photosynthesis increased strongly from 2,000 to 10,000 lux and slowly up to 30,000 lux. Bright sunlight provides about 80,000 lux, so we may assume that when one quarter of that intensity was reached, the curve was already flattening out. The temperature and the age of the leaves were not stated.

In an unpublished MSc thesis Gietema-Groenendijk (1970) reported that young banana leaves at 32 °C could reach a photosynthetic rate of 500 μg CO_2/m^2/second; twice as much as Brun found. Her findings are shown in Table 6.3. Light saturation would probably have been reached at slightly over 210 W/m^2, or less than half the irradiance of full sunlight. No significant effects of daylength on banana have been reported.

Table 6.3   *Influence of irradiance on photosynthesis of banana*

W/m^2	Photosynthesis	Duration
0–28	Gradual increase	50 minutes
28–70	Fast increase	20 minutes
70–105	Increase less fast	20 minutes
105–161	Increase less than before	20 minutes
161–209	Very little increase	20–30 minutes

Note: maximum irradiance under cloudless sky is 1,100 W/m^2, half of which is photosynthetically active.
Source: Gietema-Groenendijk (1970)

Purseglove puts the minimum amount of water needed for good growth at 25 mm/week, he considers an average of 2,000 to 2,500 mm/year satisfactory, if well distributed. Simmonds (1966), on the other hand, has made certain assumptions about levels of rainfall and temperature and regards 100 mm rain/month and a monthly average of 27 °C as ideal. On that basis he defined 'effective rainfall' as 'counting all months that have four inches or more of rain as having four inches only and all months which have less as having the amount actually recorded'; the maximum effective rainfall is therefore 1,200 mm. Similarly, 'effective temperature' was defined, with 27 °C as basis. From data on 27 regions he constructed a scatter diagram, which is reproduced in part and with metric units in Fig. 6.5; subtropical countries have been omitted but Surinam was added. In the upper right part of the diagram we find the warm and wet regions, in the lower right the warm and dry ones. The other regions are dry to moderately wet and rather cool. In Simmonds' definition temperatures above 27 °C count as optimal; furthermore

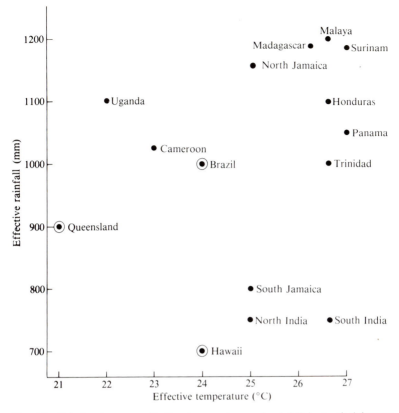

Fig. 6.5   Relation between effective temperature and rainfall in tropical banana regions. *Source*: Simmonds (1966)

one 100 mm downpour on the first day of the month, followed by 30 dry days, would be on a par with ten showers of 10 mm at three-day intervals. Therefore, Purseglove's weekly amount is agronomically more correct, but it would still be better to work with E_p and the water-holding capacity of the soil, as set out in Chapter 2.

How much water does a banana plant need for sustained growth? In bright sunlight the transpiration amounts to 50 mg per dm^2 of leaf per minute ($= 300$ g m^{-2} h^{-1}) according to Champion (1963), who quotes examples from Israel and Brazil. A 'Dwarf' plant used 25 litres water on a clear day, 18 litres on a cloudy day and 9.5 litres when the sky was overcast. If we may generalize these figures, we find 1,875 m^3 water in a sunny month for a full-grown plantation at a density of 2,500 trees/ha. This is equivalent to 187.5 mm rainfall.

and conforms to experiments in Guinea where the best result was obtained with 180 mm irrigation per month in the dry season.

In Surinam 200–300 mm rain per month was considered optimal, but this was not based on experiments. Van Emden (1967) found a correlation between bunch production and rainfall 16 weeks earlier. Van Sloten and van der Weert (1973) used the stomata as an indicator of moisture stress. When the soil contained ample water, the stomata opened before 8 a.m. and closed between 12 and 2 p.m. They closed much earlier when the soil moisture tension in the root zone reached a value equivalent to pF 2.7; irrigation at this pF was recommended.

Meyer and Schoch (1976) have reviewed the literature on water needs of banana and found the results diverging, even contradictory; they cannot be simply applied elsewhere. Measurements in lysimeter tanks on Martinique and Guadeloupe showed maximal E of an adult banana plant to be 1.6 times that of a lawn and 1.4 times the evaporation from a class A pan. The last datum is probably the simplest approximation of Ep. Bovée (1975) found an ET to class A pan relation of 0.82 in Lebanon (33° N); the yearly requirement was 1,200 mm, with a peak of 6 mm per day in August.

According to Champion it seems that the banana can easily take up 30 per cent of the available water from a soil at field capacity; but when 60 per cent has been used, the plant will show signs of wilting. This causes the stomata to close, which leads to a decrease in photosynthesis, slower growth and formation of fewer leaves, even under irrigation. Continuation of drought brings about shrivelling of leaves and breakage of stems; the corm, on the other hand, is drought-resistant and can resume growth later. As a consequence we may also assume that low relative humidity is deleterious to the banana.

Simmonds (1966) describes the effects of wind on banana in detail. Tearing of blade halves is usual, but a wind velocity of 25–30 km/h gives rise to crown distortion, breakage or uprooting of pseudostems and root damage (the result of which may not become evident until months later). A wind speed of 65 km/h will cause considerable loss and at 100 km/h banana fields are competely destroyed. Yet, other crops suffer even more from hurricanes and preference may then be given to a fast-growing crop such as banana over slow-growing crops like cacao and nutmeg. This happened in the Windward Islands after Hurricane Janet in 1955 (Haarer, 1964). The following areas are affected by cyclones: Central America, West Indies, Canary Islands, Mozambique, North Madagascar, Peninsular India, middle Burma, Indochina, South China, Taiwan, Philippines, Queensland, Solomon Islands to Marquesas. Not affected are equatorial areas: Ecuador, Colombia, Panama, Brazil, Guiana, Guinea, Cameroon, Kenya, Tanzania, Malaysia and Indonesia. Eastern South Africa, New South Wales and Hawaii are also free of hurricanes.

Effects of cold, frost and hail are not considered in this book, although they may occur if the crop is grown in mountainous areas. Lightning too can do severe damage to single plants and its effect could be mistaken for disease.

What kind of soils are needed for banana culture? According to Purseglove 'a wide range of soils is suitable, provided there is good drainage and adequate fertility and moisture'. The clay content should be below 40 per cent and the water table below 1 m (Stover, 1972). The question should not be 'is this a good banana soil?', but 'can this soil be made to grow good bananas?' (Simmonds, 1966). Three factors determine this: structure, depth and absence of toxic substances. Nutrients can come out of a bag and pH is not of itself essential (except that low pH favours the spread of Panama disease).

Good drainage is the crucial factor. Champion points out that the banana root is fragile, cannot stand stagnant water and must grow in a well-aerated, yet moist environment; a ground-water table below 80 cm, absence of rocks or impenetrable layers, good structure and porosity are also essential features. Although the banana is tolerant of pH in the range of 4.5–8.0, the best growth is found from pH 6 to 7.5.

Concerning topography, Simmonds states that flat terrain is preferred for commercial banana production, for reasons of transport, irrigation and prevention of erosion; however, there is not enough flat land to go round and a fair proportion of the world's bananas are grown on slopes. A high humus content is regarded as very favourable. Simmonds mentions the 'low flat Tabatinga of the Brazilian coast, consisting of 12–30 cm of a fertile, humic top-soil on acid, blue waterlogged clay' and the Poto-poto of Guinea. These soils probably resemble the 'pegasse' (peat) soils of Surinam that

Table 6.4 *Surinam 'pegasse', compared to 'normal' banana soil*

Properties	'Pegasse'	'Normal' soil
Topography	Flat	Flat
pH (water)	4.8	4.5–8.5
Organic matter %	5	3–8
Base saturation %	58	'High'
Salt content in mg/100 g dry soil	59–88	Less than 50
Chemical fertility	Good	Good
Horizontal drainage	1–3 m/24 h	Good
Vertical drainage	Low	Good
Horizontal root growth	1 m	Up to $4\frac{1}{2}$ m
Vertical root growth	50 cm	Up to $\frac{1}{2}$ m

Source: van Amson (1961)

Fig. 6.6   Aerial view of a banana plantation on pegasse soil in Surinam

have a high production potential (Fig. 6.6) provided they were reclaimed without burning and are well drained. Van Amson (1961) compared such a soil to one generally preferred for banana growing (Table 6.4).

There is much variety in pegasse soils, especially in thickness of the organic layer. The soil of Table 6.4 has three critical points: the salt content of the subsoil, the vertical movement of water through

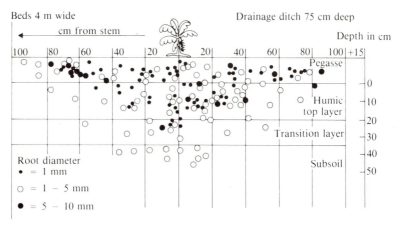

Fig. 6.7   Root profile of 'Robusta' banana on pegasse soil in Surinam

the profile and root growth. Most roots were found to grow in the pegasse layer (see Fig. 6.7). Lassoudière and Martin (1974) recommend intensive drainage for an organic (peaty) soil in the Ivory Coast. The drains should be 10 to 15 m apart to a depth of 80–100 cm. Thus a good root system develops and this influences aerial growth significantly.

Champion (1963) has reported that nutrients immobilized in 'Dwarf' plants from two locations were present in the proportion N:P:K = 10:1:35. As for the amounts taken out with the harvest, the averages were: N 1.5; P 0.20; K 4.6; Ca 0.18; and Mg 0.21 kg/tonne of fruit. A yield of 40 tonnes/ha would therefore correspond to an export of 184 kg K, a heavy taxation on chemical fertility; apart from this, however, no great demands on fertility were made.

De Geus (1973) has formulated the following requirements: 'a soil that is not too acid, rich in organic material, with a high nitrogen content, an adequate P level and plenty of potash'.

Cultivars

The number of musa cultivars is not known exactly, but is estimated at 100–300. All are propagated vegetatively and may therefore be called clones. We have already mentioned the genome groups and pointed out that the triploid groups AAA, AAB and ABB are the most important ones. We shall now discuss the main cvs of these groups and then give a few examples of the other groups. Names like Apple, Fig and Lady's finger have been omitted as they are loosely applied to many different cvs and are therefore confusing.

'Gros Michel' was the leading banana cv. in world trade for nearly a century and there are good reasons for this. It produces heavy, symmetrical bunches that permit unpacked transport; the fruit has an attractive colour and appearance and is long and slender (five times as long as broad). The old bunch grading system also worked in favour of this cv.; a nine-hand bunch received the full price, bunches with fewer hands progressively less. Bunches are now generally paid for by weight and fruit is shipped in cartons as detached hands; 'Gros Michel' has therefore, lost two important advantages over its competitors: the high 'count' and the easy shipping.

One disadvantage of the cv. is that it is big. Planting density must, therefore, be low, resulting in a lower yield per ha than from the Cavendish group. Its height also makes it very vulnerable to strong winds. But the greatest handicap is that 'Gros Michel' is susceptible to Panama disease and has therefore largely been replaced by members of the Cavendish group. Small (1961) reports that it was still the preferred cv. in Central and South America in 1960, although large areas had been abandoned because of Panama disease. As long as new land could be put into use at relatively low cost, he said, 'Gros Michel' would continue to be planted.

On the other hand, Bigi *et al.* (1969a) report that in 1956 'Gros Michel' had disappeared on the Atlantic coast of Costa Rica to be substituted by Cavendish cvs. It is not known exactly how big the areas for each cv. are, but we assume that Ecuador is the only country where 'Gros Michel' is still grown on a large scale. Even there Bigi *et al.* (1969b) reported that the area planted to 'Gros Michel' was expected to drop from 147,000 ha in 1967 to between 120,000 and 80,000 ha in 1970, whereas the Cavendish cvs were predicted to rise from 12,000 to 40,000 or 60,000 ha at the same time. It seems likely that the trend will continue. All over Africa and the West Indies and in large parts of Central America 'Robusta' and other Cavendish cvs are now in favour, as they always have been in the Canary Islands and Brazil. Several mutants of 'Gros Michel' are known and two of these, called 'Highgate' and 'Cocos' have been planted to some extent; 'Highgate' is shorter than 'Gros Michel', has about one hand per bunch more, but the fruits are shorter and less compact. 'Gros Michel' is known as Pisang ambon in Indonesia and Malaysia.

The name Cavendish is often used for one cv. only, the 'Dwarf'. Here we prefer to use the name Cavendish for a subgroup, of which 'Dwarf' is the smallest member. At the other end of the scale is a tall cv. called 'Lacatan' in most banana regions; this is really a misnomer, for a quite different cv. in the Philippines rightly bears that name, but it has become common so it will be used here. Between 'Dwarf' and 'Lacatan' there is a practically continuous transition of Cavendish forms, some of which are recognized as cvs. Mutations regularly appear.

All Cavendish cvs are resistant to Panama disease and have fruit with blunt tips, in contrast to 'Gros Michel' which has a bottle-necked fruit tip. The bunch is irregularly shaped and must be packed in paper or polythene for transport; this disadvantage has lessened because detached hands are now shipped in cartons. 'Dwarf Cavendish', or for short 'Dwarf', is better adapted to a cool climate than any other cv.; it is grown extensively in the Canary Islands (Fig. 6.8) and Brazil, but only scattered in tropical countries. In India it is called 'Basrai'.

'Giant Cavendish' is only a giant compared to 'Dwarf'; it is but slightly taller and the French name 'Grande Naine', big dwarf, is nearer to the truth. In Australia the names 'Mons Mari' and 'Williams Hybrid' are used, in India one of its names is 'Harichal'. Bracts and male flowers are persistent on the lower part of the male

Fig. 6.8   Flowering 'Dwarf' banana in Canary Islands

axis in 'Dwarf' and 'Giant Cavendish'. The 'Paz' of Israel is also in this group.

'Robusta' or 'Poyo' has taken over the role of leading cv. from 'Gros Michel', It is grown extensively in the West Indies, Central and South America and Africa. Unlike the other two, the male bracts fall off. 'Lacatan' is the tallest Cavendish cv. and thus shares some of the disadvantages of 'Gros Michel'. The Malayan name Pisang masak hijau means 'green, but ripe' [fruit].

Champion has set out some properties of the main cvs in a table, to which have been added planting density and reaction to pathogens (Table 6.5). Other Cavendish cvs are 'Valery' and 'Americani'.

Table 6.5 *Comparison of five banana cultivars*

Properties	Gr. Michel	Lacatan	Robusta	Giant Cav.	Dwarf
Height (m)	4–8	4–5	2.8–4	2.5–3	1.8–2.1
Stem colour	Green	Green-reddish-brown _____			
Blade length (cm)	400	300	220	164	156
Blade width (cm)	110	80	80	72	78
Proportion l/w	3.65	3.70	2.75	2.30	2.00
Bunch type	Cylindric _____			Truncate	Conic
Internal fr., hand 1	¾ straight _____		½ straight	⅓ straight	Curved
Fruit top	Bottleneck	Rounded _____			
Fruit stalk	Thick	Thin	Thin	Medium _____	
Cycle (months)	13–15	13–14	12	11	11
Male axis	Bare _____			Partly covered _____	
Density, stems/ha	1,000	1,500	2,000	2,200	2,500
Panama disease	Susceptible	Resistant _____			
Leaf spot	Susceptible	Very susceptible _____			
Burr. nematode	Resistant	Susceptible _____			

Source: partly from Champion (1963)

'Valery' closely resembles 'Robusta'. When Walker (1970) compared them with 'Lacatan' he found 'Valery' to be more wind-resistant and much higher yielding than the others; it was more erect than 'Robusta' and the male flower bracts were persistent during the later stages of the maturation period. Stover (1982) compared cvs Valery and Grande Naine: the latter was 78–98 cm shorter, had a smaller leaf area, suffered less damage from wind and produced

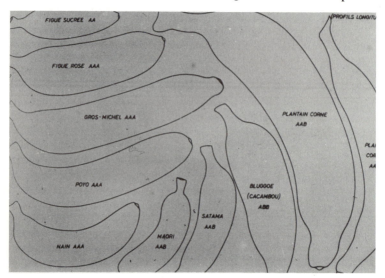

Fig. 6.9 Comparative shapes and sizes of *Musa* cvs
   Source: Champion (1968)

more fruit. It was more resistant to Black Sigatoka, nematode rootrot and wilt.

'Americani' is smaller than 'Robusta' and has a leaf length to width proportion of 2.2–2.5, which places it close to 'Giant Cavendish'. It has been grown in Madagascar since about 1966 (Moreau, 1976) and has successfully been introduced into Cameroon and other French-speaking countries. It is drought-resistant and capable of a production of 60 tonnes/ha in the second year, with many bunches of 40 kg. In Queensland three types of 'Mons Mari' are recognized: Dwarf, Queensland and Maroochy. The mutation rate of bananas is about two per million plants per generation; say 1/125 ha/year. This makes it likely that new forms will keep on appearing, some of which may become cvs.

Before we turn to the AAB and ABB groups, it is necessary to make a comparison between *Musa acuminata* and *M. balbisiana*. Simmonds gives a list of 15 characters used in the scoring of musa cvs. For each character corresponding with the wild *M. acuminata* a score of 1 was given; characters corresponding to *M. balbisiana* got a score of 5 and intermediate characters were given intermediate scores. Thus a cv. got a total score of at least 15 and at most 75. The AA and AAA cvs obtained scores of 15–24, AAB cvs 25–54 and ABB 55–64. Table 6.6. summarizes seven of these characters. The AAB group includes the plantains, but also 'Mysore' and 'Silk' which can be eaten raw. On the other hand, not all cooked musa fruits are plantains; some, especially in East Africa, come from AAA cvs.

Table 6.6 *Seven characters of musa cultivars as expressed in the parent spp.*

Character	Musa acuminata	Musa balbisiana
Stem colour	Black blotches	Green
Petiolar canal	Open, U-form	Closed, O-form
Fruit stalk	Short	Long
Ovules	2 regular rows	4 irregular rows
Bracts	Roll back	Do not roll back
Bract shape	Narrow	Broad
Bract tip	Sharp	Blunt

Source: Simmonds (1966)

The plantain subgroup contains the French plantains (Fig. 6.10), which have persistent relics of male flowers and bracts along the male axis and the Horn plantains where the male axis is absent or degenerates early. Horn plantains have small bunches with very large fruits and are often greatly preferred over other types. They are grown on a large scale in association with coffee in Colombia (Stover, 1983b). In both plantain groups apical dominance is so strong that only the higher eyes grow out, producing peepers (de

Fig. 6.10   French plantain in a variety collection at Azaguié, Ivory Coast

Langhe, personal communication). Three groups of plantains are known in Ghana: Apem (French), Apantu (Horn) and an intermediate group (Karikari and Abakah-Gyenin, 1976).

Simmonds describes two cvs of the ABB group: 'Bluggoe' and 'Pisang awak'. The first was important in Grenada and Samoa but suffers from Moko and Panama disease; it has to be cooked. The second can be eaten raw and is popular in Thailand.

Plantains and cooking bananas are among the most important food crops of the tropics. 'Because these crops are not exported, their importance to the economy is often underestimated' said E. de Langhe, president of the recently formed International Association for Research on Plantain and other Cooking Bananas (IARPCB) in the first issue of *Paradisiaca*, its Newsletter (1976). The second of these Newsletters (1977) is a bibliography of 391 articles on plantain and cooking banana.

Plantains are often intercropped with other food crops; the combination with cocoyam (Fig. 4.3) gives a high return and has a low labour input (Devos and Wilson, 1979).

Cultivar collections of plantains exist in Jamaica, Honduras, Colombia, Venezuela, Ivory Coast, Ghana, Cameroon, Tanzania, Philippines and India (*Paradisiaca*, 1976). I have seen the collections of Cameroon and Ivory Coast and was struck by the great diversity of plantain types. Among the many interesting cvs were 'Monthan' (ABB), 'Njock Korn' and 'Popoulu' (AAB); the latter

Table 6.7 *Comparision of plantain cultivars, first cycle, in intensive, irrigated culture in Cameroon*

Cultivar	Njock Korn	Njock Korn	French Sombre
Altitude (m)	80	550	550
Density (stems/ha)	1632	1666	1449
Yield (tonnes/ha)	45.4	53.9	31.3
Duration (days)	413	511	461
Yield /day (kg/ha)	110	105	68

Source: Melin *et al.* (1976)

has small round fingers and is suitable for making chips. Two issues of *Fruits* (**31**, 11 and **38**, 4) were dedicated to the plantain and its culture in Tanzania, Ghana, the Philippines and other countries. Data on two plantain cultivars are presented in Table 6.7.

The yield of cv. Njock Korn at 550 m was also reported as 45.5 tonnes/ha, which looks like a mistake; if it is not, then the yield per day would come to 89 instead of 105 kg/ha. Even so we may conclude that the semi-dwarf 'Njock Korn' is capable of high yield, both at low and middle altitude; in contrast to 'French Sombre' it makes few suckers.

'Sucrier' or 'Pisang mas' (AA), 'Red' and 'Green red' (AAA), 'Ney poovan' (AB), 'Pisang raja' (AAB) and 'Klue teparod' (ABBB) are widely distributed, but are nowhere grown to a large extent. Descriptions and colour illustrations of many of the cvs named can be found in Haddad and Borges (1973). Saba (BBB) is the most important cooking banana of the Philippines (Valmayor, 1976). It has some resistance to black leaf streak (Laville, 1983). For synonyms, see Valmayor (1980).

Finally two tetraploids (AAAA), bred in the West Indies, must be mentioned: 'IC$_2$' and 'Bodles Altafort'. These clones, obtained after years of intensive research, are a triumph of science but have little practical significance. Breeding bananas is exceedingly difficult. The plant breeder first has to get seed from a seedless parent, then must raise plants from those to produce seedless fruits from them. It is an onerous and thankless job. During those years many mutations arise, one of which may be the answer to our problems.

Changes in cultural practices and handling methods may render much breeding work obsolete, as happened when transportation changed from bunches to detached hands. It is true, as Simmonds (1966) observed, that no crop better illustrates the dangers inherent in mono-clone cultivation, but it seems to me that selection works faster and cheaper than breeding; yet Simmonds (1976) states that a mood of cautious optimism prevails. More particulars are provided by Menendez and Shepherd (1975) and Anon. (1978).

Cultivation measures

Propagation

Let us quote from Simmonds (1966): 'The material preferred for planting varies widely in different parts of the world. In Jamaica, for example, bits of large corms, maidens and sword suckers are all regarded as satisfactory planting material, water suckers and peepers being rejected. By contrast, in Martinique, heads of plants which have flowered are preferred to maidens, and sword suckers are regarded as a last resort.' Peepers are the preferred planting material in West Australia and water suckers are recommended in Israel. He concludes that: 'for the tropics, the type of sucker used is, within wide limits, unimportant'. However, 'the larger, the better'.

IFAC (1957) prefers 'the corm from a plant that has already produced a bunch of fruit; the next best is the maiden sucker that is about to shoot a bunch' and recommends one to 'plant a field with the same type so as to obtain uniform bunches'. The corm should be trimmed and all eyes other than those on the neck must be removed, before it is disinfected.

Champion (1963) prefers a big conical sucker, 60–150 cm high, with narrow leaves (sword sucker) or a maiden sucker 1.50 to 2 m high, when starting from a central bud; the two types must be planted out separately. When a side bud is to be used, the corm must be trimmed and after four to six weeks one of the suckers arising from it is chosen. Bits are not advisable as they may rot; if used, they are divided into three or four parts . The relative values of different propagules are summarized in Table 6.8.

A great advantage of corms or bits is that they can be disinfected easily by a hot water treatment, which kills nematodes. Before suckers are treated, they must be cut back to 25 cm of pseudostem. The best results are obtained on trimmed corms when held in

Table 6.8 *Agronomic value of different propagules (first cycle)*

Propagule	Duration – First cycle – Yield	
From central bud:		
small size	Average	Low
big size	Short	Average
From side bud:		
trimmed corm	Long	Average to high
corm with buds	Rather long	Average to high
corm with selected bud	Average	Average to high
corm with sucker	Short	High
corm with 1 m pseudostem	Short	Average to high

Source: Champion (1963)

water of 65 °C during 15 minutes, or at 60° during 15 to 20 minutes. In the field a metal tank of 120 × 60 × 60 cm can be used, heated by three braziers fed from a propane gas cylinder. Two smaller cages of wire netting holding the corms are lowered in the tank, and this causes the temperature to drop 3–4 °C. In an eight-hour day 1,200 corms (which must be at least 12 cm wide) can be treated this way. After this, they must be dried in a single layer for 24 hours in order to prevent rot. The survival rate is over 90 per cent (Small and Bomers, 1962).

The cleaned and disinfected material is set out in nurseries on nematode-free soil. After multiplication it is planted out in the field. Ndubizu and Obiefuna (1982) improved plantain peepers in Nigeria by growing them in shaded polythene bags during the dry season. Ezumah *et al.* (1977) in India got 3,000 plants from 18 corms in 16 months. Ké and Ké (1980) in Taiwan compared suckers to corms: survival was 72 and 95 per cent respectively, but yield per plant was higher in the first case (29 against 25 kg). Plants obtained from suckers fruited earlier, had more hands per bunch and more fingers per hand than those from corms.

Where and how is planting material obtained? One can get it from producing plants, but there are certain objections to this. Pruning aims at maintaining a balance between yield and growth; to achieve this only one follower is allowed every five months. All others are destroyed periodically, so they cannot compete with 'mother' and 'daughter'. This pruning system is interfered with when suckers are allowed to grow for propagation purposes. Taking them off damages the corm far more than pruning, as they are bigger and have to be cut off deeply. Yet Simmonds (1966) quotes an experiment by Gregory in which suckers were cut back to 60 cm above the ground; this gave five suckers per plant and affected fruit production but slightly.

A better method of getting material is from plantations being cleared after three ratoon crops; roughly one quarter of the area passes through that stage each year, so there should be enough to go round. However, when a *new* banana industry is being established other methods are required. Three means of rapid extension can be used: importation from elsewhere, special nurseries and a combination of the two. Importing corms or suckers is expensive and risky: pests and diseases are introduced and the material is often inferior in quality or wrongly labelled. Yet importing is practically unavoidable in the early stages. Strict supervision is necessary. Nurseries should be established on clean land, with clean planting material. All plants with borer or nematode symptoms must be thoroughly trimmed and disinfected. In the nursery a density of about 4,000 plants/ha is maintained, at a spacing of 1.60 × 1.60 m. As most suckers are produced during the fruiting stage, the plants should be left alone for at least six months before suckers are taken. These are set out in another part of the nursery and treated as described previously. This system permits a tenfold increase per year.

A regional banana industry is bound to a certain minimum size. Suppose that a ship calls at the harbour every ten days. It has to be loaded with 1,000 tonnes of banana, which means a yearly production of at least 36,000 tonnes, or an area of not less than 1,000 ha. No banana ship will call before that volume is reached unless a combination can be organized, as was done in the Windward Islands (where the ship calls at four ports on successive days). Likewise in Surinam the ship first goes to Nickerie, where a bar prevents the exit of a fully loaded ship, and then to Paramaribo.

At a multiplication rate of ten we need four years to increase the planting material: from one ha in the first year to ten in the second, 100 ha in the third and 1,000 in the fourth. The production of these four years is largely lost as the local market is quickly glutted. It is therefore understandable that efforts have been made to speed up multiplication.

Barker (1959) describes such a system. Vigorous plants, at least five months old, are used. The outer leaf sheath is divided lengthwise and both halves are pulled carefully sideways and upwards until a bud is exposed. This bud is ringed with a sharp knife to prevent damage to it when the next sheath is removed. One or two more sheaths are removed in the same way. The last exposed bud must be at least 2 cm^2 in size, about as big as a thumbnail. After the operation, which may be repeated every two weeks, soil is heaped against the stem, just covering the buds. The suckers are left on the plant until they are at least 60 cm high, or weigh 700 to 900 g and are then removed. The rate of multiplication has gone up to about 20.

Small (1961a) describes the same method, but suggests a multiplication rate of 50; the suckers are set out at a spacing of 1 × 1 m. Hamilton (1965) arrives at an even faster rate, 150, but his plants are small and require constant supervision; once enough plants are obtained the stripping method should be used for further reproduction. Mbumba (1980) has done research on *in-vitro* techniques for fast propagation.

Planting

Clearing is preferably done without burning to preserve organic matter. The land is cultivated and ploughed, suckers or corms are set in furrows. Where slope or roughness prevent this, holes are dug. The heavier the land, the bigger the hole, but a cube with sides of 30 to 40 cm is generally considered big enough. No ploughing or

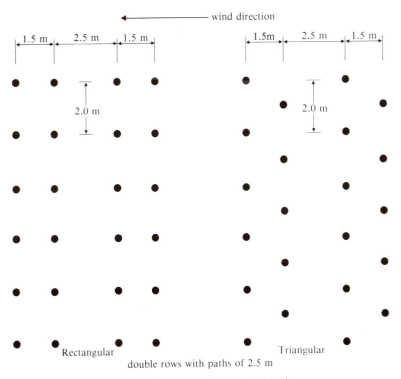

Fig. 6.11   Two ways to plant banana at a density of 2,500/ha

cultivation is necessary on organic soils and only shallow holes are made. Lime and phosphate are applied to the hole. The tendency is toward closer planting if bunches are sold by weight and not by count (number of hands). Tall cvs like 'Gros Michel ' and 'Lacatan' need wider spacing than dwarf cvs. Thus the density varies between 1,000 and 3,000/ha, with a distance of more than 3 to less than 2 m between plants.

Contours are followed on sloping land, while on flat land the planting can be on the square, rectangle or triangle. Where ground spraying is to be used, double or triple rows are advised. IFAC (1957) recommended twin rows for 'Robusta' and triple rows for 'Giant Cavendish', but Champion (1963) sticks to double rows; Figure 6.11 gives two examples. Row direction should be perpendicular to the prevailing wind to facilitate spraying.

The best planting time is at, or just before the beginning of, a rainy season. In Surinam the small rainy season is preferred. Sword suckers were found less suitable there for use in the dry season, whereas corms could rot in the long rainy season; maiden suckers were good at all times (Small, 1961b). In some places efforts are made to time planting in such a way that the main harvest occurs when it is likely to fetch the highest price on the market. However, producing countries do not always know where their fruit is to be sold; that is often decided after the departure of the ship.

Four to six weeks after planting a careful check of all plants should be made; casualties are replaced, preferably by big maiden suckers. As a rule no more than 10 per cent of the plants set out have to be replaced.

Soil management

We must remember that the root system of the banana is superficial and easily damaged by cultivation. The use of a cover of *Pueraria* or other climbing plants is not advisable in a high density crop. Cover crops like *Centrosema* and *Indigofera* have been found to be toxic to banana (Mien-Chun, 1976). However, weeds must be controlled until the musa plant is big enough to shade the ground. Broad-leaved weeds do little damage and may even be useful, yet they may harbour pathogens, e.g. nematodes in *Commelina* and virus in Cucurbitaceae. Grasses, on the other hand, must be ruthlessly suppressed. Slashing seems to be the best solution, if the cost of labour is not unduly high. It must be done above soil level, as bare spots will promote erosion, and should be repeated every six to eight weeks until weeds are shaded out.

Chemical weed control is increasingly adopted. Kasasian (1971) reports the use of atrazine, diuron, simazine or linuron in the West Indies at a rate of 1.5–3 kg/ha as a pre-emergence herbicide. Ametryne and chlorbromuron have given promising results as they appear to combine pre- and post-emergency activity. For post-emergence treatment paraquat is in general use at rates up to 4 kg/ha; there is no residue problem. Against perennial grasses dalapon and glyphosate are recommended. Against broad-leaved weeds, 2,4-D may be used, but only in non-volatile form and on older plants where no drift can settle on the leaves. For more details, see *Pest control in bananas* (1971).

Irrigation

This may be necessary even in areas with a short dry season. If rainless periods of more than two weeks occur, banana yields will be affected and the installation of irrigation equipment is likely to pay for itself. Melin and Marseault (1972) report that irrigation was profitable in Cameroon, even where yearly precipitation reached 2,750 mm. The usual methods of flood, furrow, basin, overhead and

Fig. 6.12 Rain gun mounted on a tracked vehicle, Surinam

under-tree irrigation are followed; the common amount is 25 mm per week. Ghavani (1974) in Honduras, had better results with 44 mm (inclusive of rain) than with 25 mm per week; higher applications had no effect and there was no difference between one and two applications per week.

In Surinam rain guns were mounted on old rice combines (Fig. 6.12); this seemed a very efficient system at first, but later several disadvantages became apparent: a high water table had to be maintained in the canals and leaf spot was furthered by sprinkling (van Emden, 1967).

Self-propelled rain guns are now used in many places. These may consist of a big wheel on which 100 m rubber hose is rolled back in 11 hours; thus 2½ ha can receive 30 mm of water. This can be repeated every ten days in the dry season.

Drip irrigation of banana has become common practice in Israel, Australia and other subtropical countries (Fig. 6.13). In the tropics this system would be useful in hilly areas or regions with a long dry season. Meyer (1976) used 1,200 nozzles per ha in Martinique, dispensing the equivalent of 0.48 mm rain per hour, A yield of 50 tonnes/ha was achieved and four or five cycles were possible, instead of the usual three, before replanting was necessary.

Crop fertilization

Although bananas are commonly grown on fertile soils, manuring

Fig. 6.13 Drip irrigation of young banana suckers in Israel, note the very wide spacing

is indispensable, except in very exceptional cases. The pegasse soils of Surinam provide such a case, for Small (1964) found no significant results from applications of 225 kg N, 90 kg P_2O_5 and 135 kg K_2O/ha for three successive years. If such an experiment is continued long enough, one day N deficiency will appear in the zero treatment, probably to be followed by K deficiency. The grower must be familiar with the deficiency symptoms. Briefly these are:

N   pale colour, reduction of leaf size, stunting
P   dark colour, slow rate of leaf production, marginal chlorosis
S   chlorosis of youngest leaves
K   reduction of growth, rapid yellowing of older leaves
Ca   field symptoms not known
Mg   purple blotches on petiole ('le bleu')
Fe   interveinal chlorosis of young leaves
Mn   marginal interveinal chlorosis of young leaves
Zn   narrow leaves, stunted growth
Cu   drooping, umbrella-like leaves.

A little booklet published by BASF (1974) contains beautiful colour photographs of deficiency symptoms, as well as of pests and diseases. To replace the amounts removed by the crop, fertilizers should supply at least 2 kg N, 0.5 kg P_2O_5 and 6 kg K_2O/tonne of bananas. For a yield of 40 tonnes this is 80 kg N, 20 kg P_2O_5 and 240 kg K_2O/ha (de Geus, 1973). N is applied in many small doses, sometimes as a 1 per cent urea solution in the sprinkler; P and K are given twice a year. To cover the requirements of a *new* plantation the amounts become respectively 250–60–1,000, plus 200 kg CaO and 100 kg MgO. 'Gros Michel' needs the same amounts although it produces far less than Cavendish cvs; this is explained by the fact that the fruits constitute only 25 per cent of the total bunch weight in the first and 50 per cent in Cavendish cvs.

Farmyard manure and mulch may be applied to increase the amount of organic matter in the soil. These are rather expensive, but sometimes very effective measures in banana growing. Where there is less than 1 per cent organic matter in the soil, applications are absolutely necessary and their effect may be noticeable up to levels of 3–4 per cent. On certain soils yearly applications of 20 to 50 tonnes of manure are given. Mulching is a common measure in banana growing on small farms. Giant plantain in Nigeria produced 18 tonnes/ha with fertilizer but no mulch, 17 tonnes/ha with mulch alone and 31 tonnes/ha with both. For a medium high cv. these figures were 17, 16 and 20 tonnes/ha respectively (IITA, 1982). Fertilizer had to be applied two months after planting in South-West Nigeria; yield dropped 5 to 8 per cent for each month of delay (Obiefuna, 1984).

As in other crops, leaf analysis is used to assess the nutritional status of the banana crop. In Jamaica adequacy levels of 2.9 per cent (N), 0.20 per cent (P) and 3.3 per cent (K) were found. Several methods of sampling are currently employed, but according to Martin-Prével (1977) they cannot be compared to each other because of various interactions (cultivar, age, leaf number, longitudinal and transversal gradients in the blade, moisture, climate, parasites). Generally the third fully expanded leaf is used, but leaf 2 is probably as good and less torn by wind (Boland, 1980). Messing (1978) reports fair to good agreement between various methods of sampling for most elements; only for P was agreement poor. Lahav *et al.* (1981) found the petiole to be the best part for analysis of P, Mg and Mn in suckers. Samuels (1977) compared responses of a banana and a plantain cv. to potassium uptake: 'Highgate' had a much higher percentage of fruit-bearing plants than 'Maricongo' (a Horn cv.); this was also true for number and weight of fruits. Some differences in uptake of N, P and Mg were observed too.

A relative excess of K over Mg causes 'le bleu': mottled blueish colours on the underside of petioles and midribs. Other imbalances between Ca, Mg and K have been described. In the Ivory Coast a proportion of $CaO:MgO:K_2O$ in the soil of 20:10:1 was considered correct; if the ratio $MgO:K_2O$ went below 4 blueing would appear, but a ratio higher than 25 caused K-deficiency. K was said to give better quality fruit as sulphate than as chloride in the Windward Islands.

Pruning

As was said before, pruning aims at maintaining a balance between growth and yield; all unwanted suckers must be removed. This also diminishes internal competition within a 'stool', so that bigger bunches of better quality are produced. The system of one plant about to bear a bunch (the 'mother') and one follower (the 'daughter') has been generally adopted. A third plant, which must be a follower of the daughter, not of the mother, is allowed to remain after the inflorescence has appeared; we may call this the granddaughter. If these three plants were left standing in a straight line, the original spacing would soon be disrupted. Therefore, a rotating succession is preferred (Fig. 6.14). The time lag between mother and daughter fluctuates from four to six months, depending on the growth of the crop (which itself depends on climate, fertilization and general care). The wish to deliver fruit at a certain period may also influence the choice of followers.

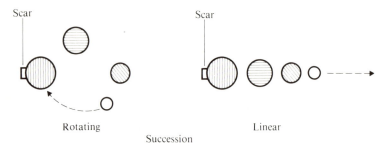

Fig. 6.14 Two pruning systems of banana. *Source*: Champion (1963)

According to Small (1964), the formation of suckers in cv. Robusta is delayed from the third ratoon on; he therefore suggests leaving two suckers unpruned in the third and fourth ratoon, one of which must not be propagated further. Schematically this is represented by:

$$\begin{array}{ccccc} & & & R4a & R5a \\ P \rightarrow R1 \rightarrow R2 \rightarrow R3 \rightarrow & R4 \rightarrow & R5 \end{array}$$

Two methods of killing unwanted suckers are used. In the first the sucker is cut off and the cutlass is pushed in to break the connection of sucker with motherplant. In the second the sucker is cut back to ground level and the heart is removed with a special knife; as this is not always sufficient, kerosine may be poured into the cavity. Applications of 2,4-D and 2,4,5-T have also been tried successfully.

Fig. 6.15 A banana bunch in need of deflowering

Suckers must not be allowed to become too big before they are cut off, so regular checks, every six to eight weeks, should take place.

Special operations

A number of measures specific to the banana must be mentioned.
1. Deflowering consists of the removal of the withered style and perianth. It is unnecessary in 'Gros Michel', as they fall off normally within a few days after fruit set. In Cavendish cvs flower rests often remain and then have to be removed during packing. This may damage the fruit and it is therefore better done just after flowering, when light brushing removes them easily. Israeli *et al.* (1980)

Fig. 6.16   A banana plantation in the Ivory Coast; note props and stumps

have proved that the flower parts fall in dry air and remain on the fruit in wet air.

Leaves rubbing against bunches cause unsightly scars on the fruit. The plants should be checked regularly and all rubbing leaves must be removed, or turned aside. In some places it is also customary to remove collapsed old leaves, but this is unnecessary and possibly even harmful.

2. Propping (Fig. 6.16) serves to protect bearing plants from falling over and from wind damage. It is necessary because the weight of the bunch is apt to pull the plant out of balance. Two props, forming a triangle, are better than one. Forked branches or bamboo poles are placed against the stem on the side where it is leaning over. Care must be taken that they do not damage the bunch. However, if nematodes are under control the loss incurred by falling over is small and propping may not be necessary.

3. Earthing-up also protects the plants against wind damage; it is reported to reduce losses by 30 to 50 per cent.

4. Removal of the male bud is said to promote fruit development; bunch weight would increase 2 to 5 per cent. However, Small (1962) found the following bunch weights (kg) in cv. Robusta if the male bud was removed after the stated number of flower clusters: 18.1 (5–7), 19.0 (11–15) and 18.4 (not removed). He concluded that it was better not to cut the male bud off, as it is an indication of the development of the bunch. Yet it is often done. It has also been

reported that it decreases the incidence of Moko disease, which is spread by insects visiting flowers.

5. Bagging of bunches protects the fruit against cold, sunburn, dust, spray residue, insects and birds. The bag is left open on the underside. Even in the tropics the effect is considerable; in Guadeloupe bagging increased the average temperature of the bunch by $\frac{1}{2}$ °C which advanced the harvest five days. Generally blue plastic bags are preferred, although it is not clear why other colours should be less desirable. In India temperature increases of 1.1–1.6 °C were found and the bunch weight was 1 kg higher.

6. In dehanding the last, so-called false hand of a bunch is removed. It is usually an incomplete hand, not fit for export; cutting it off soon after bloom is said to increase the weight of the other hands. Sometimes the hand just above the false hand is also removed. Meyer (1975) also cut off two hands above the false hand; the resulting higher quality did not compensate for the loss. The last three measures: removal of the male bud, dehanding and bagging, are often carried out in one operation.

It may happen in outside rows that bunches are exposed to excessive sunlight; fruits then turn yellow even though they are not 'full', become unsuitable for export and may even rot. Such bunches can be protected by bending down one or two of the older leaves and tying them firmly to the fruit stalk, just below the last hand. Care must be taken that they do not rub against the fruit.

Yield

Banana yields were shown in Table 6.2. As in citrus, some high yielding countries are not fully tropical: Argentina, Paraguay, Brazil, Azores, Sudan and Israel. Duration of the cycle, from planting to harvest, must be taken into account; this takes under a year in favoured tropical locations to two years far from the equator or high above sea level. Wilson (1981) noted 85 tonnes/ha for 'Dwarf' and 139 tonnes/ha for 'Williams' in the Zimbabwe lowveld. If the cycle lasts 2 years, we must halve Wilson's figures to 43 and 70 tonnes/ha. Even so, the second figure is remarkably high. Still higher yields are reported in carefully supervised trials.

Diseases and pests

In his first bulletin of the newly-founded Surinam Agricultural Research Station, van Hall had very little to say on diseases of the banana (1904). This situation drastically changed a few years later. A contract was signed between the government of Surinam and the United Fruit Corporation, stipulating that Surinam was to plant 1,000 ha of 'Gros Michel' in each of the three succeeding years. As there was not enough planting material in the country, it had to be imported. One year later Panama disease made its appearance and only one year after that the plantations, and hence the country, were in serious trouble.

'Panama disease, also known as banana wilt and vascular wilt, caused by *Fusarium oxysporum*, forma *cubense,* is one of the world's most catastrophic plant diseases' (Purseglove, 1972). The causal organism lives in the soil and penetrates into the roots, from where it slowly spreads until the corm is reached (this takes about two months). From then on the disease develops very rapidly. Purplish stains appear in the vessels which are blocked, outer leaves turn yellow and collapse. Soon only a few of the youngest leaves remain functional. Later the pseudostem splits; it smells strongly of rotten fish when cut. Not all cvs react in this way; in some there is no collapse of leaves and the fishy smell is absent (*Pest control in Bananas,* 1971).

Panama disease is spread by infected soil, tools and water. All chemical and sanitary efforts to contain the disease have failed; the only thing one can do is use resistant cvs. 'Gros Michel' and 'Silk' are highly susceptible, the Cavendish group and the plantains are resistant. Therefore, 'Gros Michel' has been replaced practically everywhere by Cavendish cvs; as a result of this Panama disease is no longer important.

Sigatoka disease, also called leaf spot, is now by far the most serious disease of banana. It has been known for a long time in Indonesia and Fiji, but was first seen in Surinam and Trinidad in 1933. It is now found in all banana regions of the world, with the exception of the Canary Islands, Egypt and Israel. The causal organism was previously known as *Cercospora*, but when the sexual spores were found the name had to be changed, to *Mycosphaerella musicola.* Stover (1975) has proposed another change of nomenclature, based on the form of the conidia: *M. fijiensis* causes black leaf streak, *M. fijiensis,* var. *musicola* causes Sigatoka and *M. fijiensis* var. *diformis* causes Black Sigatoka. It is up to the taxonomists to decide on this matter.

Sigatoka causes premature death of large areas of the plant's leaf surface; in some cases the entire leaf is affected. The photosynthetic area is drastically reduced and in severe cases the fruit does not mature at all. In less severe outbreaks, the size of bunches and individual fingers is reduced, the fruit ripens prematurely before or after harvest and has an abnormal flavour and smell (*Pest control in Bananas,* 1971). The earliest leaf symptoms seen in the field are small lesions; they are yellowish-greenish and run parallel to the veins. Symptoms become evident on the third and fourth leaf and get gradually worse farther down. The fungus produces two kinds of spores, vegetative and sexual. The vegetative spores are produced daily in wet weather or dew and are spread by drops of water to other leaves; the lesions form the so-called *line spotting*. The sexual spores are released periodically, when a leaf containing fruiting bodies is wetted. They are spread by the wind and the lesions lie along leaf margins; this pattern is called *tip spotting*. A surface film of water on the leaves is essential for the formation of both kinds of spores. In the dry season dew may form at night and if this stays on long enough, infection will take place. Plants growing under shade are less affected as the dew forms not on the musa, but on the shade tree (just as a car parked outside in a clear night becomes covered with moisture, while in the garage the car remains dry). In many places attempts have been made to forecast the appearance of the disease from (micro)climatic conditions, but this is not easy. Persistently damp weather and temperatures of 25 to 28 °C are optimal for the fungus. However, as we saw above, infection can also occur during dry periods as a result of dew. Anyway, forecasts can only have local significance. Tezenas (1976) made a calendar for six localities in Cameroon.

Certain cultural measures help to reduce humidity and thus infection within the plantation. Good drainage, weed control and pruning are *conditiones sine qua non* in banana growing and in disease control. Removal and destruction of badly spotted leaves is sometimes advocated, but this labour-intensive practice is only feasible in regions of low wages and may not be very effective. Plant breeders are searching for resistant cvs, but so far with little success: only cvs IC2, Silk and Fig have exhibited some resistance; Horn plantains, especially 'Pelipita', are fairly resistant (Laville, 1983; Stover, 1983a).

Another possibility is chemical control; for a long time this could only be done by spraying large quantities of Bordeaux mixture. Pipelines were run through the plantation to facilitate spraying. In Surinam spraying was done from boats in the drainage canals. After the harvest the spray residue had to be washed off before the bunch could be exported. All this changed when Guyot and Cuillé introduced the 'West Indian Control Method'. It consisted of a low-volume application of a fungicide, e.g. copper oxychloride or zineb, in an oil-emulsion. Later it was shown that an oil-emulsion alone had practically the same effect (IFAC, 1957).

Tollenaar (1960) argues that oil works fast but disappears quickly; it cures the disease, but cannot prevent subsequent attacks. Copper does not cure, but prevents infection and should therefore be used once control has been gained. Gonesh (1964) compared oil to maneb; oil cost only half as much and worked better. The difference was small as far as the fifth leaf, but became noticeably bigger on older leaves.

Nowadays, oil alone or mixed with a fungicide is applied from aeroplanes in 2-week cycles at a rate of 12 litres/ha (Fig. 6.17). The

Fig. 6.17 Plane spraying oil in Surinam; speed 160 km/h, height 30 m above the ground, swath width 29 m

capacity is about 100 ha/hour. Under very unfavourable circumstances the treatment has to be stepped up to once a week; in drier periods once every four to six weeks may be enough. Altogether, there are 15 to 20 cycles a year. The field must be marked clearly so that the pilot knows exactly where to spray and leaves no strips unsprayed. The best time is early in the morning, or late in the afternoon, when there is little wind. The cost is small, not much more than $1/ton of fruit (Stover, 1972.).

Ground spraying costs are much higher and control is not as effective. The amount is increased to 18 litres/ha and the nozzle must be pointed straight up; the oil spray should not be directed towards the leaves, but must drift onto them; a light wind is favourable. Spacing should be adapted to this treatment, with plants closer together but with more space between rows. The capacity is 4 ha/day for two men with a motorized knapsack sprayer (Fig. 4.6).

Oils effective in the control of Sigatoka are also phytotoxic; they reduce the yield by about 10 per cent. Consequently, there is a tendency to use as little oil as possible and to add a fungicide for better control; maneb is frequently used. A new era of Sigatoka control was initiated by the use of thiabendazole and benomyl in oil in 1970. However, these systemic fungicides are still expensive (Stover, 1972). Ganry (1978) reports that TBZ and benomyl at 125 g active ingredient/ha gave good control in aerial spraying. When resistance against these chemicals developed, imazalil replaced them; at 300 g active ingredient in 10 litres of oil per ha, it gave good control (Melin, 1975). Tilt (propiconazole) was also effective at 100 g active ingredient in 20 litres of oil per ha (Mourichon and Beugnon, 1982).

Black Sigatoka is caused by *Mycosphaerella fijiensis*. It was long known in the Far East and appeared in Honduras (1972), Venezuela (1981) and in other countries of Central and South America. It has also been found in Africa (Gabon), but not in Australia. The plantain is very sensitive to this disease. The symptoms resemble those of Sigatoka but control is far more difficult and expensive (Darthenucq, 1978; Frossard, 1980; Tarté and Presa, 1981).

A third, but less virulent form of *Mycosphaerella* causes Black leaf streak. Within all three fungus species (or varieties) races of different virulence exist. Several other leaf spots may infest banana, but none are so dangerous to the crop as the Sigatokas. In certain cases, however, a serious defoliation of plantain may occur.

Among the fruit rots, pitting disease is the most important. It can be serious in Mexico, Central America, Taiwan and the Philippines and may cause up to 50 per cent rejection of the hands at the fruit packing station if not controlled. The disease occurs especially during periods of heavy rain. Sanitation is important for control: all collapsed leaves must be removed. Another disease of consequence is called cigar-end rot. In transport too, many kinds of fruit rot appear. Good disinfection at the packing station is essential.

Bacterial wilt or Moko disease is a major disease in the western hemisphere, especially on small farms. It is transmitted from plant to plant by infected tools and by insects visiting the flowers. Later it spreads from root to root. The symptoms resemble those of Panama, but without the smell; furthermore fruits blacken internally. Tools must be disinfected with formaldehyde to prevent the spread of the disease. Removal of the male bud also helps to control Moko. Infected plants, their neighbours and all weeds must be destroyed, after which the area must be kept free from plants which could serve as a host for 6 to 12 months, depending on the strain. All commercial cvs are susceptible and 'Bluggoe' is highly susceptible; it is now being replaced by another ABB cv. called 'Pelipita', which is resistant.

Bunchy top virus disease, which is transmitted by the banana aphid *Pentalonia nigronervosa,* occurs in scattered places but not in the western hemisphere. Dark green streaks appear on the leaves and later the plant becomes stunted. Infected plants must be destroyed immediately. The disease is serious in southern India (Nat. Semin., 1980).

Cucumber mosaic is another virus disease, with different aphids as vectors. It is seldom a serious problem, but may occur where cucumbers and other host plants are grown as catch crops between musa. Destroying the virus hosts is the only measure that need be taken.

The burrowing nematode *Radopholus similis* is now the major banana root pathogen (Stover, 1972). The same species, but a different strain, is a serious pathogen of citrus in Florida. All triploid banana cvs are susceptible, but 'Gros Michel' less so than the Cavendish group. Corms must be pared and treated with warm water before planting. They can remain free of nematodes for six to eight months. 'Pralinage', i.e. coating with a mud containing nematicides, will prolong this period to 18 months; 1 g active ingredient/corm is used (Guérout, 1975). Soil can be freed of nematodes by keeping all host plants out for six months. A flood fallow is used in Panama, Honduras and Surinam. Sarah *et al.* (1983) report success from these measures in Ivory Coast.

Soil can also be disinfected by treatment with nematicides, but this is more expensive and less reliable. However, Roman *et al.* (1976) report that fenamiphos and several other compounds could prevent a reduction in yield, so that it was possible to grow banana for three years in succession. Aldicarb and UC 21865 were shown to control nematodes in Ecuador; these materials are not transported to the fruits (Hasing-Lama *et al.* 1976). Granular Miral, in three doses of 2.5 g active ingredient per plant, greatly improved yields in Cameroon and Ivory Coast (Sarah *et al.* 1983).

The banana weevil, *Cosmopolites sordidus,* is – as the name suggests – a true cosmopolitan: it occurs in nearly all banana-growing regions of the world. It also leaves the plant in a sordid mess: the larva (grub) feeds and tunnels in the corm (Fig. 6.18) which is reduced to a blackened mess of rotten tissue. Tunnels may go one metre up into the stem. Leaves turn yellow, wither and die prematurely. Other pathogens attack the plant, which eventually dies or is blown over. The adult weevil feeds on dead or dying banana plants and lives in newly cut stems; it can be recognized by the snout. Eggs are laid in these stems, or on the corm, just above the ground. The weevils stay hidden during the day, but come out at night to feed and lay eggs. When disturbed they feign death. They rarely fly and the pest is spread mostly by infested corms.

Control is achieved by using clean planting material and by preventing the weevil from entering. Therefore stems must be cut close to the ground and the cut face must be covered with a layer of earth. Harvested stems must be cut up in small pieces and the

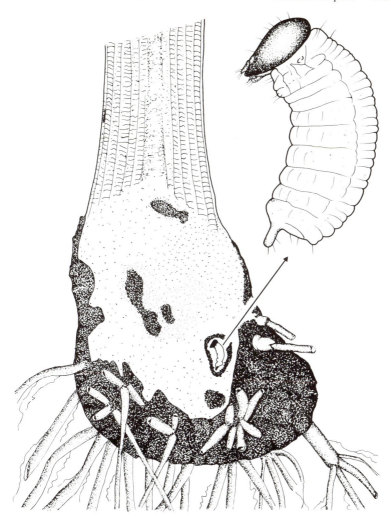

Fig. 6.18   Longitudinal section of banana corm infested by *Cosmopolites. Source*: van Dinther (1960)

base of the stem must be kept free of weeds. In some regions (Indonesia and Surinam, for example) predator beetles effectively control the pest. Other control methods consist of trapping, dipping and field treatment. In trapping, old stem pieces are cut and placed face down between the plants; beetles can be collected by hand from these pieces, or they can be poisoned with diazinon. A 0.1 per cent solution is used for dipping suckers.

A fruit-scarring beetle, *Colaspis hypochlora,* occurs in Mexico,

Central America, Colombia and the Guianas. The adult beetle feeds on the skin of young fruits rendering the fruit unsaleable. A stingless bee, *Trigona amalthea,* causes similar damage. A rust thrips is responsible for corky spots on the fruits. These pests rarely demand much attention; some are controlled fairly well by predators and bagging of bunches also lessens the damage. In Surinam a sugarcane borer, *Castnia licus,* attacks banana and plantain. One or two larvae are present in each infested plant, from which a jelly-like sap oozes. Control is difficult. Scab moth (*Nacoleia octasema*) occurs in Indonesia, Queensland and several Pacific islands; it causes serious damage to fruit. Each newly emerged bunch has to be sprayed with DDT to control this pest. See BASF Guide (1974) for colour pictures of pests and diseases.

From harvest to consumption

Under optimal conditions it takes nine months from planting to the harvest of the crop; this period may be as much as 18 months, depending on climate, cultural practices and cultivar. From bloom to harvest takes 80–95 days under good conditions and up to 120 under suboptimal conditions (in the subtropics it lasts even longer). Ganry (1978b) has shown that the sum of temperatures (above 14 °C) required, is 900 degree-days from shooting to harvest. In other words: it takes 90 days at an average day temperature of 24 °C and 100 days at 23 °C; the warmer it is, the less time is needed. But whatever the circumstances, bananas and plantains are always harvested green.

For home consumption the bunch is left hanging until the fingers are 'full' (rounded). The weight of a bunch increases considerably during the last two or three weeks and it is to the farmer's advantage to postpone cutting as long as possible. On the other hand, a full bunch is very vulnerable in transport; the longer this will take, the sooner the bunch must be cut. For export and transport overseas the bunch is therefore cut in a stage known as 'three-quarters full', when the fingers are still clearly angular; an intermediate stage is called 'full three-quarters', 'high $\frac{3}{4}$' or 'heavy $\frac{3}{4}$' and this is the preferred stage when the voyage is not a long one.

There is little agreement on the use of these terms; much depends on local experience. French researchers have studied this problem and proposed a 'fullness index', which is the weight divided by the length of an internal fruit of the first or second hand. For instance, at the correct cutting stage for 'Dwarf' the fruit weighs 133–140 g and the length is 16.3 to 17.7 cm; the quotient is from 7.9 to 8.3. Such values have to be determined for each cv. separately (Champion, 1963). According to Purseglove (1972) the outer central finger of the second hand is measured with calipers in Central America; the optimum diameter is about 3.37 cm.

The first task in harvesting is to remove the props. In a 'Dwarf' plantation a man can harvest a bunch alone, but for taller cvs at least two men are required: a cutter, a helper and a line of carriers. The cutter selects the bunch; if it is in the right stage he nicks the pseudostem with his cutlass in such a way that it comes down gently

Fig. 6.19 Harvest of a bunch by one cutter and one carrier

(Fig. 6.19). The helper then puts it on the head of a carrier, after which it is cut off from the plant. Enough stalk should be left on both sides to allow easy handling.

The pseudostem is then cut back to about 1 m above the ground. Nutrients remaining in it are supposed to flow back to the follower. Simmonds states that there is no proof of this, and if borers cause trouble it is better to remove the stump completely and chop it up in small pieces.

The carrier takes the bunch to a loading platform, from where it is transported by mules, carts, trucks, train or cableway. The cableway shown in Figs 6.20a and b is rather expensive. A simple conveyor was designed for the Windward Islands; it is inexpensive, usable on slopes and can be made locally (Kemp and Matthews, 1977). The bunches should always be handled gently to prevent damage; they should stand on the thick end of the stalk, or lie down in a padded cradle. Banana trash and foam rubber are used as padding.

Until 1960 bananas were exported in bunches in nearly all banana regions. 'Gros Michel' could be shipped as naked bunches, all other cvs had to be packed in parcels of paper and straw, or in perforated plastic tubes. In many regions bunches were paid for by 'grade', not by weight. A nine-hand bunch received the full price (count), an 'eight' got $\frac{3}{4}$, a 'seven' $\frac{1}{2}$ and a 'six' only $\frac{1}{4}$ of the fixed count price.

In 1960 the Standard Fruit Company began to pack detached hands in boxes and soon this became common. The fact that 'Gros Michel' had to be replaced by Cavendish cvs speeded up the conversion, but it may also have worked the other way round: boxing speeded up the elimination of 'Gros Michel'. Simmonds (1966), quoting Cadillat, sums up the advantages of boxing: it frees the grower from the need to produce a particular size of bunch; the fruit is handled less; more labour is needed in the exporting and less in the importing country; no waste material is exported; good parts of damaged bunches can be sold; mechanized bulk handling is made possible. The relatively small disadvantages are the increased cost of packing material, the need to wash and disinfect the hands and the need for better ventilation aboard ship.

Bunches arriving at the packing station are hung up and dissected with a knife, or with a piece of nylon string between two wooden handles by looping it over the cluster stalk. The hands are washed for eight minutes in a tank to remove latex (Fig. 6.21), then dipped in maneb or TBZ and (after this has drained off) packed in cardboard boxes which generally hold 18 kg of fruit. The fruit is now ready for transport, possibly cooled, to the harbour.

Bananas must be carried by specially built ships which can be loaded and unloaded quickly, make a fast trip, have a large cargo

Figs. 6.20a and 6.20b Cable transport of bunches to receiving station, Surinam

Fig. 6.21 Washing and disinfecting hands in tank, packing house in Surinam

space and are refrigerated. Only when the distance is short, as from the Canary Islands to Spain, can the voyage be made without cooling.

The holds must be pre-cooled before loading (Fig. 6.22) starts. The temperature should be brought down as soon as possible to 13 °C, or lower if the cv. can stand it. Below 11 °C chilling occurs in all cvs, so this is regarded as the minimum temperature. The boxes are stacked up to eight layers high. Ventilation must be provided.

Fig. 6.22   Boxed bananas being loaded on board ship

In the port of destination the fruit is unloaded and taken to specially constructed banana ripening rooms. Here they are subjected to intermediate temperatures, usually 18–20 °C, high humidity and an ethylene treatment. The exact conditions depend on the cv., the picking stage, outside temperature, the day of arrival and the day when the fruit is required for consumption. In ripening the skin changes colour from dark to light green and greenish yellow to bright yellow. Meanwhile the pulp softens outwards from the core and from tip to stalk. If left too long the pulp becomes watery, the skin turns brown and finally the whole fruit rots away. Even in the tropics the quality of bananas is improved if they are ripened properly at the right temperature and humidity.

References

Amson, F. W. van (1961) 'De bodem' in: *Bacovencursus 1961*, 5–7, Landb. Proefst. Suriname.

Anon. (1978) *Genetic resources of bananas and plantains*, IBPGR 77/19, Rome.

Anon. (1981) 'Le marché bananier dans le monde en 1980' *Fruits* **36**, 479–80.

Barker, W. G. (1959) 'A system of maximum multiplication of the banana plant', *Trop. Agric.* (Trinidad) **36**, 275–84.

BASF (ca. 1974) *Guide to diseases, pests and principal deficiencies in banana*, Ludwigshafen.

Bigi, F. *et al.* (1969a) 'Il relancio della bananicoltura sul versante atlantico della Costa Rica', *Rivista Agric. subtrop. e tropicale* **63**, 88–113.

Bigi, F. *et al.* (1969b) 'Aspetti tecnico-economici della riconversione varietale nella bananicoltura ecuadoriana', *ibidem*, 181–211.

Boland, D. E. (1980) 'Some aspects of banana leaf analysis in Jamaica', *Fruits* **35**, 355–9.

Bovée, A. C. J. (1975) 'Lysimeteronderzoek naar de verdamping van bananen in Libanon', *Landbk. Tijdschrift* **87**, 174–80.

Brun, W. A. (1961) 'Photosynthesis and transpiration from upper and lower surfaces of intact banana leaves', *Plant Physiology* **36**, 399–405.

Champion, J. (1963) *Le bananier*, Maisonneuve et Larose, Paris.

Champion, J. (1968) *Les bananiers et leur culture, Tome I: Botanique et Génétique*, Paris.

Darthenucq, A. *et al.* (1978) 'Notes d'un voyage d'étude dans quelques zones bananières d'Amerique latine', *Fruits* **33**, 157–65.

Devos, P. and **Wilson, G. F.** (1979) 'Intercropping of plantains with food crops: maize, cassava and cocoyam', *Fruits* **34**, 169–74.

Dinther, J. B. M. van (1960) *Insect pests of cultivated plants in Surinam*, Landb. Proefst. Suriname, Bull. 76.

Emden, J. van (1967) *Rapport over de bacovencultuur in Suriname*, LVV, Suriname.

Ezumah, H. C. *et al.* (1977) 'Increasing sucker production in Giant Cavendish banana', *Indian J. of Hortic.* **34**, 93–9.

FAO (1976) *Production Yearbook 1975*, Rome.

FAO (1977) *Production Yearbook 1976*, Rome.

FAO (1978) *Production Yearbook 1977*, Rome, etc.

Frossard, P. (1980) 'Apparition d'une nouvelle et grave maladie foliaire des bananiers et plantains au Gabon', *Fruits* **35**, 519–27.

Ganry, J. (1978) 'Etude comparée de fongicides à longue durée d'action, pour la lutte contre la cercosporiose du bananier aux Antilles', *Fruits* **33**, 149–55.

Ganry, J. (1978) 'Recherche d'une méthode d'estimation de la date de récolte du bananier à partir de données climatiques dans les conditions des Antilles', *Fruits* **33**, 669–80.

Ganry, J. and **Meyer, J. P.** (1975) 'Recherche d'une loi d'action de la température sur la croissance des fruits du bananier', *Fruits* **30**, 375–92.

Gietema-Groenendijk, E. (1970) *Fotosynthese metingen bij banaan*, Internal rpt. V. 70/7, Dept. Trop. Crops, Agric. Univ. Wageningen.

Geus, J. G. de (1973) *Fertilizer guide for tropical and subtropical farming*, Centre d'étude de l'azote, Zürich.

Ghavani, M. (1974) 'Irrigation of Valery bananas in Honduras', *Trop. Agric.* (Trin.) **51**, 443–6.

Gonesh, D. J. (1964) 'De bestrijding van Cercospora musae' in: *Voordrachtencyclus*, Versl. LVV Suriname, 64.

Guérout, R. (1975) 'Banana corm coating with nematicidal mud: a preplant treatment', *Nematropica* **5**(2), 22.

Haarer, A. E. (1964) *Modern banana production*, Leonard Hill, London.

Haddad, O. and **Borges, O.** (1973) *Los bananos en Venezuela*, Fundación Shell para el agricultor, Cagua.

Hamilton, K. S. (1965) 'Reproduction of banana from adventitious buds', *Trop. Agric.* **42**, 69.

Hasing-Lama, M. *et al.* (1976) 'Effects of Aldicarb, UC 21865 and DBCP on *Radopholus similis* (Cobb) infestations and on yields in a banana plantation in Machala, Ecuador', First Int. Symp. on trop. and subtrop. fruits, *Acta Horticulturae* **57**, 213–21.

Holder, G. D. (1982) 'Effects of irrigation at critical stages of ontogeny of the banana cv. Robusta on growth and yield', *Trop. Agric.* (Trin.) **59**, 221–6.

IFAC (1957) *Practical guide to banana cultivation in the French West Indies*, Caribbean Commission, Port of Spain.

IITA (1982) *Research highlights 1981*, Ibadan, Nigeria.

Israeli, Y. *et al.* (1980) 'Influence of relative humidity on the type of flower in the Cavendish banana', *Fruits* **35**, 295–9.

Karikari, S. K. and **Abakah-Gyenin, A. K.** (1976) 'Some guidelines in the classification of Ghanaian plantains (*Musa* group AAB)', *Fruits* **31**, 658–60.

Kasasian, L. (1971) *Weed control in the tropics*, London.

Ké L. S. and **Ké D. F.** (1980) 'Effect of different planting materials on the growth of the Taiwan banana', *J. Agric. Ass. of China* **111**, 47–53

Kemp, D. C. and **Matthews, M. D. P.** (1977) *Banana conveyor*, Nat. Inst. Agric. Engineers (UK), O.D. techn. Bull. 7.

Klasen, J. and **Mulder, H. R.** (1975) 'De bacovencultuur in Suriname', *Surin. Landb.* **23**, 91–108.

Lahav, E. *et al.* (1981) 'The suitability of the blade, vein and petiole for determination of nutrients in the banana sucker', *Fruits* **36**, 417–20 and 485–8.

Lassoudière, A. (1978) 'Quelques aspects de la croissance et du développement du bananier 'Poyo' en Côte d'Ivoire; le système radical', *Fruits* **33**, 314–38.

Lassoudière, A. and **Martin, Ph.** (1974) 'Problèmes de drainage en sols organiques de bananeraie (Agneby, Côte d'Ivoire)', *Fruits* **29**, 255–66.

Laville, E. (1983) 'Les cercosporioses du bananier et leurs traitements', *Fruits* **38**, 147–51.

Loesecke, H. von (1950) *Bananas*, New York.

Martin-Prével, P. (1977) 'Echantillonage du bananier pour l'analyse foliaire: conséquences de différences de techniques', *Fruits* **32**, 151–66.

May, S. and **Plaza G.** (1958) *The United Fruit Company in Latin America*, Nat. Planning Assoc., Wash. DC.

Mbumba, N. (1980) *Recherches in-vitro en relation avec la multiplication végétative accélérée du bananier (Musa sp.)*, Cath. Univ. Leuven.

Melin, Ph. and **Marseault, J.** (1972) 'Intérêt de l'irrigation en bananerie au Cameroun', *Fruits* **27**, 495–508.

Melin, Ph. *et al.* (1975) 'Activité comparée de l'imazalil sur la cercosporiose du bananier au Cameroun', *Fruits* **30**, 301–6.

Melin, Ph. *et al.* (1976) 'Potentiel de productivité de deux cultivars de French plantain', *Fruits* **31**, 655–7.

Menendez T. and **Shepherd, K.** (1975) 'Breeding new bananas', *World Crops*, **27**, 104–12.

Messing, J. H. L. (1978) 'A comparison of diagnostic sampling methods in banana', *Fruits* **33**, 167–81.

Meyer, J. P. (1975) 'Influence de l'ablation de mains sur le rendement en poids des régimes de bananes par catégorie de conditionnement aux Antilles', *Fruits* **30**, 663–8.

Meyer, J. P. (1976) 'Premières indications sur l'irrigation localisée en culture bananière en Martinique', *Fruits* **31**, 349–52.

Meyer, J. P. and **Schoch, P. G.** (1976) 'Besoin en eau bananier aux Antilles. Mesure de l'évapotranspiration maximale', *Fruits* **31**, 3–19.

Mien-Chun, L. (1976) 'The evolution of soil conservation practices in Taiwan', *J. Agric. Assoc. China* New Series XCVI, 52–4.

Moreau, B. (1976) 'Développement du cultivar 'Americani' dans les conditions de la côte est de Madagascar', *Fruits* **31**, 83–92.

Mourichon, X. and **Beugnon, M.** (1982) 'Efficacité comparée du Tilt (GCA.64250) sur la cercosporiose des bananiers en Côte d'Ivoire', *Fruits* **37**, 595–7.

National Seminar on banana production technology (1980), Coimbatore.

Ndubizu, T. O. C. and **Obiefuna, J. C.** (1982) 'Upgrading inferior plantain propagation material through dry-season nursery', *Scientia Hortic.* **18**, 31–7.

Obiefuna, J. C. (1984) 'Effect of delayed fertilizer application on the growth and yield of plantains in South-West Nigeria', *Fertilizer Research* **5**, 309–13.

Paradisiaca (1976/1977) *Newsletter* of the Intern. Assoc. for research on plantain and other cooking bananas, IITA, Ibadan.

Pest control in bananas (1971) Pans Manual no. 1, London.

Purseglove, J. W. (1972) *Tropical crops, Monocotyledons*, Longman.

Roman, J. *et al.* (1976) 'Chemical control of nematodes in plantains', *J. Agric. Univ. Puerto-Rico* **60**, 36–44.

Rowe, P. R. (1984) 'Recent advances in breeding dessert bananas, plantains and cooking bananas', *Fruits* **39**, 149–53

Samuels, G. (1977) 'The response of bananas and plantains to common fertilizer levels', *Nouvelles Agron. Antilles et Guyane* **3**, 3/4, 204–8.

Sarah, J. L. and **Vilardebó, A.** (1979) 'L'utilisation du Miral en Afrique de l'ouest pour la lutte contre les nématodes du bananier', *Fruits* **34**, 729–41.

Sarah, J. L. *et al.* (1983) 'La jachère nue et l'immersion du sol: deux méthodes intéressantes de lutte integrée contre *R. similis* dans les bananeraies des sols tourbeux de Côte d'Ivoire', *Fruits* **38**, 35–42.

Simmonds, N. W. (1966) *Bananas*, Longman.

Simmonds, N. W. (1976) Bananas, in *Evolution of crop plants* (ed. N. W. Simmonds), Longman.

Sloten, D. H. van and **Weert, R. van der** (1973) 'Stomatal opening of banana in relation to soil moisture', *Sur. Landb.* **21**, 80–6.

Small, C. V. J. (1961a) *Bacoven studiereis*, Med. 28 Landb. Proefst. Suriname.

Small, C. V. J. (1961b) 'Plantmateriaal' in: *Bacovencursus 1961*, 8–9, Landb. Proefst. Suriname.

Small, C. V. J. (1962) 'Musa' in: *Jaarverslag 1961*, Landb. Proefst. Suriname.

Small, C. V. J. (1964) 'Musa' in: *Jaarverslag 1963*, ibidem.

Small, C. V. J. and **Bomers, H. B. O.** (1962) 'Warmwater-behandeling van bacoven plantmateriaal tegen wortelaaltjes', *Surin. Landb.* **10**, 109–17.

Stover, R. H. (1972) *Banana, plantain and abaca diseases*, Kew.

Stover, R. H. (1975) 'Variation de pathogénie et de morphologie chez *Mycosphaerella fijiensis* (*M. musicola*)', *Fruits* **30**, 306.

Stover, R. H. (1982) 'Valery and Grand Nain: plant and foliage characteristics and a proposed banana idiotype', *Trop. Agric.* (Trin.) **59**, 303–5.

Stover, R. H. (1983a) 'Effet du *Cercospora* noir sur les plantains en Amérique Central', *Fruits* **38**, 326–9.

Stover, R. H. (1983b) 'The intensive production of horn-type plantains (*Musa* AAB) with coffee in Colombia', *Fruits* **38**, 765–70.

Tarté R. and **Presa, J.** (1981) 'Avanza la Sigatoka negra', *Informe Mensuel* UPEB (Panama) **5**(38/39), 29–32.

Tezenas du Montcel, H. (1976) 'Observations sur la cercosporiose du bananier au Cameroun en 1974', *Fruits* **31**, 437–58.

Tollenaar, D. (1960) 'Effects of copper and oil in the control of sigatoka banana leaf spot', *Neth. J. Agric. Sci.* **8**, 253–60.

Turner, D. W. and **Lahav, E.** (1983) 'The growth of banana plants in relation to temperature', *Austral. J. Pl. Physiol.* **10**(1), 43–53.

UPEB (1979) *Programa coordinado de investigaciones; proyectos*, Panama.

UPEB (1980) *La flota mundial de barcos refrigerados y el transporte de banana*, Panama.

Valmayor, R. V. (1976) 'Plantains and bananas in Philippine agriculture', *Fruits* **31**, 661–3.

Valmayor, R. V. *et al.* (1980) 'Philippine banana cultivar names and synonyms', *IBPGR Reg. Comm. S.E. Asia* **4**(4), 4–11.

Vorm, P. D. J. van der and **Diest, A. van** (1982) 'Redistribution of nutritive elements in a Gros Michel plant', *Neth, J. Agric. Sci.* **30**, 286–96.

Vos, L. de (1953) 'Bacovenjam', *Surin. Landb.* **1**, 287–90.

Walker, L. A. (1970) 'A study of the growth and yield of the Valery, Lacatan and Robusta cultivars of banana in Jamaica', *Trop. Agric.* (Trinidad) **47**, 233–42.

Webster, C. C. and **Wilson P. N.** (1966, 1980) *Agriculture in the tropics*, Longman.

Wilson, J. H. (1981) 'Phenology and yield of bananas in the Zimbabwe lowveld', *Zimb. Agric. J.* **78**, 201–6.

Chapter 7

Pineapple

Taxonomy and morphology

The pineapple, *Ananas comosus*, belongs to the Bromeliaceae, a large family of the American tropics (one species originated in Africa). Most bromeliads are epiphytes, they live on trees. The pineapple and some relatives, however, live on the ground. There are many *Ananas* cvs, but the predominant one is 'Smooth Cayenne'; the following description is based on that cultivar.

It is a perennial, monocarpic herb. As in musa each stem flowers only once and dies after fruiting, a side shoot then takes over. Viewed from the side, it has the shape of a spinning top, about 1 m high and 1.5 m wide (Fig. 7.1). The leaves are long and narrow, arranged in a spiral on a short stem, forming a 'rosette'. From 70 to 80 leaves are formed and there is a bud in the axil of each one; some buds grow out into shoots or suckers, all others remain dormant. Shoots are between leaves, a ratoon or sucker appears to come out of the ground and has roots.

In cv. Cayenne the leaf has smooth edges, except for some 'spines' just below the top. Actually, the word spine is botanically incorrect but we shall conform to general usage. Most other cvs have spines all along the leaves. The tip is elongated, ending in a fine point. The leaf blade is shaped like a shallow trough and conducts water to the base of the plant. The upper leaf surface is green while the lower one is silvery due to the presence of trichomes: stalked, many-celled hairs which absorb water. Trichomes and stomata lie at the bottom of sunken longitudinal ridges. Inside the leaf there are water-storing tissue and air canals. All these features contribute to the pineapple plant's ability to withstand drought.

The number of stomata has been reported by Collins (1960) at 180 per/mm^2 in 'Cayenne' and less for triploid and tetraploid hybrids (132 and 105). On the other hand, Py and Tisseau (1965) give 70–80 and Purseglove (1972) mentions 70–85. Whichever is

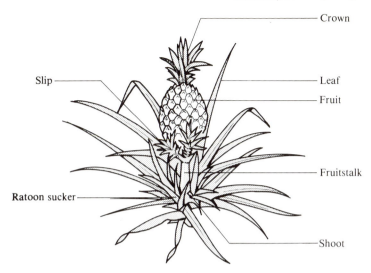

Fig. 7.1 Sketch of a pineapple plant with different kinds of propagules. After Figueroa *et al.* (1970)

correct the number is low compared to banana ($220/mm^2$), citrus (around $500/mm^2$) and other crops. This may be connected with the type of photosynthesis found in pineapple.

The form of the leaves varies and depends on the position on the stem and therefore on the age. It is important for the grower, as well as for the researcher, to know the different leaf forms. They can be placed in the following classes (Py and Tisseau, 1965):

A. Outer leaves, already fully developed when the shoot (or sucker, slip, crown) was planted; they have a 'neck', a zone of restricted growth, near the base and occupy a practically horizontal position.

B. Leaves present, but not fully developed at planting; the neck lies higher and above it we see some spines (this, incidentally, occurs after each cessation of growth).

C. Oldest leaves developed after planting; there is no clearly observable neck.

D. Young, but fully developed leaves, standing at an angle of about 45°; these leaves are generally taken for leaf analysis and measurement. Their weight (which can reach 100 g) is closely related to yield.

E. Leaves in full development, not completely green yet.

F. Leaves standing straight up inside the rosette; small and lightly coloured.

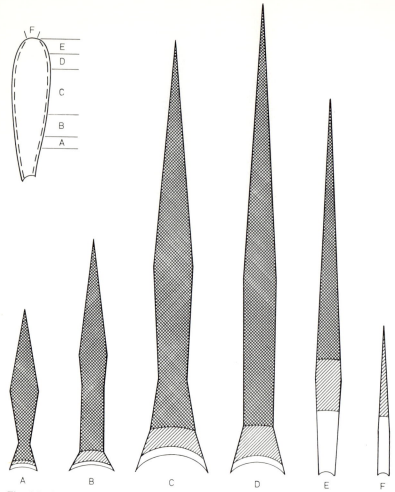

Fig. 7.2 Leaf forms of 'Cayenne' pineapple and their position on the stem. *Source*: Louvel (1975)

The stem is 20 to 30 cm long, narrow at the base (about 2 cm) and wider on top (about 6 cm). The base is curved in slips, but straight in other propagules. The meristem produces 70 to 80 leaves, unless it is prematurely induced to bloom. The time between planting and formation of the inflorescence varies between 6 and 16 months, and depends on cultivar, size of propagule, date of planting, climate and soil. The flowers, 100 to 200 in number, are hermaphrodite, each sitting in the axil of a bract; they secrete nectar. 5–10 flowers open every day, from the base up, over a period of 10–20 days. Both pollen and ovules are functional, yet they set no seed unless cross-pollinated. Humming birds are the principal pollinating agents and their importation into Hawaii is therefore forbidden. Bees visit the flowers, but cannot pollinate them. Fruits of compatible strains may contain up to 3,000 very hard seeds (Pickersgill, 1976).

Fig. 7.3 A 'Queen' pineapple coming into bloom

The compound fruit is formed by fusion of the parthenocarpic fruitlets with the bracts and the central axis of the inflorescence; it takes 5–6 months to ripen. On top of the fruit is a crown of leaves, which continues to grow until the fruit is mature and may be used for propagation. Likewise, the slips below the fruit can be used for propagation. Slips are really the crowns of small fruits, which are sometimes visible (see Fig. 7.4). A slip can be recognized by a bend or curve at the bottom.

The root system of pineapple is shallow and limited. Even in the best growing media roots go no deeper than 50 cm and in soil they rarely extend below 30 cm depth, or outwards beyond the drip area of the plant. This makes it possible to grow pineapples at very high planting densities.

Fig. 7.4   Slip of a 'Queen' pineapple with scar of fruit stalk

Uses and composition

By far the greater part of the world's pineapple production is canned. Slices are the most valuable product, then juice, chunks and dice; other products are fruit salads, sugar syrup, alcohol and citric acid. Recently the export of fresh pineapples to markets in temperate countries has increased tremendously. In general smaller fruit, in an earlier state of maturity, are chosen for this purpose.

Sixty per cent of the fresh pineapple is edible. The fruit contains 80–85 per cent water, 12–15 per cent sugars (of which two-thirds is in the form of sucrose, the rest glucose and fructose), 0.6 per cent acid (of which 87 per cent is citric, the rest malic acid), 0.4 per cent protein, 0.5 per cent ash (chiefly K), 0.1 per cent fat, some fibre and several vitamins (mainly A and C). The vitamin C content varies from 8 to 30 mg/100 g. Pineapple leaves yield a strong white fibre, from which the silky piña cloth is made in the Philippines and Taiwan. Special cvs are used for this purpose, grown at high density in shade. The fruits have to be removed, so it is not possible to combine fruit and fibre production.

Origin, distribution and production

Columbus and his crew were the first Europeans to taste pineapple. In 1493 they landed on the island of Guadeloupe where they found a fruit that 'astonished and delighted them'. Later explorers saw pineapple growing everywhere near tropical American coasts.

It was formerly assumed that Tupi-Guarani Indians, in the region where the frontiers of Brazil, Argentine and Paraguay now meet, domesticated the pineapple (Collins, 1960). Indeed, several *Ananas* spp. and related genera have been found growing wild there. However, Brücher (1977) is of the opinion that what he calls *Ananas sativus*, var. "Cayenne", originated in the highlands of Guiana. On a map he indicates that related spp. and genera arose near the mouth of the Amazon river, in the NE corner of Brazil, near Sao Paulo and in Paraguay. Some of these were cultivated for fibre. He does not comment on the remarkable fact that a crop so adapted to drought should have originated in such a perhumid area.

Pineapple was soon spread to other parts of the world by travellers. This proved to be easy as the crown (and other plant parts) can withstand drying up very well. Nowadays, the crop is grown everywhere in the tropics and in some favoured parts of the subtropics. A special case is the cultivation of pineapple in greenhouses in the Azores (38° N).

It is worthy of note that large scale cultivation of pineapple generally occurs at a distance from the equator and preponderantly along the east coasts of continents (west coasts are too cool) or on islands just inside the tropics (Hawaii, Taiwan). Where we find pineapple cultivation near the equator, as in Kenya, it is at an altitude of 1,500 m above sea level, or at 500 to 800 m above the sea in the Philippines (5–20° N). Humid tropical lowland cultivation is practised in Malaysia and Thailand on peat soils; the cv. grown is not 'Cayenne', but 'Singapore Spanish'. The Ivory Coast is the only major producer of 'Cayenne' in the humid tropics. The reasons for

Table 7.1 *The major pineapple producers (×1,000 tonnes)*

	1961–5	1975	1980	1981	1982	1983	1984
Thailand	327	500	1,372	1,673	1,824	1,439	1,650
Philippines	148	360	901	896	871	1,300	1,250
Brazil	281	515	566	620	667	841	956
India	81	102	549	593	613	660	691
USA (Hawaii)	829	653	596	577	549	549	544
Mexico	203	371	551	560	550	400	400
World	3,660	5,842	7,843	8,594	8,864	8,665	8,752

— No data available
Sources: Py and Tisseau for 1935–8; *FAO Production Yearbooks*
NB: FAO indications for estimates have been omitted

this will be examined later. Hawaii once produced more than half the world's pineapple crop (see Table 7.1). By the early 1960s, however, its share had dropped to less than one quarter, although the actual production had increased; it now holds fifth place. There is a rapid increase in output in most producing countries, especially in Thailand and the Philippines. However, it is impossible to judge by statistics alone. As Py and Tisseau (1965) have pointed out, only export figures are known exactly. It is difficult, if not impossible, to establish just how much is consumed locally, which part goes to canning, what proportion is exported as fresh fruit, etc. Furthermore, one large producer (Taiwan) no longer appears in the FAO statistics. Its production was reported by Naville (1976) to be 358,000 tonnes in 1971 and 294,000 tonnes in 1972.

According to Akamine (1976) the decrease in production of pineapple in Hawaii took place so that land could be released to the more lucrative real estate business. Naville (1976) also reports a decrease: canning went down from 361,000 tonnes in 1972 to 284,000 tonnes in 1975; he ascribes this to the investment by two large American companies in Taiwan, Philippines and Thailand, where labour is less expensive. He also gives an example of the influence of prices on the market: when $80/tonne was paid for fresh fruit in Mexico and only $48 for cannery fruit, a big drop in processing ensued.

Naville has estimated that 28 per cent of the world's pineapples went to the canneries in 1972 and 5 per cent was exported as fresh fruit. Lately, fresh fruit exports are growing fast. For instance, imports into the European Economic Community rose from 1970 to 1975 as follows (×1,000 tonnes): slices 121 to 171, fresh fruit 30 to 70 and juice 20 to 28. Most fresh pineapples exported to the EEC, came from the Ivory Coast. Malaysia and Kenya have also become important exporters.

Growth and development

Three phases can be distinguished in the life of a pineapple plant: the vegetative phase of leaf growth, the generative phase of fruit growth and another vegetative phase, that of shoot growth (*Études sur Ananas*, 1977). These phases overlap to a certain extent since leaf growth continues after bloom has been induced and only stops after the inflorescence has appeared; shoot growth may start before the fruit is harvested.

The total weight of the plant increases at a regular rate. Growth of plant parts starts rather slowly, increases to a maximum and slows down again. If we plot these weight increases against time we get a graph resembling an oblique S (Fogg, 1970). The initial phase of slow growth lasts longer in a crown than in a shoot or slip. However, for the plant as a whole one hardly notices different growth rates as they appear successively. Yet one should be able to recognize them, in order for example to apply fertilizer at the right time.

On average a new leaf appears once a week, but the rate is slower in the beginning and more rapid later. As in the banana the leaf stands up straight at first, is pushed aside and gradually hangs down. However, the pineapple leaf remains functional much longer than the banana leaf. Growth can be accelerated by well-timed applications of fertilizer and by efficient control of pests and diseases. The maximal leaf elongation is 9 cm per week. It is about four months before a new leaf reaches stage D (Fig. 7.2). Each successive leaf is bigger than the previous one, until a maximum leaf weight of 80–100 g is reached. Soon after, and sometimes before this maximum is attained, floral induction takes place. It has become customary to force the bloom by the use of 'hormones', but we shall first discuss the natural course of events.

Collins (1960) stated that pineapples, being native to the tropics, show almost no photoperiodicity. The inference that all tropical plants are insensitive to daylength, is incorrect. Py and Tisseau (1965) state that 'Smooth Cayenne' is a short-day plant, or rather that a succession of long dark periods will induce bloom. Other cvs react only weakly to daylength but may flower when the temperature falls (Bourke, 1976). As low temperature usually coincides with short days, these effects are generally confused. 'Cayenne' plants, if sufficiently big, initiate bloom close to the shortest day and flowers appear 60 days later; February or March in the northern hemisphere. If, however, the plants are too small to bloom, induction is postponed for six months. The growth cycle becomes shorter the closer one approaches the equator or sea level.

It has been shown in Martinique that the duration of the growth cycle depends on the size of the planting material and that cycles may overlap. Let us suppose that on a certain day in June one field is planted with shoots weighing 600 g and another field with crowns of 100 g. In the first field induction will take place in December of the same year and harvest 7 months later, in July; the length of the cycle is 13 months. In the other field induction will be postponed until May or June of the second year, to be followed by bloom in July or August and harvest in December or later; the cycle has lasted 17 to 19 months. The bigger the propagule, the shorter the growth cycle (Fig. 7.5). The cycle, as shown in Fig. 7.5 can take place in a region where the rains are well distributed throughout the year (in this case Martinique). In the Ivory Coast and Guinea similar cycles were observed under irrigation. But without irrigation one cycle became much longer and the other practically disappeared due to the long dry season.

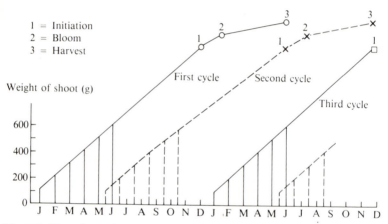

Fig. 7.5 Natural growth cycles of pineapple in relation to planting date and shoot weight. *Source*: Py and Tisseau (1965)

Wide variations in weight of shoots (e.g. 500–1,000 g) occur, while in slips the weights are grouped much closer, say 300–500 g. Crowns are also quite homogeneous (75–200 g). This conforms to the Hawaiian experience that it takes 16 to 18 months for a shoot to complete a cycle, 20 to 22 months for a slip and 24 months for a crown. Near the equator the differences remain, but all cycles are about four months shorter.

The natural cycle is rather unreliable: induction, bloom and harvest may range over many months. Treatment of pineapple plants with 'hormones' causes an abrupt change from the vegetative to the generative phase, irrespective of time or plant size. Thus it is possible to concentrate the harvest of each field in a short time; one can also distribute harvests of separate fields evenly over the whole year. This also applies to the phase of shoot growth.

Ecology and physiology

The pineapple, being tropical, perennial and herbaceous, cannot tolerate frost. The first requirement must therefore be that the climate is practically frostless. This limits the distribution of the crop to latitudes between 25° North and South, with a few exceptions such as Assam (30° N) and South Africa (33° S). The upper altitude of 1,800 m on the equator (Kenya) is not determined by the minimum, but by the average temperature. An absolute maximum temperature has not been established, but it is known that over-heated fruits suffer from 'sun scald'. Leaves and roots grow best at 32 °C and 29 °C respectively and their growth practically ceases below 20 °C and above 36 °C. A high temperature at night is deleterious and a difference of at least 4 °C must be maintained between day and night, as is clear from an experiment made in a phytotron in North Carolina (Bartholomew and Kadzimin, 1977).

Py and Tisseau (1965) report 25 °C to be optimal for fruit ripening in Guinea and Hawaii; at higher temperatures the acid content of the fruit and its quality diminish. The daily amplitude was 12 °C in Guinea and somewhat less in Hawaii. The best quality 'Cayenne' fruits in Kenya are obtained at altitudes of 1,400–1,800 m (Purseglove, 1972). The sugar to acid ratio there is 16, while it is 38 at 1,150 m.

Generally speaking, pineapple needs a sunny climate, though there are no exact figures on hours of sunshine or of solar radiation required. In Hawaii and Martinique a 20 per cent reduction in light caused a 10 per cent reduction in yield. Fruit colour and the content of malic (but not of citric) acid were also found to be influenced by light. We have already mentioned the effect of daylength on bloom.

Pineapple is a true xerophytic crop: it has many adaptations to drought. One of these is the behaviour of the stomata. They are kept closed during the hot hours of the day, when much moisture might otherwise be lost. On the other hand, this prevents an ordinary system of photosynthesis and the plant uses the CAM (crassulacean acid metabolism) system in which CO_2 is not reduced all the way to sugars, but only to organic acids (mainly citric and malic). They accumulate in the leaves at night and the reduction to sugar is completed during the day with closed stomata.

Most authors regard 1,000 to 1,500 mm rain as optimal for pineapple. Potential evaporation figures are provided by Bartholomew and Kadzimin (1977); they range from 1,742 mm in Malaysia to 839 mm in South Africa. In *Études sur Ananas* (1977) a potential evapo-transpiration of 3.5–4.5 mm/day is reported for Yamoussokro at 7° N in the Ivory Coast. There are few data on *Ep* from which irrigation needs could be calculated. High rainfall occurs in some pineapple regions, e.g. 3,000 mm in Cameroon and Costa Rica, 2,600 to 2,900 mm in Malaysia, nearly 2,000 mm in Taiwan. In such cases there is an excess of 700 mm to over 1,000 mm a year. As the pineapple root is very sensitive to waterlogging, drainage is necessary, but this causes leaching of fertilizers. Most pineapples are grown near the sea, where the atmospheric humidity is fairly high. Dew collects in leaf bases and is absorbed by the trichomes; Teisson (1979) found a maximum of 40 mlitres per plant per day. This is the equivalent of at most 0.24 mm of rain, which is insignificant. Salty sprays may cause leaf burn; this can be reduced by using trees such as *Casuarina* as windbreaks.

'Smooth Cayenne' is chiefly grown on sandy loam soils with a pH of 5 to 6. Red-brown oxisols are used for pineapple; some contain much iron oxide or manganese. Coral sand from beaches is often used for roads and the lime it contains causes poor growth of nearby pineapple plants. In Malaysia and other parts of South-East Asia peat soils with a low pH are in use for production of 'Singapore Spanish' pineapples (Driessen and Sudewo, 1977; Wee and Tay, 1968). For 'Red Spanish' a pH of 5 is about right.

The pineapple plant is efficiently built: leaf area continues to increase for 12–14 months and remains constant for another five months. The LAI of a closed crop is between 9 and 12, which is exceptionally high. Although the net assimilation rate is fairly low (0.4–2 g m^{-2} day^{-1}), the crop growth rate can be high: 15 g/m^2 or 150 kg ha^{-1} day^{-1} during several months. In one Hawaiian experiment 390 tonnes/ha of fresh plant material and 62 tonnes/ha of dry matter was produced in a cycle of 656 days (Bartholomew and Kadzimin, 1977).

Cultivars

The pineapple cultivars are classified in five groups: Cayenne, Queen, Spanish, Brazilian and Maipure. The description given at the beginning of this chapter was based on the Cayenne group, or rather on its main cv. 'Smooth Cayenne' or in French 'Cayenne Lisse', the most commonly grown pineapple. It combines nearly spineless leaves with high production, high fruit quality and resistance to gummosis. No other cv. has fruit of such good, cylindrical shape, ideal for canning. Table 7.2, assembled from data by

Table 7.2 *Characteristics of pineapple cultivars*

Cultivar	Spiny	Fruit form	Prod.	Fruit kg	Flesh colour	Slips	Exp.	Can.	Gum.	Wilt
Sm. Cayenne	No	Cyl.	High	2.5	Yellow	Few	G	VG	R	S
Red Spanish	Yes	Glob.	Fair	1.5	White	Many	VG	F	S	R
PR 1–56	No	?	High	2.5	White	Many	G	F	R	R
PR 1–67	Yes	?	High	2.7	Pale	Many	VG	F	R	T
Queen	Yes	Con.	Fair	1	Golden	Few	VG	No	?	?
Sing. Spanish	No	Cyl.	Fair	1.2	Golden	Many	No	VG	?	?
Abacaxi	Yes	Pyr.	Fair	1.5	Pale	Many	No	No	?	T

Exp. = export, Can. = cannery, Gum. = gummosis
G = good, VG = very good, F = fair,
R = resistant, S = susceptible, T = tolerant
PR = Puerto Rican clones

Samuels and Gonzáles-Tejera (1976), Purseglove (1972), Wee and Tay (1968) and Py and Tisseau (1965), compares some characteristics of the main cultivars.

Although 'Smooth Cayenne' is regarded as one cv., it is very variable. Other members of the Cayenne group are 'Hilo', which does not produce slips and 'Baronne de Rothschild', which has spiny leaves. In the Queen group the leaves are rather short and have many spines which curve back. The yield is only moderate and the fruit shape is conical, which makes it unsuitable for canning. On the other hand the flesh colour and aroma make it popular for consumption as fresh fruit. Well known cvs are 'Natal Queen', 'Z Queen' and 'Ripley Queen'. The slips and shoots are rather small and the plant is fairly resistant to most diseases.

Cv. Red Spanish has long, narrow, spiny leaves and is rather variable. The fruit is roundish with few, but large, flat 'eyes'. This cv. is very good for fresh fruit export, but only fair for canning. It will probably be replaced by 'Smooth Cayenne' wherever that is possible. 'Singapore Spanish' has smooth leaves and is good for canning; it grows well on peat soils. A mutant is called 'Selangor Green'.

Brazilian cvs, e.g. Abacaxi and 'Pernambuco' (= 'Perola'), have long spiny leaves and a long fruit stalk. The white-fleshed fruit is mainly for local consumption and not suitable for canning nor for export as fresh fruit. For a description of pineapple growing in Brazil see Giacomelli and Py (1981).

Leal and Soule (1977) added a new group of pineapples: Maipure; the completely spineless cvs Maipure, Perolera, Monte Lirio and Bumanguesa belong to this group. The fruits are cylindrical to oval, somewhat fibrous, with a small core and yellowish-white flesh. The quality is only fair for canning and export but good for local consumption.

Although 'Smooth Cayenne' can be regarded as ideal in many respects, it could be improved. Apart from selection within this very variable population, breeding work is also necessary. To this end it has been crossed *inter alia* with 'Perolera', a cv. with a high vitamin C content. Two Puerto Rican (PR) clones, derived from crosses with 'Red Spanish', are shown in Table 7.2.

Breeding aims are summarized as follows: a vigorous plant, having a short cycle, resistant to disease (especially mealy bug wilt); with broad, short and smooth leaves and cylindrical fruit, well coloured, with flat eyes, on a short but strong fruit stalk; a small ratio of leaf to fruit; flesh firm, well coloured, not fibrous, with high dry matter content, moderate acidity, high vitamin C and a narrow axis; early formation of one or two shoots and presence of one or two slips at least 2 cm below the fruit base (*Études sur Ananas*, 1977).

Selection in the 'Singapore Spanish' population in Malaysia has led to a new cultivar, 'Masmerah'. It is more vigorous, has more leaves, which stand more erect and has heavier fruit than the parent cultivar (Wee, 1974).

Cultivation measures

Once the decision to plant pineapple is taken, one must also decide on: which cultivar to plant, how large an area, at what density and to what purpose. It makes a big difference whether one grows fruit for the local market, for canning or for export as fresh fruit. The real choice is between the latter two, as the local market is commonly supplied from their remainders. Market greatly influences the choice of cultivar, acreage and planting density. A cannery needs a very large catchment area, several thousand hectares, and will favour the cv. Smooth Cayenne. Only big fruit will be accepted, indicating a long growth cycle and a relatively low planting density. If, on the other hand, the fruit is to be exported fresh, then the area can be smaller, the growth cycle shorter and the planting density substantially higher.

A second decision concerns the number of ratoon crops. In Malaysia ten or twenty may be taken with ease; presumably nematodes give no trouble on peat soil. In tropical West Africa even one ratoon is not advisable, because of nematodes, symphilids and so on. In Hawaii and most other countries at least one, and sometimes two ratoons are taken before the land is replanted. Pineapple can also be grown as an intercrop, e.g. with coconut (Nair, 1979).

Ground preparation

Clearing of forest or scrubland proceeds according to principles already discussed in Chapter 4. In replanting, old pineapple plants must be removed, chopped up and incorporated into the soil. If this is not carried out well, seats of infection remain from which nematodes will quickly reinfest the plantation. It is a major operation, as more than 200 tonnes/ha fresh material may be involved. Special machinery has been developed for this purpose.

Next, the field is ploughed and drains are laid down in such a way that erosion is checked. On sloping land the rows must follow the contours. A network of paths and roads is constructed, enabling machinery to be used efficiently: the distance between two roads is usually twice the range of the machines, so that all work can be done from the road. Then lime, fertilizers, herbicides and nematicides can be applied to the soil. Heavy infestation with nematodes, especially *Pratylenchus*, can be expected after a cassava or maize crop (Lacoeuilhe and Guyot, 1979).

Propagation

The different propagules (suckers, shoots, slips, crowns) have already been described. When an industry is just starting it may be necessary to maintain special nurseries for fast propagation. Here, stripped stems of plants that have borne fruit are also used. They are cut lengthwise and are sectioned. In principle each bud can grow out to a new plant, provided their nutrition is assured and the apical dominance is broken. Thus Louis (1969) succeeded in growing 80 plants from one stem. Tissue culture and similar techniques have been described in *Études sur Ananas* (1977). The different groups should be kept separate and must be sorted for size, so that a field is planted with uniform material. Slips and shoots are dried for one to two weeks and may be kept in storage for several months; crowns are dried for a few days but must not be stored. They can be used as such, or be split lengthwise in six to eight parts.

Planting

When planting pineapples, the rows are staggered; i.e. each second row is displaced by half a space. Two rows make a bed and between beds a path is maintained. A spacing of 30 × 60 × 90 means: 30 cm in the row, 60 cm between rows and 90 cm between beds. A simple way to figure out the planting density for this spacing is: each plant takes up 0.3 m × (0.6 + 0.9): 2 m = 0.225 m^2; the density therefore is 10,000:0.225 = 44,444 plants/ha. Three- or four-row beds are possible, but are never used for canning. Not only is maintenance

Fig. 7.6   Sorting propagules at Azaguié, Ivory Coast

more difficult, but the fruits from inner rows weigh appreciably less (10–14 per cent) and produce less shoots. Generally a density of 40,000 to 50,000 plants/ha is chosen for cannery fruit and 60,000 to 70,000/ha for fresh export; in both cases with Cayenne. The tendency is to lower the distance between rows from 60 to 45 cm and the row interval to 25 or 20 cm (Guyot *et al.*, 1974; *Études sur Ananas*, 1977).

Lower densities are usual with other cvs, especially when they have spiny leaves. For 'Red Spanish', e.g., a distance of 1.20 to 1.30 m was deemed necessary between beds, but Treto *et al.* (1974) report significant yield increases at densities up to 41,600/ha, while the fruit weight did not change.

A large number of density experiments have been carried out. They all show great increments in yield with greater densities, but lower fruit weight and quality. Waithaka and Puri (1971), for example, report that grade I fruit production increases up to a density of 57,500 plants/ha and then slowly decreases, with a combined fast rise in grades II and III up to 127,000 plants/ha. The economic optimum of density is thus decided by the price ratio for the grades. However, at very high densities the field becomes unmanageable. Lee (1977), with cv. Serawak, harvested 44 tonnes/ha at 24,000 plants/ha and 50 tonnes/ha at a density of 72,000 plants/ha; a top yield of 66 tonnes/ha was obtained with 54,000 plants/ha. Ramirez and Gandía (1982) compared nine planting distances of cv. PR 1–67 and found no effect on fruit weight or juice quality. The highest yield was 86 tonnes/ha at density 51,625 for the plant crop; the ratoon crop brought in 45 tonnes/ha. Slip production decreased at higher densities.

Mulching

Mulching is widely practised in pineapple growing. Green manure and organic refuse, such as sugarcane bagasse, may be used. In heavy rainfall areas this provides an environment for parasitic fungi that attack pineapple. A 20 cm thick layer of grass has given good results, especially with *Anadelphia arrecta*.

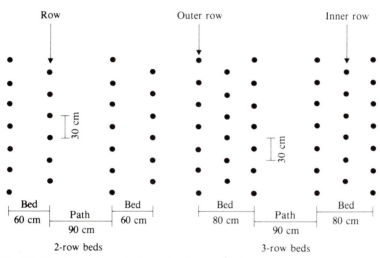

Fig. 7.7 Spacing systems of pineapple. *Source: Études sur Ananas* (1977)

Table 7.3 *Density and spacing in relation to fruit weight and production of shoots (until 5 months after harvest)*

Plants/ha	Rows	Spacing	Fruit weight (g)			% Shoots		
			av.	inner	outer	av.	inner	outer
63,500	2	90 × 40 × 25	1,207			154		
65,900	2	91.5 × 30 × 25	1,180			153		
70,500	2	83.5 × 30 × 25	1,197			186		
65,900	3	90 × 30 − 30 × 30	1,164	1,061	1,214	141	129	146
70,500	3	90 × 26 − 26 × 30	1,150	1,074	1,185	138	131	142

Source: Études sur Ananas (1977)

Mulching paper has also been used with success for a number of years. However, it has now largely been replaced by black polythene which lasts longer and has the same advantages: it warms the soil, protects against erosion, prevents weed growth, reduces leaching and increases yields. The plants are inserted into the ground through holes made at marked spots in the mulch. If replanting is necessary for each crop, as in West Africa, there is little sense in using mulching paper or polythene.

Control of weeds

Weed control is essential. In Guinea the average fruit weight without it was only 609 g, as against 1,553 g in plots with adequate weed control. Not only are weeds competitors for water, light and nutrients, but they also harbour nematodes and other pests. The most notorious weeds, which must be destroyed at all cost, are nut grass (*Cyperus rotundus*) and lalang (*Imperata cylindrica*).

Bromacil is very active as a pre-emergence herbicide against nut grass; paraquat and glyphosate are among the best of the contact herbicides. IRFA (page 10) constructed a small machine for application of herbicides, called *l'herbinet*, which has given good results on small farms. This proved to be a paying proposition as it raised the yield by 10 per cent, multiplied the number of acceptable fruits by five and halved the number of working hours (*Études sur Ananas*, 1977).

Irrigation

Potential evapotranspiration, as we already saw, can go up to 4.5 mm per day in pineapple. As the water-holding capacity of the soil rarely surpasses 100 mm, this means that the supply will be exhausted in three to four rainless weeks. The plant may be drought resistant, but growth and fruiting will be delayed under such conditions. It is therefore not surprising to read that irrigation in the Ivory Coast has increased yields by 14 to 22 tonnes/ha; the cost was equivalent to a yield of 5 tonnes/ha. Some less favourable side effects have to be acknowledged: more weeds, leaching and premature flowering. Sprinkling is the most efficient method, since some water is retained in the leaf bases. Irrigation may be used at the same time for application of fertilizer, herbicide and insecticide.

Crop fertilization

Data from West Africa are shown in Table 7.4. The plant crop needs a good supply of all nutrients and for each harvest at least 68 kg N, 24 kg P_2O_5, 174 kg K_2O, 27 kg CaO and 16 kg MgO/ha

Table 7.4 *Minerals in kg/ha in the pineapple crop*

	N	P_2O_5	K_2O	CaO	MgO
Yield (55 tonnes) takes up	205	58	393	121	42
Harvest removes	43	16.5	131	17	10
One shoot/plant removes	24.5	8	43	10	6.2

Source: de Geus (1973)

have to be provided extra. In the Ivory Coast the following formula is used with fresh fruit for export: 4–2–11–2 g/plant of $N-P_2O_5-K_2O-MgO$. At a density of 60,000 this comes to 240 kg N/ha. N is given as urea and K as sulphate. For processing from 8 to 14 g N and 10 to 20 g K_2O/plant is applied. In each locality the exact fertilizer formula must be determined from data obtained by soil and leaf analysis. And, of course, the plants have to be watched constantly for signs of deficiency.

Deficiency symptoms conform to those of other crops. N deficiency delays growth, stunts plants and causes yellowish leaves. The production of fruit, slips and shoots is affected. Too little P is seldom observed, but too much depresses yields. K deficiency becomes apparent by yellow dots on the leaves. The basal part of the D leaf should contain at least 3.2 per cent K in the dry matter; in air-dry soil 140 ppm exchangeable K is adequate. However, the nature of the cv. exerts some influence too, as is evident from Table 7.5. Much lower figures were found in Ghana for cv. Sugarloaf: N 0.35 to 0.40; P 0.04; and K 0.44 per cent (de Geus, 1973).

Table 7.5 *Optimum leaf nutrient levels (% dry matter) for two cvs in Puerto Rico*

	Cv. Red Spanish	Cv. Smooth Cayenne
N	1.7–2.2	1.6–1.9
P	0.20–0.25	0.16–0.20
K	3.5–4.0	1.8–2.5

Source: de Geus (1973)

High yields are associated with Mg contents of 0.32–0.35 per cent in leaves of 5-month-old plants and of 0.19–0.20 per cent in 9-month-old plants, but much depends on the K/Mg ratio. The N/K ratio is important as too much K causes acid fruits with big cores and pale, firm flesh. Iron deficiency occurs at a pH above 6.5, high Ca content and an excess of manganese in the soil. An Mn/Fe ratio above 2 produces chlorosis of leaves, due to iron deficiency. It can be controlled by low volume spraying with iron sulphate or a chelate. Zinc and copper deficiencies may also occur, and can be corrected by spraying a chelate.

Fertilizer can be applied in solid form to the soil, or in solution

Fig. 7.8　A pineapple field at Azaguié, Ivory Coast

to the lower leaf axils; the latter has given better results. It must be split in small monthly portions for N and in a few applications for K. Nitrogen fertilizer must be stopped about two months before flower induction.

Floral induction

The natural cycle of flowering in pineapple is rather unreliable. It has become customary to induce bloom by:
1. the natural hormones NAA and ethylene
2. products that release ethylene, e.g. ethephon and BOH
3. calcium carbide, which releases acetylene.
All these products are effective in Hawaii and Queensland, Australia, but problems arise near the equator. In Ivory Coast NAA has undesirable side-effects: fruit becomes pointed, ripens unevenly and over a longer period; it is now only used on young fruits, to make them bigger and to strengthen the fruit stalk.

Ethylene gas must be sprayed at night; expensive equipment and much water (6,000 litres/ha) are needed for this. The small farmer uses calcium carbide, which can be applied by hand; this must be done at night and may burn the leaves.

The best product for floral induction is ethephon. It splits, on contact with water, into ethylene and phosphoric acid at a pH of 4.5 or higher. However, when the pH of the liquid in the rosette of pineapple plants was found to be 2.5–3.8, Dass *et al.* (1975)

added 2 per cent urea and 0.04 per cent lime to 25 ppm ethephon and got more than 90 per cent flowering in 50 days; each plant received 50 mlitres of liquid on a sunny day.

A disadvantage of this method is that urea causes fruit stalks to grow out excessively and that young leaves may become burned. On the other hand, bentonite produced very good results. This cheap absorbent clay does not damage the plant and raises the pH of the rosette liquid to ten. Two mg ethephon in 100 to 400 mg bentonite gave 90 to 100 per cent floral induction. Furthermore, this product can be applied by day, either early in the morning or late in the afternoon, when the stomata are open. Application to the rosette gave better results than to the leaf bases and bright weather was favourable; however, rain within a few hours could practically eliminate the effect. For further details, see *Études sur Ananas* (1977) and Teisson (1979). Norman (1982) has investigated the role of growth regulators in cv. Sugarloaf.

Fig. 7.9 An ornamental pineapple (*Ananas bracteatus*)

For canning one commonly waits with an induction treatment until at least 60 per cent of the plants show signs of natural induction. For fresh fruit export the treatments are applied a fixed number of months after planting. Induction treatment also makes it possible to control certain pests (e.g. *Thecla*) more efficiently.

Other measures

Pineapple fruits in the field that are excessively heated by the sun on one side may suffer from sun scald, which renders them useless. This occurs in ripening fruits on slopes exposed to the sun, or in fruits which lean over because the fruit stalk is weak. The best protection is to tie leaves around the fruit, a rather labour-intensive operation; one man can do from 700 to 1,500 plants a day.

Reduction of the crown is necessary for 'Smooth Cayenne' fruits intended for export; they would become too big if nothing was done about it. When the crown is about 8 cm long, the meristem is gouged out with a chisel. From 1,000 to 2,000 plants can be treated per man-day. Fruits of Queen cvs have smaller crowns that do not require reduction.

Fig. 7.10 'Smooth Cayenne' fruit with crown reduced for export

Diseases and pests

Mealy bug wilt is the most widely distributed pineapple disease and probably also one of the most damaging, particularly to 'Smooth

Cayenne' (Collins, 1960). A rapid spread takes place from a focal point; as soon as symptoms appear the mealy bugs move to healthy plants. Roots cease to grow, collapse and rot, causing the plant to wilt. This starts at the leaf tips and a reddish-yellow colour develops. Control of the mealy bugs is essential, but this can only be achieved if the attendant ants are destroyed. A border area of six to eight beds must be sprayed regularly to keep the ants outside the field. Several insecticides e.g. parathion are used. 'Smooth Cayenne' is very susceptible to mealy bug wilt, but some resistant Cayenne clones exist. 'Red Spanish' and 'Singapore Spanish' are resistant and so are some wild relatives that are used in breeding. The real cause of the disease seems to be a virus, but this has not yet been proved.

Yellow spot is certainly caused by a virus. It is transmitted by a thrips from hosts such as *Emilia sonchifolia*, a well-known composite weed (Collins wrote Emelia, a mistake that has been copied in most books on pineapple). Eradicating the weed hosts is the only practical control measure.

Nematodes may number over 100,000/dm³ soil and are extremely dangerous enemies of pineapple. They attack the roots, causing rootknots (*Meloidogyne*) and lesions (*Pratylenchus*) or penetrate partially into the roots (*Rotylenchus*). Other genera have been found too. Generally they prefer light soils, but some also thrive on heavier soils. *Pratylenchus brachyurus* is a dangerous pest of pineapple in Ivory Coast.

During the first three or four months after clearing and planting the nematodes are few in number, but then they increase dramatically and slow up the growth of the pineapple plants significantly; a chlorosis appears on the leaves. This is followed by an abrupt decrease in nematode numbers. The same happens during a drought. These 'waves' are particularly marked in the short cycles (for export). All plant material that may harbour nematodes must be destroyed before planting.

Two weeks before planting the soil is fumigated with D-D (300 litres/ha), which is very toxic to the plant. Collins (1960) advises waiting only three days in Hawaii, probably because of the very light soil type. EDB (ethylene dibromide), used at 100 kg/ha, is less toxic. Fumigation gives yield increases from 3 to 32 per cent (*Études sur Ananas*, 1977), depending on the abundance of nematodes at that time. It is advisable to fumigate only when strictly necessary, as it is a dangerous and expensive operation.

New nematicides, with systemic action, are fenamiphos (Nemacur), carbofuran (Furadan, Curaterr) and ethoprophos (Mocap). Plants soaked in a 1,500 ppm Nemacur solution and treated with 0.2 g/plant every two months were free of nematodes all year. Dusting with 0.5 g/plant was very effective, as the dust could be mixed with fertilizer and other pesticides. Application of pellets to leaf bases is another possibility (*Études sur Ananas*, 1977). Another promising nematicide is oxamyl (Dupont 1410, Vydate) which can be sprayed; unlike other nematicides it is transported downwards, from the leaf to the roots. However, not all chemicals mentioned in this paragraph have been found safe for use yet, and the danger of residue in and on the fruit must be kept in mind. Sarah (1980) got excellent results from spraying with carbofuran (0.15 g/plant) and fenamiphos (0.2 g/ plant) twice a year. Later (id. 1981) some phytotoxic effects were noticed. A rest period, to eliminate nematodes, is not mentioned in articles on this subject; it is probably too expensive to leave the soil unused for six months.

Symphilids are myriapods that can be destructive on pineapple. They are about 4 mm long and feed on roots. The reaction of the plant causes a sort of 'witches' broom' in the roots; a much reduced root system that is easily attacked by fungi, and growth stagnation of the plant are the result. The above mentioned fumigants are active against this pest, but addition of lindane (2 kg active ingredient/ha) reinforces the action. Fenamiphos and ethoprophos may be used too.

Other pineapple pests are scales, mites, fruit flies, the moth *Castnia licus* (the same as in the banana) and the butterfly *Thecla basilides*. The last named can only be controlled properly after hormone treatment, so that no untreated fruits remain to maintain the pest population.

Heart rot is caused by *Phytophthora cinnamomi* and *P. parasitica*; the latter is adapted to warmer regions. Their zoospores are drawn towards the trichomes chemically and penetrate young leaf cells. Resistance is based on structure of the rosette and the tissue below the epidermis; selection programmes may now make use of these properties (Louvel, 1975). To control the disease plant material is soaked in difolatan but Allen *et al.* (1980) prefer metalaxyl and Aliette. Sprays of 0.2 per cent captafol at 3,500 litres/ha immediately after planting, one month later and one week after the treatment for flower induction are recommended by Frossard (1978). *Thielaviopsis paradoxa* causes rot of planting material and of fruits after harvest. To prevent this, the stalk is dipped in benzoic acid, Shirlan, or imazalil. Fruit gummosis occurs in 'Red Spanish', but 'Smooth Cayenne' is resistant. It seems to be caused by the feeding of a caterpillar and can be controlled by spraying with insecticides.

From harvest to consumption

A yield of at least 40 tonnes/ha may be expected in a well-managed pineapple plantation; under optimal conditions the yield goes up to 70 tonnes/ha or even more. To compare yields one must know the number of months the crop took to ripen. This may range from less than 12 to more than 24. It would therefore be more correct to state yields in terms of tonnes/ha. month. It is also necessary to record whether a plant crop or a ratoon is involved.

The point at which a fruit is considered ready for harvest depends on its ultimate destination: cannery, export or local market. As the local market is often supplied as a by-product of the export and cannery trades, we may ignore it here: fruits that are too big, too small or otherwise unfit for the other outlets go to the local market. Fruit for the cannery must be picked ripe. This is judged by the colour and sometimes also by the sound made when the picker knocks on the fruit. Formerly harvesting was largely a hand operation (Collins, 1960). It still is in most places. Bags, baskets or sleighs take the fruits to the roadside. There they are trimmed (crowns removed, bases cleaned) and placed in wooden boxes. These are loaded on flat-bed lorries and sent to the cannery. The work is mechanized on large plantations. Lorries with an endless conveyor belt of 18 m, i.e. half the distance between two roads, are used. The pickers walk on the paths, remove crowns and put fruits on the belt. Smaller units have movable slat floors. Crowns are left on the fruit as protection if the cannery is far away.

Fruit for export is picked in a half-ripe state. The maturity is judged on the basis of external colour, but this depends on cultivar, fruit size and weather. Generally grade II (coloured halfway) is exported. Internal colour is also taken into account: flesh must be translucid when cut transversely at one-third of the height. A minimum of 12 per cent total soluble solids is required in Hawaii (Akamine, 1976).

In the Ivory Coast fresh fruit is graded in four classes: D(700–900 g), C(900–1,100 g), B(1,100–1,500 g) and A(1,500–1,800 g). The fruit is packed in boxes holding 3 × 2 of the biggest to 5 × 4 of the smallest fruits, and shipped in holds kept at 8 °C. Sizes D and C were greatly preferred in France until a TV advertising campaign boosted the sales of bigger fruit.

Pineapples are attacked by fruit flies, but they do not survive in the fruit; therefore quarantine restrictions have been lifted. SOPP (7.2 g/litre) is used to control *Thielaviopsis* decay. A physiological malady called endogenous brown spot caused big losses in cooled pineapples until a practical method of control was discovered: dry heat is applied at 35 °C for 24 hours, either before or immediately after cooling (Akamine, 1976). Blackheart (Smith, 1983) is probably identical with endogenous brown spot.

In the cannery, Ginaca machines remove shells, ends and cores at a rate of 125 fruits/minute. Then the fruits are sliced in three sizes: 4/4, 3/4 and 2/4 in the French terminology, and manually packed in cans. Other products are: chunks, spears, cubes, crush, tidbits, juice, sugar-syrup, alcohol and cattle feed. For details on the world market for these products, see Harman (1984).

References

Akamine, E. K. (1976) 'Problems in shipping fresh Hawaiian tropical and subtropical fruits', *Acta Horticulturae* **57**, 151–61.

Allen, R. N. *et al.* (1980) 'Fungicidal control in pineapple and avocado of diseases caused by *Phytophthora cinnamomi*', *Austral. J. of exp. agric.* **20**, 119–24.

Anon. (1980) *Action report Hawaii, 1976–1980*, Haw. Agric. Exp. Sta.

Bartholomew, D. P. and **Kadzimin, S. B.** (1977) 'Pineapple' in: *Ecophysiology of tropical crops*, Academic Press, 113–56.

Bourke, R. M. (1976) 'Seasonal influences on fruiting of rough leaf pineapples', *Papua New Guinea Agric. J.* **27**, 103–6.

Brücher, H. (1977) *Tropische Nutzpflanzen*, Springer Verlag, Berlin.

Collins, J. L. (1960) *The pineapple*, Leonard Hill, London.

Dass, H. C. *et al.* (1975) 'Flowering in pineapple as influenced by ethephon and its combination with urea and calcium carbonate', *Scientia Hortic.* **3**, 231–8.

Driessen, P. M. and **Sudewo, P.** (1977) *A review of crops and crop performance on South-East Asian lowland peats*, Bogor, SRI.

Études sur Ananas (1977) *Fruits* **32**, 447–522.

Figueroa, R. *et al.* (1970) *El cultivo de la piña en el Perú*, La Molina, Boletin tecnico 75.

Fogg, G. E. (1970) *The growth of plants*, sec. ed. Pelican.

Frossard, P. (1978) 'Lutte contre la pourriture de coeur à *Phytophtora* de l'ananas', *Fruits* **33**, 183–91

Geus, J. G. de (1973) *Fertilizer guide for the tropics and subtropics*, Zürich.

Giacomelli, E. J. and **Py, C.** (1981) 'L'ananas au Brésil', *Fruits* **36**, 645–87.

Guyot, A. *et al.* (1974) 'L'ananas en Côte d'Ivoire', *Fruits* **34**, 85–117.

Harman, G. W. (1984) *The world market for pineapple and pineapple juice*, Trop. Dev. and Res. Inst. (London), G 186.

Lacoeuilhe, J. J. and **Guyot, A.** (1979) 'Les techniques culturales de l'ananas en Côte d'Ivoire', *Fruits* **34**, 159–68.

Leal, F. J. and **Soule, J** (1977) 'Maipure, a spineless group of pineapple cultivars', *Hortsci.* **12**, 301–5.

Lee, S. A. (1977) 'Influence of planting density on the yield and fruit quality of pineapple cv. Serawak', *Malaysian Agric. J.* **51**, 223–7.

Louis, A. M. (1969) *Vegetatieve vermeerdering van ananas met hoog rendement*, Internal report V 69/2, Dept. Tr. Crops, AU Wageningen.

Louvel, D. (1975) 'Étude des relations entre l'ananas et le Phytophthora parasitica', *Fruits* **30**, 669–80.

Nair, P. K. (1979) *Intensive multiple cropping with coconuts in India*, Parey.

Naville, E. (1976) 'Production et commerce de l'ananas dans le monde', *Fruits* **31**, 789–806.

Norman, J. C. (1982) *Growth, flowering and fruiting of 'Sugarloaf' pineapple, as influenced by growth regulators and cultural practices*, Diss. Bonn.

Pickersgill, B. (1976) Pineapple, in *Evolution of crop plants* (ed. N. W. Simmonds), Longman.

Py, C. and **Tisseau, M. A.** (1965) *L'ananas*, Maisonneuve et Larose.

Purseglove, J. W. (1972) *Tropical crops, Monocotyledons*, Longman.

Ramirez, O. D. and **Gandía, H.** (1982) 'Comparison of nine planting distances on the yield of pineapple variety PR 1–67', *J. Agric. Puerto Rico* **66**, 130–8.

Samuels, G. and **Gonzáles-Tejera, E.** (1976) 'Recent research findings on pineapple growing in Puerto Rico and the Caribbean Islands', *Acta Horticulturae* **57**, 241–6.

Sarah, J. L. (1980–1) 'Utilisation de nématicides endothérapiques dans la lutte contre *Pratylenchus brachyurus* en culture d'ananas', *Fruits* **35**, 745–57; **36**, 275–83 and 491–500.

Smith, L. G. (1983) 'Cause and development of blackheart in pineapples', *Trop. Agric.* (Trin.) **60**, 31–5.

Teisson, C. (1979) 'A la recherche d'un traitement d'induction florale de l'ananas par voie solide', *Fruits* **34**, 515–23.

Treto, E. *et al.* (1974) 'Étude de différentes densités de plantation chez la variété d'ananas Española roja à Cuba', *Fruits* **29**, 279–84.

Waithaka, J. H. G. and **Puri, D. K.** (1971) 'Recent research on pineapple in Kenya', *World Crops* **23**, 190–2.

Wee, Y. C. and **Tay, T. H.** (1968) 'Pineapple cultivation in West Malaysia', *Research paper* 11, Pineapple research station, Johore.

Wee, Y. C. (1974) 'The Masmerah pineapple', *World Crops* **26**, 64–7.

Chapter 8

Mango

Taxonomy and morphology

The botanical name for the mango plant is *Mangifera indica*. It belongs to the Anacardiaceae family, to which cashew nut and some other fruit crops (see Appendix 1) also belong. The genus is native to South-East Asia and consists of 62 species. About 16 of these have edible fruits but apart from mango, only *M. caesia*, *M. foetida*

Fig. 8.1 'Golek' tree in Surinam

and *M. odorata* are regularly eaten, although they strongly taste of turpentine. *M. verticillata* of the Southern Philippines is of interest; it may be a variety of *M. caesia*.

Mango forms an erect, well-branched evergreen tree with a dense crown (Fig. 8.1). The leaves are spirally arranged and come out in reddish flushes that initially hang straight down. Later they take on a more horizontal position and turn green; they stay on the tree for one to three years. There are two to five flushes a year, depending on the climate. The leaf blade is elliptic or oval, 15–40 cm long and 2–10 cm wide. The midrib is prominent, with 30 pairs of lateral veins. Stomata are present on both sides, but mainly on the lower surface. The leaf stalk is swollen at the base and 2–10 cm long.

The inflorescence (Fig. 8.2) is a widely branched panicle, 10–60 cm long, with a thousand or more male and hermaphrodite flowers. The proportion of bisexual flowers ranges from 1–100 per cent, depending on cv., climate and weather. The flowers are small, 5–8 mm, usually with five sepals, petals and stamens (only one of which is fertile) and a pistil with an oblique style.

The fruit is a fleshy drupe with edible pulp (mesocarp) and a woody stone (endocarp) around the seed. A fruit may weigh from

Fig. 8.2 Inflorescence of mango

100 g to 2 kg. There is a double seed coat consisting of two papery layers around the seed. Inside are two fleshy cotyledons and one to several embryos: the mono-embryonic cultivars contain one zygotic or sexual embryo, polyembryonic cvs contain two or more; one may be zygotic, all others are nucellar. It is therefore possible to propagate mango vegetatively by seed (see Fig. 3.3), just like citrus.

Uses and composition

Mango is the most popular fruit of the Orient and has been called 'king of the fruits', but also 'a ball of tow soaked in turpentine' or 'fit to be eaten in the bathtub only'. Fruit from seedling trees may be stringy and watery or may have an unpleasant aroma. On the other hand, fruit from the better cvs has melting yellow flesh, good flavour and a fine aroma. People tasting it for the first time often compare it to the peach.

Ripe mangoes are eaten for dessert, are canned or used for making juice, jams and other preserves. Pickles and chutney are prepared from unripe fruits, or a powder is ground from them, after slicing and drying. The wood is utilized for boats. Seeds and leaves can be eaten or fed to cattle in times of food shortage, but prolonged feeding may result in death (Purseglove, 1968).

The edible portion takes up 60–75 per cent of the fruit weight. It contains water (84 per cent), sugars (15 per cent) and protein (0.5 per cent), plus fair amounts of vitamins A, B and C. The seed consists of carbohydrate (70 per cent), protein (6 per cent) and fat (10 per cent).

Origin, distribution and production

The name *Mangifera indica*, assigned by Linné, suggests an Indian origin but this is not at all certain. It is more likely that the origin lies in the Burma–Malesian region. Yet the name is appropriate, as this fruit has been cultivated in India for more than four thousand years. From here it spread to other countries in the region. Persian sailors took it to East Africa, probably in the tenth century A. D., and Portuguese travellers of the 16th century brought the mango to West Africa and South America. Since then, the mango has been introduced into every tropical and subtropical country in the world.

India has the largest mango cultivation area by far: about one million hectares. Cultivation is also widespread in Pakistan, Bangladesh and other countries of South-East Asia. The southern Sahel

Table 8.1 *The major producers of mango (× 1,000 tonnes)*

	1975	1980	1981	1982	1983	1984
India	8,500	8,363	8,516	8,500	8,700	8,919
Mexico	389	581	620	663	665	670
Brazil	615	530	560	600	600	520
Pakistan	605	550	550	552	683	683
Philippines	250	374	380	390	550	550
Indonesia	—	345	444	340	344	360
China	203	276	341	338	353	387
Haiti	278	326	330	335	340	340
Bangladesh	284	207	203	203	182	185
World	12,664	13,091	13,507	13,508	13,954	14,213

Source: FAO Production Yearbooks
NB: FAO indications for estimates have been omitted

is well suited to mango culture and it is grown commercially in Transvaal, Egypt, Israel, Florida, Mexico and Queensland.

Insufficient information is available on the production, export, import, processing and consumption of mangoes. The market for mango juice, pickles and chutney is small, but is growing at a rate of 5 per year (Cadillat, 1976 a+b). Most EEC countries do not itemize mango separately in their statistics, but it probably constitutes 90 per cent of the 'other fruit' figures.

The major producers are shown in Table 8.1. Principal exporters to the EEC (European Economic Community) are India, Kenya and Mali. In 1975 they exported 515, 514 and 320 tonnes respectively. In that year the United States imported 8,050 tonnes of mango from Mexico and Haiti, in addition to the 6,500 tonnes produced locally in Florida. It therefore seems likely that mango consumption is ten times higher in the United States than in Europe.

It is clear that there is much room for expansion in Europe. Only a few years ago it was almost impossible to buy mangoes on a Dutch market, but they are now becoming increasingly popular. Table 8.2 shows the amounts and seasons of import into Holland from some countries.

Growth and development

Mango seeds lose their viability rapidly, unless kept in moist charcoal. In germination, the taproot grows straight down; it may reach a depth of 5 m. However, its growth is often checked by a hardpan or a water bearing layer. The taproot provides good anchorage for the tree, together with other vertical roots. There is a dense mat of feeding roots too, but altogether the rooting volume is surpris-

Table 8.2　*Importation of mangoes into the Netherlands (tonnes)*

From	1975	1980	1981	1982	1983	Main period
Mali	223	361	418	660	863	April–June
Upper Volta	5	20	120	231	116	April–May
Mexico	6	53	137	119	241	Aug.–Sept.
Guinea		2	13	105	8	April–June
Senegal	75	45	38	66	66	April–June
Israel	9	44	59	57	43	Aug.–Oct.
Kenya	48	75	74	50	38	All months
Total import	531	1,415	1,300	1,534	1,864	

Source: Produktschap voor groenten en fruit, The Hague

ingly small. Moutounet *et al.* (1977) found a rooting volume, in a deep brown ferrallitic soil, of only 26 dm^3 for a five-year-old tree on 'Maisonrouge' rootstock and 64 dm^3 for a similar tree on 'Carotte'; considerably less than a guava tree (370 dm^3).

Zygotic embryos do not breed true, so mono-embryonic types must be propagated vegetatively if one wants to preserve their properties. Nucellar seedlings have the same genetic constitution as the mother plant, but is is not easy to recognize them.

A seedling tree passes through a juvenile period which may last seven years, but a grafted tree commonly flowers in the third year. However, Singh (1960) advises the removal of all inflorescences during the first four years, in order not to weaken the tree. Bloom initiation occurs after a rest period in autumn and winter in the subtropics; a dry period of three months or longer has the same effect in the tropics. Apical buds need at least two months of dormancy before they can be differentiated into flower buds.

Pollination is effected mainly by flies and thrips. Singh (1960) asserts: 'no honey bee is ever observed visiting mango flowers', but this is incorrect. Bijhouwer (1937) put bee hives in his orchard on Java and found that bees regularly flew to some cvs (e.g. Alphonso) but rarely to others. Purseglove's remark: 'bagged panicles set no fruit' is also incorrect: Bijhouwer found up to 64 fruits per 50 bagged panicles and five times as many in open ones. Free and Williams (1976) report that mango is relatively unattractive to honeybees in Kenya and Jamaica.

Only a small percentage of the flowers set fruit and again a small percentage of these reach maturity. Once a fruit has attained a

Fig. 8.3　Mango fruits in Israel

diameter of 25 mm, its chances of survival improve. Alternate bearing, a phenomenon discussed in Chapter 3, is one of the most pressing problems in mango cultivation. Popenoe (1920) reported that a good mango crop in Saharanpur (northern India) was had in only 8 out of 27 years; there were light crops in 7 and bad ones in 12 years. From 1949 to 1955 'Langra' produced 484–5–653–0–635–0–417 fruits in northern India: an extreme case of alternation.

Mango is characterized by heavy fruit drop at all stages; several authors point out that less than one out of a thousand perfect flowers develops into a mature fruit. Rain, high humidity, attacks by fungi and insects, a low C/N ratio, mineral deficiencies, a low per-

centage of perfect flowers and the hormonal balance have been blamed for alternate bearing. As a result of one or more of these causes, not enough vegetative growth takes place in the on-year to support bloom in the off-year. In order to correct this situation, the following measures have been proposed:
1. Avoid cultivars that are known alternate bearers.
2. Plough, manure and irrigate during the on-year.
3. Double the amount of N during on-years.
4. Irrigate immediately after fruit setting.
5. Deblossom partly in on-years.
6. Ring branches, prune and apply salt to roots.
7. Light fires and produce smoke during 1 to 4 weeks.

None of these measures is entirely successful and the reasons why they sometimes work are little understood. It is possible that the smoke contains ethylene, which would promote flowering. Far more surprising is that a 1% KNO_3 spray also worked: bloom occurred 10 to 14 days later (Anon., 1977).

Ecology and physiology

The optimum climate for mango ranges from subhumid-equatorial to subarid-subtropical, provided there is a marked dry season. Even a light frost at ground level kills young trees, but older trees tolerate some frost when dormant. Inflorescences are damaged by frost but may bloom again later. Subtropical growers should leave the lower slopes of hills unplanted as cold air accumulates there. The average temperature must be at least 21 °C and the optimum is 25 °C. High temperature, especially when combined with low humidity and strong winds, is injurious to mango; the plant cannot cope with the increased transpiration rate and dries out. Leaf scorch, seed abortion and fruit fall follow; above temperatures of 42 °C a collapse, called sunstroke, occurs.

Single mango trees are occasionally seen growing quite well at an elevation of 1,000–1,200 m, but it is bad policy to plant an orchard higher than 600 m above sea level: the average temperature would be too low. Mango requires much heat, but an exact number of degree-days has not been established.

Mango's light requirement and reaction to daylength are not mentioned in literature. Perhaps it is regarded as self-evident that mango needs much light and that daylength has no effect. I can say that in our greenhouse at Wageningen (52° N) bloom usually occurred in November.

How much rain does the mango need? This varies from a small amount where irrigation is available, to very much on permeable soils with a long dry season. Rainfall distribution is more important than total precipitation. If the rainless period lasts four months, we must know the Ep for the other eight months. Unfortunately, such data are hard to find. I have seen but one indication: Sizaret (1970) mentions an E_p of 2,276 mm in a nursery at Niamey (Niger). As a provisional figure, I propose 150 mm per month; this may serve as an indication for the irrigation needs of the mango.

Terra (1949), on Java, found that mango needs at least three dry months and the wet season should not last more than seven months. Laroussilhe (1980) tells us that mango grows well at Tamatave (Madagascar) with more than 3,000 mm of rain per year. At any rate, mango is not a good crop for equatorial regions: the rest period is too short and if bloom occurs at all, it is likely to be rained upon and perish because pollination will fail. In drier areas bloom appears during the dry season and fruit is set before the rains begin.

High humidity, mist and heavy dew promote fungal attack and are therefore harmful to the mango. Low humidity and drying wind increase evapotranspiration unduly and cause wilting. Strong wind impedes pollination and may destroy the crop; where it occurs, windbreaks are required.

Like citrus, mango makes no high demands on the soil. It can be sandy or loamy, lateritic or alluvial, as long as it is deep and free-draining. Very poor, shallow, rocky and alkaline soils should be avoided. Chemical fertility is not a necessity but good structure and aeration are essential; pH should be between 5.5 and 7.5. At a higher pH, deficiencies of iron and zinc appear. The soil must have at least a moderate water-holding capacity and a ground water table at a depth of 3–4 m is an advantage. A light slope furthers drainage but steep slopes are unsuitable.

Cultivars

There are hundreds of cultivars. Gangolly *et al.* (1957) list some 210, all from India. Singh (1960) lists 77 from India and 24 from other countries. Raghava Kurup *et al.* (1967) limit their choice to 40. From Java 42 are listed by Bijhouwer (1937); the best were 'Golek', 'Arumanis' and 'Manalagi'. Florida has produced many valuable cvs such as Haden, Irwin, Tommy Atkins and Keitt (Malo, 1972). The best West Indian cvs are Julie and Peter (Purseglove, 1968). The Indian and Florida cvs are mainly mono-embryonic, those from South-East Asia and Hawaii are predominantly polyembryonic. Several attempts were made to classify the cvs, none of them very enlightening. Examples are:
1. Round-fruited, long-fruited, indefinite.
2. Mulgoba, Alphonso, Sandersha and Cambodiana group.
3. Seedling races, polyembryonic types and horticultural varieties.
It frequently occurs that one cv. has more than one name, e.g.

'Totapuri' = 'Bangalora'. Or, one name may refer to two distinct cvs, e.g. 'Safeda'. We need accurate cv. descriptions, according to a list of characters, internationally approved, including:
1. Morphology of tree, leaves and flowers.
2. Shape, size, weight, taste and quality of fruits.
3. Agronomic characters, such as preferred climate, bearing habit, harvest season, resistance to diseases and pests, etc.

Good photographs, preferably in colour, of tree, leaf, panicle and fruit should go with the description. Aubert (1975) has described bloom, fruit setting, vegetative growth and harvest period in Réunion (20° S), for 17 months, compared to temperature, rainfall, irrigation and incidence of cyclones. Larousilhe (1980) presents simple but clear drawings of 12 leaf, 18 fruit and 3 panicle types. Singh (1972) has drawn up a list of merits and demerits of the commercial cvs and concludes that no cv. is ideal in all respects. There is a need for a dwarf cv. with fruits of medium size (300 g), of good quality, yielding a crop every year and tolerant to malformation.

Most cultivars arose from chance seedlings. Breeding work is difficult because of low success rates in pollination, a long life cycle and other problems. Certain cvs, e.g. Dusehri, are self-incompatible. Even self-compatible cvs are better planted mixed, as fruit setting is generally poor in the centre of a monoclonal plot (Singh, 1969).

Table 8.3 shows the main properties of mango cvs in India and Table 8.4 does the same for Florida and some other regions. In the Philippines 'Carabao' is the most popular cv. and rootstock (Anon., 1977). Hawaii's main cvs are Momi K, Gouveia, Pope and Ah Ping (Nakasone, 1982). Fiji has some local cvs but mostly uses Hawaiian cvs and 'Haden' for commercial production. 'Eldon' rootstock increases dwarfing (Perez and Cedeño-Maldonado, 1984).

Table 8.3　*Main properties of seven mango cultivars of southern India*

Cultivar	Plant characteristics			Season	Fruit characteristics			
	climate	habit	yield		size	form	weight	quality
Alphonso	Humid	Reg.	Med.	Mid	Med.	Ovate	241 g	Exc.
Pairi	Humid	R–A	Heavy	Early	Med.	Ovate	213	Good*
Banganpalli	Dry	Reg.	Heavy	Mid	Large	Oval	626	Good
Mulgoa	Dry	Irr.	Poor	Late	Large	Round	922	Exc.
Neelum	Dry	Reg.	Heavy†	Late	Med.	Ovate	364	Good
Suvarnarekha	Dry	Reg.	Heavy	Early	Med.	Ovate	294	Med.
Bombai	Subtr.	Reg.?	Good	Early	Med.	Ovate	227	Med.

Reg. = regular, R–A = regular to alternate, Irr = irregular,
* = poor keeping quality, † = 2 crops a year

Source: Gangolly *et al.* (1957)

Fig. 8.4   The 'Arumanis' mango of Java. *Source*: Ochse (1931)

Cultivation measures

Propagation

Mono-embryonic cvs must be propagated vegetatively; this is generally done by grafting on polyembryonic rootstock. Some nurs-

Table 8.4 *Main properties of mango cultivars outside India*

Cultivar	Tree season	yield	Fruit weight (g)	form	qual.	Remarks
Florida group						
Irwin	early	high	300	ovoid	good	R, rb, small tree
Haden	early	low	450	oval	good	A
Tommy Atkins	mid	med.	580	ovoid	good	R, rb
Smith	mid	high	800	obl.	med.	A
Kent	mid	high	740	ovoid	exc.	rb
Palmer	mid	med.	560	obl.	med.	A
Zill	mid	high	290	ovoid	VG	rb
Eldon	mid	high	380	ovoid	exc.	rb
Ruby	mid	med.	225	oval	med.	A, fr in clusters
Keitt	late	med.	560	oval	VG	R. rb
Sensation	late	high	310	oval	med.	rb
Brooks	late	high	675	obl.	med.	A, rb, needs support
Indian group						
Alphonso	early	high	350	obl.	exc.	R, rb
Pahiri	early	high	med.	ovoid	G	A
Mulgo(b)a	late	low	big	round	VG	
West-Indian group						
Amélie	mid	?	340	round	VG	exp. from Mali
Julie	?	?	210	obl.	med.	
Polyembryonics						
Cambodiana	early	low	250	obl.	VG	
Ameliorée du Cameroun	early	high	275	ell.	G	R
Sabre	mid	high	?	?	G	R
Peach	late	med.	?	?	VG	R, exp. from S. Afr.

Sources: Laroussilhe (1980); Campbell and Malo (1974)
A = very sensitive to anthracnose, R = resistant to anthracnose, G = good,
VG = very good, exc. = excellent, rb = regular bearer, ell. = elliptic,
med. = medium, obl. = oblong

erymen prefer to use mono-embryonic seedlings as rootstock because they grow faster, having more reserves at their disposal. However, they are much more heterogeneous and growers do well to insist on polyembryonic stock.

A sowing of polyembryonics is around 75 per cent nucellar and has to be rogued to remove aberrant seedlings. Van der Meulen (1971) advises to retain only the strongest seedling of each seed, but this may very well be a zygotic. Tammes (1965) has shown that the seedling at the micropylar end must be removed. Well known polyembryonic rootstocks are: Bappakai, Pahutan and Goa in India; Golek and Arumanis in Indonesia; Carabao and Pico in the Philippines and Sabre in Israel and South Africa. Carotte and Maison-rough perform well in Réunion (Aubert, 1975) but their rooting volume is rather poor (Moutounet, 1977).

A good rootstock must be uniform, grow vigorously, possess tolerance to soil-borne diseases and induce regular bearing. Mango seeds can be stored in moist charcoal for up to 100 days, but it is better to sow them fresh. They are washed, and dried in the shade for a few days; then they are sown at a spacing of 15 × 30 cm, 5 cm deep, with the convex side up. Germination takes at least 10 days, the average is 18 days for peeled seed and 1 month for seed with a stone. The cotyledons remain in the ground and the shoot has a reddish colour. A seedling is ready for transplanting when it has grown to a height of 10 cm: the seedbed is wetted and the seedlings are carefully removed. Every precaution must be taken to prevent breakage of roots and stems. The spacing in the nursery bed is 40 × 100 cm, the density being 25,000/ha.

Our previous statement that mango needs a rest period, does not hold for young plants: they should be kept growing continuously. They must be irrigated in dry periods and fertilized regularly, i.e. every 6 weeks. Weeds, pests and diseases must be strictly controlled. A mulch of straw or grass may be applied. It has recently become common practice to transplant young plants into plastic bags; they should be wide and at least 15 cm deep.

Sowing can also be done at stake (in the field) to avoid transplanting but it is easier and cheaper to look after plants in a nursery. Approach grafting (Fig. 8.5) was, and probably still is, the most common method in India: stocks 6–18 months old are placed on scaffolds around a good tree and joined to shoots of pencil size. Modified methods, with six weeks' old seedlings, have been described (Garner *et al.*, 1976).

Outside India budding is generally preferred, it requires less labour. Several methods are in use and the age of the stock varies from a few weeks (in the rainier climates) to over a year. In Florida chip budding on very young rootstocks has become popular (Malo, 1972). Cuttings, marcots and stooling have also been tried. A useful description of recent nursery work is given by Lichou (1977). More detail is provided by Garner *et al.* (1976).

Planting

Grafted trees are usually set 8–10 m apart, while seedling trees need more room; a density of 100 trees/ha is commonly found. Aubert (1975) has described a spacing of 6.1 × 4.8 m, or 340 trees/ha, which is gradually thinned out. Avilán *et al.* (1981) also recommend close planting in Venezuela; at an elevation of 500 m cv. Haden at 12 × 8 m was interplanted with cv. Palmer at 6 m. The 'Palmer' trees were kept to a width of 3.5 m. The overall density was 104 + 138 = 242 trees/ha. The value of this procedure is doubtful; only a very alert grower will be ready to prune, trim and thin out his

Fig. 8.5 Approach graft, made in Surinam

trees when they begin to crowd each other. It seems preferable to plant wide and to use an intercrop such as banana, pineapple, papaya or vegetables for the first five to six years. A cover crop is not advisable as it would dry the soil out. Windbreaks are needed where hot, dry or cold winds and salty sea sprays prevail.

Very big planting holes, 1 m in all directions, are often advised, but on a permeable soil a hole of 40 × 40 × 40 cm is quite large enough. About 1 kg rock phosphate and 50 kg manure may be applied in a large hole and proportionally less in smaller holes, all thoroughly mixed with the soil. The best time for planting is the beginning of the wet season. Where rains are irregular, or the wet season is short, some supplementary irrigation will be needed during the initial years.

Soil management

Weeds must be kept under control in the young orchard; this is generally done by slashing in the tropics, but shallow cultivation and herbicides, such as bromacil and paraquat, are used in the subtropics. If an intercrop is grown, it should gradually be removed as the mango trees grow bigger. Little weed control is needed in older orchards as the soil is shaded for the most part. Good drainage is required at all times; mango cannot thrive in a water-logged soil.

Irrigation

Young trees are irrigated as soon as the dry season starts. Older trees need a rest period of at least three months for bloom induction. However, when fruit hangs on the tree, an adequate supply of water must be provided. If necessary, irrigation should make up for a lack of rain.

The frequency and correct amounts of irrigation can be calculated from data on Ep, rainfall and water-holding capacity of the soil. Where such data are missing, the grower must inspect his trees for signs of wilting, or use a tensiometer. As a rule of thumb: young trees need irrigation every two weeks during the dry season. Older trees are not irrigated, except during a long drought or in order to prevent fruit drop.

Crop fertilization

A yield of 16 tonnes/ha was found by Laborem *et al.* (1979) in Venezuela to remove (in kg): N 104, P 12, K 99, Ca 88 and Mg 47, and (in g): Fe 976, Mn 871, Cu 435, Zn 375 and B 174. On this basis a yearly application of 1,000 kg of a 10–3–12–6 formula ($N-P_2O_5-K_2O-MgO$) would be needed for fruit-bearing trees. Trace elements should be added when deficiency signs become visible.

The pH is best kept at 6 to 7; when it rises to over 7.5 the roots cannot take up zinc, manganese and iron any more and in such cases those minerals must be supplied in sprays (see Table 5.20) or as chelates. Martin–Prével *et al.* (1975) have described a case of severe zinc deficiency in Senegal. The trees became unproductive, branches had shortened internodes, leaves were reduced in size and curled. The borderline was 15 ppm zinc; below that, clear symptoms of zinc deficiency developed.

In India the best ratio for fertilizers was found to be N–P–K 4–1–4 at 0.75 kg N/12-year-old tree. In Florida N–P–K–Mg 2–4–2–1, supplemented by Zn–Mn–Cu is recommended and fruit-bearing trees get 1.6 kg N/year. In the French West Indies 0.5 kg N–P–K–Mg 12–15–18–5 for each year is given, up to 6 kg at 12 years of age. Excesses of boron and chloride should be watched for (de Geus, 1973). It is clear, from these widely diverging practices, that a sensible fertilizer programme can only be arrived at after the analysis of soil and leaves is known.

Pruning

The mango needs little pruning and will grow out naturally into a symmetrical dome-shaped tree. During the first four years of its life some remedial pruning may be called for, e.g. to remove crossing branches. The rehabilitation of old trees consists of cutting them down severely and putting top grafts on the branches (see Fig. 8.6). Such trees must be protected against sunburn by a white coat. Otherwise, pruning is limited to the removal of dead wood and parasitic plants.

Diseases and pests

Several diseases and pests affect the mango. Anthracnose, caused by the fungus *Glomerella cingulata*, is especially bad in humid areas. Brown spots appear on leaves, which crinkle and die, leaving the twigs bare; fruits soften and rot. Control is achieved by frequent spraying with copper compounds, zineb and captan. 'Edward', a Florida cv. is said to be highly resistant. Other resistant cvs are indicated in Table 8.4.

A canker and a gummosis, caused by *Phytophthora cinnamomi*, was found in Ivory Coast by Lourd and Keuli (1975). Other fungus diseases are powdery mildew (*Oidium* sp.) and storage rots of fruits. Malformation is a serious disease in which flowers are replaced by leaf rudiments. The cause is uncertain. It can be cured by a 200 ppm NAA spray (Panday and Sharma, 1981). Sant Ram and Bist (1984) report success from spraying with 2.2 g/litre glutathion or ascorbic acid.

The mango-hopper or jassid is the most serious pest of mango blossoms in India. It secretes honey-dew on which sooty mould develops. At least 62 different species of scales occur on mangoes (Singh, 1960). The larvae of several fruit flies may render the fruit useless for human consumption. Butani (1975) mentions six genera

Fig. 8.6 Top graft of mango

of nematodes. Appropriate control measures against all these pests are a subject for local research.

From harvest to consumption

Mango yields average 400–600 fruits/tree in on-years, but it is well known that a big tree, after some off-years, may set a crop of 10,000 fruits or more. In Florida 8 tonnes/ha is an average yield.

The fruits are generally picked when they begin to change colour or after a few ripe fruits have dropped from the tree. For the local market one may wait until the fruit is becoming soft, but for transport by train or ship it should be picked when still green and firm. More precise standards for picking are:

1. Total soluble solids of at least 12° Brix.
2. A specific gravity of 1.01 to 1.02.
3. The ability to withstand a pressure of 1.75 to 2 kg/cm².

The reader may consult Hulme (1971) for more details on the biochemical processes in the ripening mango fruit and Caygill *et al.* (1976) for an annotated bibliography of harvesting, handling and processing of this fruit.

Picking is done by hand. The picker climbs up the tree with a bag and knife, or a special 'mango-picker' consisting of a bamboo pole with attached knife and a cloth bag held open by a ring. Chaplin (1981) advises to leave 5 mm of stalk on the fruit, in order to prevent stains from exuding sap; he also recommends to pick early in the day and to cool the fruit immediately to 12–15 °C. Anthracnose can be prevented by immersion in hot water (52 °C) for five minutes; one g Benlate per litre may be added. Gunjate *et al.* (1982) warn against sunlight: even five minutes exposure increases decay and after two hours this reaches 100 per cent!

It is customary in India to size the fruit in three classes: small

Fig. 8.7 Packed box of mangoes

(200–270 g), medium (270–320 g) and big (over 320 g). After grading and sizing fruits are packed in single layers (5 kg) (Fig. 8.7) and stored or transported at a temperature of 9 °C and a relative humidity of 85–90 per cent. Before marketing they are ripened for five or six days at 16–21 °C. Packing individual fruits in tissue paper was better than six other treatments (Mukerjee, 1972). The Working Group on Mango, Institute of Horticultural Research at Bangalore (India), publishes a relevant newsletter.

References

Anon. (1977) *The Philippines recommends for mango 1977*, Bureau of Plant Industry, Los Baños.

Anon. (1981) 'Le marché européen de la mangue', *Fruits* **36**, 723–6.

Avilán-R., L. *et al.* (1981) 'Consideraciones acerca de las sistemas de plantación en mango', *Fruits* **36**, 171–9.

Aubert, B. (1975) 'Possibilités de production de mangues greffées à la Réunion', *Fruits* **30**, 447–79.

Bijhouwer, A. P. C. (1937) *Een bijdrage tot de kennis omtrent het bloeien en het vruchtdragende vermogen van den mangga (Mangifera indica)*, Wageningen.

Butani, D. K. (1975) 'Parasites et maladies du manguier en Inde', *Fruits* **30**, 91–101.

Cadillat, R. M. (1976a) 'Importations de fruits tropicaux dans le CEE', *Fruits* **31**, 407–9.

Cadillat, R. M. (1976b) 'Chronique économique: avocats et mangues', *Fruits* **31**, 715–17.

Campbell, C. W. and **Malo, S. E.** (1974) *The mango*, Fruit crops fact sheet FC-2, Florida.

Caygill, J. C. *et al.* (1976) *The mango; harvesting and subsequent handling and processing; an annotated bibliography*, TPI, London.

Chandler, W. H. (1958) *Evergreen orchards*, Philadelphia.

Chaplin, G. R. (1981) *Post-harvest handling of mangoes*, Tech. Bull. **55**, Dept. Prim. Prod. North Terr. Australia.

FAO (1983) *Production Yearbook 1982*, Roma.

Free, J. B. and **Williams, I. H.** (1976) 'Insect pollination of *Anacardium occidentale, Mangifera indica, Blighia sapida* and *Persea americana*', *Trop. Agric.* (Trin.) **53**, 125–39.

Gangolly, S. R. *et al.* (1957) *The mango*, New Delhi.

Garner, R. J. *et al.* (1976) *The propagation of tropical fruit trees*, East Malling.

Geus, J. G. de (1973) *Fertilizer guide for the tropics and subtropics*, Zürich.

Gunjate, R. T. *et al.* (1982) 'Development of internal breakdown in Alphonso mango by post-harvest exposure to sunlight', *Science and culture* **48**, 188–90.

Hulme, A. C. (1971) 'The mango' in: *The biochemistry of fruits and their products*, vol. **2**, London.

Laborem, G. *et al.* (1979) 'Extracción de nutrientes por una cosecha de mango', *Agronomia tropical* **29**, 3–15.

Laroussilhe, F. de (1980) *Le manguier*, Paris.

Lichou, J. (1977) 'Pépinière fruitière à la Réunion; Programmes, Techniques utilisées, Perspectives', *Fruits* **32**, 197–209.

Lourd, M. and **Keuli, S. D.** (1975) 'Note sur un chancre à *Phytophthora du manguier en Côte d'Ivoire*', *Fruits* **30**, 541–4.

Malo, S. E. (1972) 'Mango culture in Florida', in: Symposium on mango and mango culture, *Acta Hortic.* **24**, 149–54.

Martin-Prével, P. *et al.* (1975) 'Un cas de carence en zinc sur manguier', *Fruits* **30**, 201–6.

Meulen, A. van der (1971) *Vegetative propagation of subtropical fruit trees*, Bull. **392**, Nelspruit.

Moutounet, S. E. *et al.* (1977) 'Étude de l'enracinement de quelques arbres fruitiers sur sol ferrallitique brun profond', *Fruits* **32**, 321–33.

Mukerjee, S. K. (1972) 'Harvesting, storage and transport of mango' in: *Symposium on mango and mango culture*, Acta Hortic. **24**, 251–5.

Nakasone H. Y. (1982) 'Fruit crops' in: *Crop improvement in Hawaii* (ed. J. L. Brewbaker), Un. Haw. Misc. Publ. **180**.

Ochse, J. J. and **Bakhuizen van den Brink, R. C.** (1931) *Vruchten en vruchtenteelt in Nederlandsch-Oost-Indië*, Batavia.

Panday, R. M. and **Sharma, Y. K.** (1981) 'Mango malformation: a disease of national importance', *Indian Hortic.* **26(2)** 9, 11, 13, 27.

Perez, A. and **Cedeño-Maldonado, A.** (1984) 'Rootstock/scion combination to reduce mango tree size, *Hortsci* **19(3)** 55.

Popenoe, W. (1920) *Manual of tropical and subtropical fruits*, Hafner.

Purseglove, J. W. (1968) *Tropical crops, Dicotyledons*, Longman.

Raghava Kurup, C. G. *et al.* (1967) *The mango, a handbook*, New Delhi.

Sant Ram and **Bist, L. D.** (1984) 'Occurrence of malformin-like substances in malformed panicles and control of floral malformation in mango, *Scientia Hortic.* **23**, 331–36.

Singh, L. B. (1960) *The mango*, Leonard Hill.

Singh, L. B. (1969) 'Mango' in Ferwerda and Wit (eds) *Outlines of perennial crop breeding in the tropics*, Wageningen.

Singh, R. N. (1972) 'An assessment of some of the existing and also potential commercial cultivars of mango in India', in: Symposium on mango and mango culture, *Acta Hortic.* **24**, 24–8.

Sizaret, A. (1970) 'Nouvelles techniques pépinières en sols sableux sous climats arides', *Fruits* **25**, 725–39.

Tammes, P. M. L. (1965) 'Vegetative propagation from poly-embryonic mango seeds', *Chron. Hortic.* **5**, 64.

Terra, G. J. A. (1949) *De tuinbouw in Indonesië*, 'sGravenhage.

Chapter 9

Avocado

Taxonomy and morphology

The avocado (botanical name *Persea americana*) belongs to the Lauraceae, a family of mainly (sub)tropical trees and shrubs; other well-known members are laurel, cinnamon, sassafras and greenheart (a timber of the Guianas). The English name derives from the Spanish word abogado, an adaptation of an Aztec word: ahuacatl. This became avocat in French and advokaat in Dutch. The Inca name palta is still used in Peru, Ecuador and Chili (Gustafson, 1976b).

Three ecological races (subspecies or botanical varieties) are recognized: Mexican, Guatemala and West Indian, that may be regarded as subtropical, semi-tropical and tropical respectively. Their main differences are set out in Table 9.1.

Table 9.1 *Properties of the avocado races*

	Leaf	Fruit				Seed		Tolerance	
	scent	*size*	*skin*	*oil %*	*months to ripen*	*size*	*cavity*	*cold*	*salt*
Mexican	Anise	Small	Thin	High	6	Big	Loose	Yes	No
Guatemalan	None	Var.	Warty	Med.	9	Small	Tight	Med.	No
West Indian	None	Var.	Leathery	Low	6	Big	Loose	No	Yes

Var. = variable, Med. = medium

The avocado tree is shallow-rooted and the leaves are arranged in spirals, coming out in flushes. Inflorescences appear by the thousand, each carrying hundreds of flowers; these are greenish, 1 cm wide and deep, with three whorls of three stamens and an ovary. The flower is thus complete, but behaves in a unique way called 'protogynous, diurnally synchronous dichogamy' (Bergh, 1969). Simply said: Each flower opens twice and is closed in between; the first time it functions as a female, the second time as a male. The openings are so timed that self-pollination is unlikely: in group A

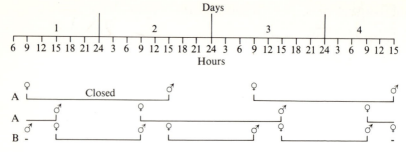

♀ = Functionally female

♂ = Functionally male

Fig. 9.1 Dichogamous flowering in the avocado

first openings take place in the morning, second openings during the afternoon of the following day; the time lapse is more than 24 hours. In group B the lapse is less than 24 hours: first they open in the afternoon, then again next morning. Therefore, every morning A-pistils can be fertilized by B-pollen, while during afternoons B-pistils are ready to receive A-pollen.

This is set out schematically in Fig. 9.1. However, not all cvs or seedling trees follow this pattern. Malo and Campbell (1967) state that synchronous dichogamy has not proven to be of practical significance in Florida. Furthermore, the rhythm is upset by cold or cloudy weather, so that overlapping occurs. Some cvs, e.g. 'Hass', are known to fruit well in monoclonal stands. Nevertheless, single trees are often unfruitful and mixed planting of cvs is desirable.

Bees are the principal pollinators and growers do well to keep hives in their grove during bloom. However, the bees much prefer citrus blossoms and will travel hundreds of metres to find them. It is therefore not advisable to plant avocado in close proximity to citrus orchards. Flies also visit avocado flowers and probably contribute to pollination. Papadimetriou (1976) reports pollination by wasps in Trinidad and overlapping of flowering in all A-type trees.

The avocado fruit is a large berry, with one seed consisting of a double seed coat, two fleshy cotyledons and a small embryo. Seedless fruits are called 'cukes', because of their resemblance to cucumbers; some people regard them as a delicacy.

Uses and composition

Shortly after the conquest of Mexico, in 1526, Oviedo described the

Fig. 9.2   First opening, female stage, of an avocado flower

avocado fruit as follows: 'the part which is eaten is a paste similar to butter and of very good taste.' It has been a popular food in Central America for thousands of years and still is: the average consumption in Mexico is about 20 kg per year. 'Four or five tortillas, an avocado and a cup of coffee – this is a good meal', say the Indians of Guatemala according to Popenoe (1920).

It is a very nutritious fruit, containing 3–30 per cent oil, similar in composition to olive oil, much vitamin A and reasonable quantities of vitamins B, and C. The calorific value is high, but the sugar content is low; it can be recommended as a high energy food for diabetics. Avocados are eaten on bread and tortillas, in salads, with lemon juice, salt and pepper; they cannot be cooked. Avocado oil is in great demand for the preparation of cosmetics, it is presssed from fruit rejected for export; the seed is converted into chicken feed. A frozen avocado spread is commercially produced in Queensland, Australia.

Fig. 9.3   Second opening, or male stage, of an avocado flower

Origin, distribution and production

The avocado is a native of the highlands of Mexico and Central America. It has been eaten there for perhaps 10,000 years, as was established by carbon dating (Sauls *et al.* 1977). The West Indian race, however, did not – as the name suggests – originate in the Caribbean Islands, but in the lowlands of Central America. From here the avocado spread to other regions of Central and South America. Purseglove (1968) informs us that it was taken to Spain in 1601, to Jamaica in about 1650 (from where it was taken to other local islands), to Mauritius in 1780 and to Asia by around 1850. The first record of avocados in Florida dates back to 1833 and 1856 in California.

Nowadays, avocado is grown in practically all tropical and subtropical countries of the world, from a latitude of 40° north to 40° south (having identical boundaries to citrus). The western hemisphere is still by far the most important producer of avocado,

Table 9.2 *The major producers of avocado (× 1,000 tonnes)*

	1975	1980	1981	1982	1983	1984
Mexico	279	424	444	468	448	440
USA	80	244	164	190	218	223
Brazil	117	107	120	140	125	115
Dominican Rep.	128	125	134	135	136	136
World	1,103	1,489	1,505	1,520	1,586	1,573

Source: FAO Production Yearbooks
NB: FAO indications for estimates have been ommitted

Table 9.3 *Importation of avocados into the Netherlands (tonnes)*

	1975	1977	1980	1981	1982	1983	Main period
Israel	160	390	263	255	984	1,294	Oct.–May
South Africa	168	294	296	298	269	290	April–Oct.
USA	—	—	387	701	138	204	
Total import	361	756	1,287	1,576	1,511	2,040	

Source: Produktschap voor groenten en fruit, The Hague

but the production and export of high quality avocado fruit now preponderantly comes from California, Florida, Israel and South Africa.

As can be seen from Table 9.2, world production has more than doubled in the last 15 years. Consumption in Europe is rising fast, as is shown for one country in Table 9.3.

Regional descriptions of avocado cultivation are found in Yearbooks of the California Avocado Society, e.g. Florida (Newman, 1976), Texas (Duke, 1978), New Zealand (Sale, 1980), Australia (Campbell, 1981), Israel (Shachar, 1982) and Hawaii (Nakasone, 1982). It is significant that descriptions from three of the largest producers listed in Table 9.2 have not yet been published.

Growth and development

There is a distinct lack of published descriptions of growth and development processes in avocado. Chandler (1958) tells us: it is a strong–growing tree, with branches that are elongated from their terminal buds. The young shoots are succulent and brittle; there are two flushes of long duration per season. The wood is spongy and breaks easily; much starch accumulates in stems and roots. Trees of the Mexican and Guatemalan race lose nearly all their leaves in winter, and a new flush starts in spring. Lateral buds are easily shed; as a result, few dormant buds remain to start growth in a severely pruned tree.

Fig. 9.4 'Fuerte' avocado in full bloom

According to Chandler (1958) avocado forms no visible root hairs. Perhaps mycorrhiza are present, but I have not seen them mentioned in literature. Between 80–90 per cent of the roots are found in the upper 60 cm of the soil (Gustafson, 1979a); in Cameroon 65–80 per cent were in the top 15 cm of soil (Anon., 1978).

In Trinidad, seed germination normally takes 30–40 days; with the seed coat removed, it takes 16–18 days. In cv. Lula the first leaves appeared at a height of 22–25 cm and the first flowers came in the third year; it lasted five to seven years in seedlings (Tai, 1969).

Ecology and physiology

Information on this subject is also hard to find. One thing is certain: the influence of climate on avocado growing can only be discussed in relation to the ecological races. Praloran (1970) has made a comparison between several areas, from Toluca, Mexico (19° N, elevation 2,675 m) to Santa Marta, Colombia (11° N, sea level).

Average temperatures ranged from 12.8 °C to 28.3 °C and rainfall was moderate (665 mm and 1,475 mm), but the dry season was well marked. In contrast to mango and cashew nut, the bloom in avocado is not harmed by rain, unless it persists for a month or more (Wolfe *et al.* 1969). More details on the influence of climate on bloom are provided by Aubert *et al.* (1972). The periodicity we

Fig. 9.5 In this tropical tree, flowers and fruits of all stages are seen at the same time

find in the subtropics is absent in tropical avocados (see Fig. 9.5). No data on optimum rainfall are available to me, but we can consider 1,800 mm as a likely figure for tropical conditions. Nobody has mentioned light specifically as an ecological factor for avocados, probably because it is considered to be self-evident; daylength also seems to be unimportant. Wind, on the other hand, is very important as avocado branches are brittle and break easily; therefore windbreaks must be provided.

'Hass' followed the normal A-pattern in a 12-hour day at a high temperature regime (day 33 °C, night 28 °C) or a moderate one (day 25 °C, night 20 °C). At low temperatures (day 17 °C, night 12 °C) the flower opened in the afternoon in the female stage and stayed open; the pistils became very long and the pollen tube grew so slowly that pollination was not achieved (Sedgley and Annels, 1981). It would be interesting to repeat this experiment with other cvs and at temperatures closer to tropical conditions.

Scholefield *et al.* (1980) report maximum photosynthesis in 'Fuerte' at one quarter of full sunlight; the stomata open rapidly at sunrise, close later in the day and open again in the afternoon. In this respect, 'Fuerte' resembles the coffee plant. Whether this is true for more tropical cvs remains to be seen. The effect of altitude on avocado is demonstrated by 'Fuerte' when grown in tropical mountains: the harvesting season moves by two months for every 500 m of altitude.

Soils for avocado cultivation must have good drainage; if this is assured the tree will grow well on clay, sand or gravel. An impervious layer or water table within 90 cm of the surface will cause trees to die in a few years, even on raised beds (Chandler, 1958). Salt is poorly tolerated, though better on West Indian stock than on the others. Where rainfall is too low, an adequate supply of irrigation water containing less than 100 ppm chloride should be available (Gustafson, 1975). In Israel, where the water may contain 250 ppm chloride, West Indian rootstock is used. The optimum pH is between 5 and 7. A sandy loam, at least one metre deep, is preferred.

Cultivars and rootstocks

There are at least a hundred avocado cultivars. They generally originated from superior seedlings that have been propagated vegetatively. 'Fuerte', the most popular cultivar, is a Mexican x Guatemalan hybrid that was found by an American grower in Mexico in 1911 (Hodgkin, 1980). It has shiny-green pear-shaped fruit that weighs 250–450 g, with a high oil content (18–26 per cent); the flower type is B. The tree is cold-resistant, has a rather

Fig. 9.6 Girdled stem of a 'Fuerte' tree in Israel

horizontal growth habit and a tendency to alternate bearing which can be countered by girdling (Fig. 9.6).

'Hass' belongs to the Guatemalan race and to flower group A, but may be planted in large blocks as it is self-fertile. The fruit is warty, medium-sized, roundish and dark purple at full maturity. These two cvs, together with Zutano (M × G, A), Bacon (M × G, B) and Nabal (G, A) are the most popular cvs of California.

The Florida cvs are better suited to the tropics. Their properties are shown in Table 9.4. In Texas, 'Lula' is the best cv.; it produces good crops (15 tonnes/ha) of high quality. The fruit is pear–shaped and has creamy, sweet flesh (Duke, 1978). Israel's main cvs are Ettinger, Fuerte, Hass, Nabal, Reed and Benik (Shachar, 1982).

Tango *et al.* (1972) give data on size, oil and flesh content of cvs grown in Sao Paulo State, Brazil. In Peru, with its great range of climates, from dry lowland to cool mountain, many cvs can be grown, e.g. 'Nabal' from 100 to 1500 m above sea level and 'Fuerte' from 800 to 1800 m elevation (Wolfe *et al.* 1969). Moreuil (1973) considers 'Collinson' and 'Simpson' to be the best in Ivoloina, Madagascar (altitude not stated). In Ivory Coast, at sea level, Perrin (1975) found the following cvs to produce well: Gottfried and Pernod (Mexican), Benik (Guatemalan), Black Prince, Simmonds and Waldin (West Indian), Fairchild and Long (G × W), Johnston, Semil 34 and Semil 44 (from Puerto Rico); Itzamna, Lula and Pollock were unsatisfactory. Around 1960 a search was made

Table 9.4   *Properties of ten Florida avocado cultivars*

Cultivar	Season	Race	Type	Weight (g)	Quality	Yield	Tree shape
Pollock	early	WI	B	800	VG	Low	Spreading
Simmonds	early	WI	A	700	VG	Med.	Spreading
Tonnage	mid	G	B	500	VG	Med.	Tall
Booth 8	mid	G×W	B	400	Fair	High	Spreading
Booth 7	mid	G×W	B	500	G	High	Spreading
Monroe	mid	G×W	A	900	VG	Med.	Med. tall
Hall	mid	G×W	B	500	G	High	Med. tall
Lula	late	G×W	A	500	G	High	Tall
Taylor	late	G	A	400	VG	Low	Tall
Choquette	late	G×W	A	900	VG	Med.	Spreading

Sources: Maxwell and Maxwell (1980); Malo and Campbell (1967)

in backyards of Surinam for seedling trees with good properties. Two clonal gardens were planted, but no evidently superior cvs were recognized. Both collections suffered from foot rot and had to be abandoned (Consen-Kaboord, 1976).

For the coffee zone of Colombia, i.e. near the equator at elevations of about 1600 m, Salazar *et al.* (1978) reported the highest yield from cvs Peterson, Trinidad, Fairchild and Trapp and the best quality from Fairchild, Trapp, Peterson and Booth 8; his recommendation was to plant Booth 8 (B) and Peterson (A) at 6 × 8 m, or Booth 8 and Fairchild (A) at 6 × 6 m. Promising cvs selected in Hawaii were Cho, Fujikawa, Kahula and Masami (Nakasone, 1982).

Selection, and possibly breeding, are going on in many countries. 'Pinkerton' in California is described as a 'medium large, green winter fruit of high quality with a small seed and a long shelf-life; it is a heavy and consistent bearer' (from an advertisement in an Avocado Yearbook). In Australia, 'Sharwil' is considered to be a very good new cv. to fill in the season between 'Fuerte' and 'Hass'. 'Ettinger' in Israel is tall, with an early but short harvesting season. Several new cvs have been selected, whose main properties are set out in Table 9.5.

Vuillaume and Moreau (1981) established the following taste preference in Réunion (20° S): 'Bacon', 'Edranol', 'Fuerte', 'Hass' and 'Peterson'.

The main qualities that breeders and selectionists look for in an avocado cultivar are: good and regular yield of high quality, fruit of medium weight with small seed in a well-filled cavity, holding well on the tree, good storage life, small- or medium-sized tree, spreading habit. Several cvs are needed with successive harvesting

Table 9.5   *Properties of four new cultivars from Israel*

Cultivar	Nordstein	Horshim	Tova*	Netaim
Season	Early	Mid	Mid	Late
Flower type	?	B	A	B
Fr. weight	270	275	250	300
Quality	G	Exc.	G	G
Yield	VG	VG	Exc.	VG
Seed %	16	13	17	18
Oil %	9–18	12–25	12–22	12–20
Tree height (m)	9	16	5	10
Altern. bearing	No	Yes	No	Yes

Source: Edgerton (1977) * good pollinizer.

seasons, to fill a period of at least six months of the year. However, as yet no single cultivar, or group of cvs, can be recommended for any location without further investigation. A grower planting an initial avocado crop is forced to make a choice and runs a big risk.

The choice of rootstocks largely depends on their resistance to three factors: cold, salt and root-rot. Mexican stocks are cold-resistant, West Indian stocks are not and the Guatemalan cvs are intermediate. In terms of salt resistance this order is reversed, West Indian being most resistant and Mexican least. In localities where the irrigation water contains more than 100 ppm chlorine, the use of West Indian stock is strongly recommended. Kadman and Ben Ja'acov (1980) have selected a dwarfing stock, the 'Maoz', that shows no leaf scorch symptoms even at 800 ppm chlorine.

'Waldin' (WI) is the common stock in Florida. Research in California to find root-rot resistant Mexican stocks has been described by Zentmyer (1978). Seedlings of 'Topa Topa' and 'Duke' were used until it was noticed that two 'Duke' trees, numbers 6 and 7, were superior to the others. Rooted cuttings from these trees were better again than the seedlings. The more vigourous 'Duke 7' came into general use and was later joined by 'G 6' and 'G 22'.

Cultivation measures

Propagation

The majority of avocado trees in tropical countries are seedlings. However, cultivars must be propagated vegetatively. The common way is to graft on seedling rootstock. Seed from mature fruit picked on the tree must be used; the tree must be registered as free from sun-blotch virus. Contamination with *Phytophthora cinnamomi*, the causal organism of root-rot, stem canker and seedling blight must be avoided at all costs.

The seeds are disinfected in water at 50 °C for 30 minutes, cooled and air-dried. The seed coat is removed and thin slices towards the bottom and top of the seed are cut away. The seed is sown with the broad side down, just as it was when the fruit was hanging on the tree. The spacing in the nursery is 30 × 60 cm and not more than 1 cm of sand should cover the tops.

Where good rootstock seed is scarce, seed may be split lengthwise into four to six parts; a part will germinate if a piece of the embryo adheres to it. Clonal rootstocks are thus produced on a small scale. Large scale production of clonal rootstocks is possible by the use of cuttings, but this is difficult. Tissue culture provides the means for clonal propagation of rootstocks but then good laboratory facilities are needed (Schroeder, 1976).

In California, seeds are commonly planted in polythene bags; these are 7 cm wide and 25 cm deep and have holes for aeration and drainage in their lower third. The potting mixture consists of three parts moss to two parts perlite, or of equal parts peat, perlite and vermiculite. In a shaded greenhouse, the seeds will germinate in about four weeks. The scions are cut from terminal growth of registered trees with plump buds. They are cleft or wedge-grafted to young seedlings, two to four weeks after germination when they have reached a height of 10 cm. Four to six weeks later, the plants are taken to a half-shaded house for transplantation into bigger containers (Platt, in Sauls, 1977).

In Florida and most tropical countries, three to four-month old seedlings are side- or veneer-grafted. In Puerto Rico, Pennock (1970) recommends cleft-grafing: the take is better than 95 per cent. In Togo, West Africa, Sizaret (1974) emphasized the need to earth–up plants in polythene bags; otherwise, the roots would die from overheating.

Planting

Planting holes are 25 cm wide and as deep as the roots of young plants go, provided the soil is permeable. The spacing depends on climate, soil and cultivar and varies from 5 × 5 m to 12 × 12 m, or even wider. In California, cv. Fuerte is initially set out at 6 × 6 m, cvs Hass and Reed at 6 × 5 m and cvs Zutano and Bacon at 5 × 5 m. These spacings are the equivalent of planting densities of 278, 333 and 400 trees/ha. The further handling of close spacings will be described under 'Pruning and thinning'.

The IRFA (see page 10) advises much wider planting for the tropics, e.g. 8 × 10 m for the erect growing cvs Collinson, Nabal and Taylor and 10 × 10 m for other cvs (Anon. 1978). During the first ten years, a cover crop, vegetables or short-lived perennials can be grown between the rows. However, tomato and eggplant should be excluded, as they frequently suffer from a fungus disease (*Verticillium*) to which avocado is also sensitive. It is bad practice to plant solid blocks of only one avocado cv. In order to promote good pollination two to four rows of an A–group cv. should alternate with some rows of a B–cultivar. It has been observed that bees seldom move more than two rows away. Even with 'Hass', which is known to be self-fertile, it is wise to include a pollinator at every fifth row. Two bee hives/ha should be placed in the orchard during bloom.

A spacing of 9 × 12 m, to be thinned later to 12 × 18 m, is the recommendation in Australia (Whiley, 1982). Whatever the spacing, a five per cent mortality must be taken into account (Duke, 1978). Open places should be filled as soon as possible. As avocado trees have brittle branches that break easily, it is necessary to plant windbreaks at least a year before the avocado trees are set out. Stems of young trees are protected against sunburn by a whitewash or a cardboard jacket, which also diminishes damage due to rodents.

Soil management

The shallow root system of the avocado is easily damaged by hoeing, disc ploughing and other forms of cultivation. Therefore, weeds are better controlled by herbicides such as monuron, diuron, simazine, paraquat and weed oil. Cover crops are recommended in the humid tropics and mulching around the tree has been shown to have a favourable effect against root-rot.

The cost of weed control in Texas was 30 per cent of the total cultivation costs, excluding picking. Other costs include: supervision and overhead 22 per cent, planting 19 per cent and irrigation 12 per cent (Duke, 1978).

Irrigation

Whenever rainfall remains below potential evapotranspiration (Ep) for an extended period, irrigation is required. Flooding is the cheapest method, it cost \$293/ha in California, but it is inadvisable as it promotes *Phytophthora* root-rot. Alternatives are drip irrigation costing \$415/ha or a permanent set of sprinklers at \$550/ha (Gustafson, 1976a).

Experiments in Israel have shown that drip irrigation represents a saving of 50 to 60 per cent in water and that 60 per cent of the class A pan evaporation is sufficient (B. Gefen, personal communication). This method only wets the soil around the tree and thus considerably facilitates weed control; the system can also be used to apply fertilizers and pesticides. The total water use in Israel amounts to 700 mm per year (Shachar, 1982). Similar conclusions were drawn in California (Marsh *et al.*, 1978).

There should be at least two emitters per tree, set closer together on sand than on a loamy soil; the soil profile should be known to a depth of at least one metre. An emitter can deliver 4 litres/hour and the maximum water use of an adult avocado tree is 150 litres/day. Drip irrigation is an intricate system that should only be serviced by experienced workers (Gustafson, 1981).

Crop fertilization

Fertilizer requirements of avocado are similar to those of citrus, with some subtle differences. Ranges of adequacy established by leaf analysis for 'Fuerte' were: N 1.6–2.0, P 0.08–0.25, K 0.75–2.0, Ca 1.0–3.0, Mg 0.25–0.80 and S 0.20–0.60 per cent; B 50–100, Fe 50–200, Mn 30–500, Zn 30–150, Cu 5–15, Mo 0.05–1 parts per million.

These ranges apply to most other cvs too, but 'Hass' needs more N: its range is 2.0–2.4 per cent. A young tree receives 25 g N in the first year, 50 g in the second and 100–200–300–400 g N in succeeding years. From the seventh to the tenth year 500 g N is given and with increasing yields more, up to 800 g N per tree per year. The application is split into three, and later two portions. On acid soils N is applied as calcium nitrate; on soils where the pH is above 6, it is applied in the form of ammonium (Gustafson, 1979a).

Slightly different figures are mentioned by de Geus (1973): 110–170 kg N/ha in California, 1.4 kg N/tree in Florida. A ½ per cent urea spray at fruit setting provides additional nitrogen. Four applications/year of a 12-6-10 mixture are given in Puerto Rico, while in Brazil trees get 90 to 180 g N-P-K when young and a 6-3-6 formula when producing. The advice in Ghana is a 15-15-15-2 mixture, 360 g/cm of stem diameter.

From these examples, it is apparent that no general advice is possible. Fertilizer needs must be determined locally from data provided by soil and leaf analysis. One conclusion seems justified: avocado needs relatively more phosphate than citrus and other fruit trees.

An average harvest of 14.4 tonnes in Venezuela removed the following elements from the soil (in kg/ha): N 40, P_2O_5 25, K_2O 60, CaO 11 and MgO 7 (Avilan *et al.* 1978). These figures may serve as an indication of the minimum amounts of fertilizer needed for bearing trees.

Zinc deficiency is as prevalent in avocado as in citrus. Apart from the common symptoms, such as narrow leaves and yellow mottling between the nerves, the fruits also tend to be more rounded. Ideas differ on how to control the deficiency. Gustafson (1979a) adheres to the traditional way of spraying with zinc sulphate (see Table 5.20); the best time is in spring, after the flush is fully expanded. A 0.2 per cent zinc oxide spray is also effective. Kadman and Cohen (1977) claim that spraying has little effect as the penetration of zinc through the leaves is very slight; they prefer a chelate (Zn-EDTA). It will move into leaves that are present, but cannot be transferred to new leaves. The dilemma is perhaps solved by an observation (in citrus) that zinc is taken up through the stomata; spraying must therefore be directed to the underside of the leaves.

Pruning and thinning

Little pruning is needed for young trees with a spreading habit. It is usually adequate to cut back some shoots or to pinch out the terminal buds of shoots growing straight up. Heavy pruning leads to low and late productivity. For trees with an erect growth habit, e.g. cvs Ettinger, Bacon, Zutano and Reed, terminal buds must be pinched out after every flush. All trees should be kept to a height of 5 to 8 m in order to keep picking costs low; this also diminishes wind damage and lowers the cost of pest control (Platt, in Sauls, 1977).

Older trees may be pruned either selectively or mechanically. In the first case every tree is treated as an individual: carefully selected branches are removed or cut back in order to prevent crowding, or to reduce the height of the tree. The best time to do this is after the harvest. In the second case hedging and topping are carried out by large machines once a year. This system has been outlined in Chapter 5. Evidently, it can only be done in large holdings on closely planted trees; the orchard should cover at least 150 ha and time spent on pruning is reduced from 40 hours to 40 minutes/ha/year (Anon. 1978). In Texas, the cost of pruning was calculated at $40 per ha (Duke, 1978).

A different way to handle close planting is thinning. When an orchard planted at 6 × 6 m is becoming crowded, for example around the tenth year, every other diagonal row is removed; a new square, turned 45° emerges, the planting distance now being $6\sqrt{2} = 8.5$ m. Approximately six years later, a second thinning removes every other row; spacing then changes to 12 × 12 m and only one quarter of the original trees are left; density has been reduced from an initial 278 to 139 trees/ha and finally to 70 trees/ha. Obviously the weak and unproductive trees are the first to go.

Diseases and pests

Phytophthora cinnamomi is the avocado's worst enemy. It causes seedling blight, stem canker and root-rot, which in turn is responsible for leaf fall, dieback and eventual collapse of the tree. *Phytophthora* causes stem cankers on more than 900 plant spp. or varieties, many of which are ornamentals (Zentmyer *et al.* 1980).

Brun (1975) reports that treatment with copper curbed the attack. Allen *et al.* (1980) state that metalaxyl (0.8 kg active ingredient/ha) and fosetyl-Al (8 kg active ingredient/ha) worked better than captafol. Metalaxyl is applied to the soil and restores seriously damaged trees to health; fosetyl-Al as a foliar spray is only effective on lightly affected trees (Bertin *et al.* 1983); Darvas *et al.* (1983) achieved very good control by trunk injection.

It is better to take preventive measures such as treatment of rootstock seed and use of root-rot resistant stocks in order to restrict the damage. Pegg and Forsberg (1982) also propose:

1. Keep nursery free from root-rot.
2. Plant on ridges in deep, well-drained soil.
3. Irrigate and fertilize carefully.
4. Use metalaxyl 5 per cent, 100 g/m^2 tree canopy every 8–12 weeks.
5. Use fosetyl-Al 80 per cent at 4 g/litre as foliar spray.

It is not easy to keep the nursery free from the fungus. Traffic must pass through a disinfecting bath, tools and utensils must be disinfected regularly. Optimal conditions for the fungus are, according to Zentmyer (in Sauls, 1977): 21–30 °C and pH 6.5. In Hawaii serious damage was seen at pH 5.2, whereas there was no damage at pH 6.2. From these observations it seems that different strains of *Phytophthora* may be involved.

Other fungal diseases of avocado are scab (*Sphaceloma*), anthracnose (*Glomerella*), *Cercospora* and *Verticillium*. They are responsible for serious yield losses in West Africa; if not controlled, fungal disease could render 90 per cent of the harvest to be unfit for export. Bordeaux mixture can only control them in combination with sulphur, but other copper compounds and carbamates are effective (Anon. 1978).

Sunblotch, a virus disease, is transmitted by pollen (Desjardins *et al.* 1979), grafting or seed. It causes white areas and streaks on fruits, leaves and stems, and reduces yield by 20 per cent or more; it also lowers quality. Some trees are symptomless carriers. Therefore, scions and rootstock seed should only be taken from indexed, virus-free trees.

Insects and mites are not serious enemies of avocado in California as they are biologically controlled by parasites. It is therefore not necessary to spray, except on heavily infested single trees

(Gustafson, 1979b). Purseglove (1968) mentions scale insects, mealy bugs and mites; in Puerto Rico the sugar-cane root-weevil causes serious damage. Some nematodes that are parasitic on avocado have been described in Florida.

From harvest to consumption

Although average yields of 14 to 15 tonnes/ha have been reported (Avilan *et al.* 1978, Duke, 1978), the general average is far lower; usually between 5 and 10 tonnes/ha.

When the fruit has reached maturity, it can be left hanging on the tree for several weeks; it will not ripen further. Opinions on

Fig. 9.7   Avocado harvest in Kibbutz Beth Haemek, Israel
Source: Mr. H. Miller, Beth Haemek

maturity differ: in California an oil content of at least 8 per cent is demanded whereas in Florida, fruit size or an assigned picking date is the criterion. Seung-Koo (1981) proposes dry weight as a maturity standard and describes a simple way to measure it. Lee and Coggins (1982b) have found a correlation: 5 per cent oil corresponds to 16 per cent dry weight and 10 per cent oil is analogous to 21 per cent dry weight.

If harvested immature, an avocado will not ripen properly. Mature fruit ripens within a week at 27 °C and in a month at 5 °C. Florida avocados ripen best at 15–24 °C and should be stored at 13 °C (West Indian cvs) or lower, but not below 5 °C for the hardier ones.

As the branches are brittle, the picking crews must not climb into the trees. Long poles with a sac and a rope-operated knife are used and several rounds of picking are necessary. The fruit must be clipped, not pulled from the tree. Pickers carry a piece of plywood with a circular cut to check the required diameter. Plastic buckets holding 12 kg fruit are emptied into bins of 119 × 119 × 61 cm for transport to the packing house (Newman, 1976). A self-propelled harvester was developed in Israel (Fig. 9.7).

In the packing house the fruit is cleaned, graded, sized and packed in flat trays (Fig. 9.8) for export, or it is stored. Tai (1969)

Fig. 9.8 Packed box of avocados in a Dutch market

reports that storing below 13 °C in Trinidad induces chilling. Whiley (1982) found that fruit in Australia could be stored for three weeks at 10–18 °C and that storage below 8 °C caused chilling. Lee and Coggins (1982a) point out that customers like to buy soft fruit. The retailer ripens part of his avocado fruit at 20–25 °C for five to fifteen days, after which it is stored for three days at 7 °C.

References

NB *CASY* = California Avocado Society Yearbook

Allen, R. N. *et al.* (1980) 'Fungicidal control in pineapple and avocado of diseases caused by *Phytophthora cinnamomi*', *Austr. J. Experim. Agric.* **20**, 119–24.

Anon. (1978) 'L'avocatier' in: Inventaire et état actuel etc., *Fruits* **33**, 587–609.

Anon. (1981) *Achievement report July 1976–June 1979*, Hawaii Agric. Exp. Sta. Misc. Publ. **184**.

Aubert, B. *et al.* (1972) 'Considerations sur la phénologie des espèces fruitières arbustives', *Fruits* **27**, 269–86.

Avilan-R., L. *et al.* (1978) 'Exportación de nutrientes por una cosecha de aguacate', *Agronomia tropical* **28**, 449–61.

Bergh, B. O. (1969) 'Avocado' in Ferwerda and Wit (eds) *Outlines of perennial crop breeding in the tropics*, Wageningen.

Bertin, Y. *et al.* (1983) 'Essai de traitement chimique des attaques de *Phytophthora cinnamomi* de l'avocatier en Martinique', *Fruits* **38**, 481–5.

Brun, J. (1975) 'Le chancre de l'avocatier provoqué par *Phytophthora cinnamomi* Rands, *Fruits* **30**, 339–44.

Campbell., A. (1981) 'The avocado industry in Australia', *CASY* **65**, 69–71.

Chandler, W. H. (1958) *Evergreen orchards*, Philadelphia.

Consen-Kaboord, M. *et al.* (1976) *Jaarverslag Landbouwproefstation 1975*, Paramaribo.

Darvas, J. M. *et al.* (1983) 'Injection of established avocado trees for the effective control of *Phytophthora* root rot', *CASY* **67**, 141–6.

Desjardins, P. R. *et al.* (1979) 'Pollen transmission of avocado sunblotch virus experimentally demonstrated', *CASY* **63**, 83–5.

Duke, D. (1978) 'Avocado feasibility study', *CASY* **62**, 73–6.

Edgerton, A. D. (1977) 'Proliferation of Israeli avocado varieties in California', *CASY* **61**, 50–6.

FAO *Production Yearbooks*, Roma.

Geus, J. G. de (1973) *Fertilizer guide for the tropics and subtropics*, Zürich.

Gustafson, C. D. (1975) *How to buy an avocado orchard*, UC 75–LE/2251.

Gustafson, C. D. (1976a) 'Avocado water relations', *CASY* **60**, 57–72.

Gustafson, C. D. (1976b) 'World avocado production', *CASY* **60**, 74–90.

Gustafson, C. D. (1979a) 'Review of average fertilizer practices in San Diego County', *CASY* **63**, 50–7.

Gustafson, C. D. (1979b) 'A review of pest control in avocados', *CASY* **63**, 58–65.

Gustafson, C. D. (1979c) 'Avocado orchard development costs – San Diego County', *CASY* **63**, 66–70.

Gustafson, C. D. (1981) 'Management of drip irrigation systems on tree crops', *CASY* **65**, 81–92.

Hodgkin, G. B. (1980) 'A pilgrimage to the parent Fuerte tree', *CASY* **64**, 63–5.

Jeanteur, P. (1970) 'Quelques caractéristiques de la culture de l'avocatier en Floride et à Porto-Rico', *Fruits* **25**, 817–23.

254 *Avocado*

Kadman, A. and **Cohen, A.** (1977) 'Experiments with zinc application to avocado trees', *CASY* **61**, 81–5.
Kadman, A. and **Ben Ja'acov, A.** (1980) 'Avocado rootstock selection' and 'Maoz avocado rootstock selection for high salinity and high lime', *Hortscience* **15**, 206–7.
LaRue, J. H. and **Opitz, K. W.** (1976) *Growing avocados in the San Joaquin Valley*, UC leaflet 2904.
Lee, S. K. and **Coggins, C. W.** (1982a) 'Feasibility of marketing soft avocado fruit', *CASY* **66**, 57–62.
Lee, S. K. and **Coggins, C. W.** (1982b) 'Dry weight method for determination of avocado fruit maturity', *CASY* **66**, 67–70.
Malo, S. E. (1971) 'Girdling increases avocado yields in South Florida', *Proc. Trop. Region ASHS* **15**, 19–25.
Malo, S. E. and **Campbell, C. W.** (1967) 'The avocado', *Fruit crops fact sheet* 3, Florida.
Marsh, A. W. *et al.* (1978) 'Irrigating new avocado orchards', *Calif. Agriculture* **32** (5), 19–20.
Maxwell, L. S. and **B. M.** (1980) *Florida fruit*, Tampa.
Moreuil, C. (1973) 'L'avocatier à Madagascar', *Fruits* **28**, 699–702.
Nakasone, H. Y. (1982) 'Fruit crops' in: *Crop improvement in Hawaii* (ed. J. L. Brewbaker), Un. Haw. Misc. Publ. **180**.
Newman, P. W. (1976) 'Harvesting and post-harvest handling of avocados in Florida', *CASY* **60**, 32–5.
Papadimetriou, M. K. (1976) 'Some aspects of the flower behaviour, pollination and fruit set of the avocado', *CASY* **60**, 106–52.
Pegg, K. G. and **Forsberg, L. I.** (1982) 'Avocado root rot', *Queensland Agric. J.* **108**, 162–68.
Pennock, W. (1970) *Plant grafting techniques for tropical horticulture*, Bull. 221, Univ. P. Rico.
Perrin, B. (1975) 'Comportement de l'avocatier en basse Côte d'Ivoire', *Fruits*, **30**, 35–43.
Platt, R. G. and **Frolich, E. F.** (1965) *Propagation of avocados*, Calif, Agric. Exp. Sta. circ. 531.
Popenoe, W. (1920) *Manual of tropical and subtropical fruits*, Hafner.
Praloran, J. C. (1970) 'Le climat des aires d'origine des avocatiers', *Fruits* **25**, 543–57.
Purseglove, J. W. (1968) *Tropical crops, Dicotyledons*, Longman.
Salazar, C. R. *et al.* (1978) 'Resultados preliminares en la evaluación de ocho variedades de aguacate en zona cafetera colombiana', *Revista del Inst. Col. Agropec.* **13**, 425–35.
Sale, P. (1980) 'The New Zealand avocado industry', *CASY* **64**, 71–4.
Sauls, J. W. *et al.* (1977) *Proc. First Intern. trop. fruit short course: The avocado*, Un. Florida, Gainesville.
Scholefield, P. B. *et al.* (1980) 'Some environmental effects on photosynthesis and water relations of avocado leaves', *CASY* **64**, 93–105.
Schroeder, C. A. (1976) 'Responses of avocado stem pieces in tissue culture', *Cal. Avocado Soc. Yearbook* **60**, 160–3.
Sedgley, M. and **Annels, C. M.** (1981) 'Flowering and fruit-set response to temperature in the avocado cv. Hass', *Scientia Hortic.* **14**, 27–33.
Seung-Koo, L. (1981) 'A review and background of the avocado maturity standard', *CASY* **65**, 101–9.
Shachar, Z. (1982) 'The avocado in Israel', *CASY* **66**, 103–8.
Sizaret, A. (1974) 'Suggestions pour la multiplication rapide de trois espèces fruitières (agrumes, avocatiers, manguiers), lors des premières introductions de matériel végétal', *Fruits* **29**, 767–70.
Storey, W. B. (1972) 'Avocado', *Capita selecta*, Wageningen (stencil).

Tai, E. A. (1969) 'Investigations on avocado in Trinidad', *Proc. Conf. trop. and subtrop. fruits*, London, 231–5.

Tango, J. S. *et al.* (1972) 'Composition du fruit et de l'huile de différentes variétés d'avocats cultivés dans l'état de Sao Paulo', *Fruits* **27**, 143–6.

Vogel, R. (1982) 'L'acclimatation en Corse de quelques fruitiers exotiques', *Fruits* **37**, 157–65.

Vuillaume, C. and **Moreau, B.** (1981) 'Bilan des premières observations sur la collection d'avocatiers de la Station de Bousin-Martin à la Réunion', *Fruits* **36**, 139–50.

Whiley, A. W. (1982) 'The avocado', *Austral. Hortic.* **80(3)** 80–9.

Wolfe, H. S. *et al.* (1969) *El cultivo del palto en el Perú*. Bol. tecnico 73, Lima.

Zentmyer, G. A. (1978) 'Origin of root-rot resistant rootstocks', *CASY* **62**, 87–9.

Zentmyer, G. A. *et al.* (1980) 'Pathogenicity of different California isolates of *Phytophthora cinnamomi* to avocado', *CASY* **64**, 131–37.

Chapter 10

Papaya

Taxonomy and morphology

Papaya, botanical name *Carica papaya*, belongs to the Caricaceae, a family mainly inhabiting South and Central America; it is the only species of economic importance in the family. It is a small, normally unbranched quick-growing soft-wooded tree – 'almost an herb' says Chandler (1958) – with latex vessels in all parts. The British call it 'papaw' or 'pawpaw', in Brazil it is known as 'mamao' and in Spanish it is called 'papaya' or 'lechosa' (Venezuela), but in Cuba (where papaya is a term of abuse) the name is 'fruta bomba'.

The stem is hollow between the nodes, except in young plants; it mainly consists of wood parenchyma and bears large triangular scars. The peltate leaves are arranged in a 2/5 spiral; they have long hollow petioles and large, deeply-lobed blades (except in one cv., see Fig. 10.3).

The plant is dioecious, but hermaphrodite (bisexual) flowers and trees also occur. The female flowers, 3–5 cm long, sit alone or in small groups in the leaf axils; the ovary is 2–3 cm long and has five fan-shaped stigmas on top. The male flowers, with ten stamens each, are found on long hanging panicles (see Fig. 10.2). Bisexual flowers have either five or ten stamens and some of these tend to become 'carpelloid' (fruit-like), in which case the fruits have a 'cat-face' appearance and are unmarketable. Different types of hermaphrodite flowers may occur on the same tree, depending on the season, or on the age of the tree.

Male trees are also variable: sometimes a fruit is found at the end of a long panicle. A complete change of sex may take place when an old male tree is cut back: sprouts bearing female flowers (and later fruits) may appear. There is a difference between pure males and sex-reverting males (Teaotia and Singh, 1967). How pollination takes place is not known with certainty; wind is probably the main agent, as the pollen is light and abundant, but thrips and moths may assist (Purseglove, 1968).

Fig. 10.1   Stem, leaves and flowers of a hermaphrodite papaya

The fruit is a large, fleshy, hollow berry (see p. 30). Fruits formed from female flowers are oblong to nearly spherical, but if formed from bisexual flowers they are pear-shaped, cylindrical or grooved. Marketable fruits weigh from 0.5 to 2 kg and are 10–20 cm long. The thin green skin turns yellow at the bottom when maturity

Fig. 10.2   Male tree with hanging inflorescences

sets in. The flesh is yellow to orange, in some cvs reddish, and has a pleasant flavour. Around the cavity lie a thousand or more black seeds, but seedless fruits occur too. Twenty air-dried seeds weigh about one g.

The root system is said to be extensive and dense (Malan, 1953) or shallow (Kasasian, 1971). Actually, one may expect a deep and well developed root system on good soil, whereas the roots will stay near the surface on a wet or compact soil.

Uses and composition

Papaya is a popular breakfast food in the tropics and is recently being sold regularly on markets of temperate countries. It is also used for fruit salads and desserts. The fruit contains about 85 per cent water, 10–13 per cent sugar, 0.6 per cent protein, much vitamin A and fair amounts of vitamins B1, B2 and C; it contains practically no starch. It is considered to have a mild laxative action and the seeds are used medicinally against worms.

Processed papaya fruit has a neutral taste that is greatly improved by the addition of passion fruit juice to make soft drinks, jams and various preserves. Unripe fruit can be fermented into 'sauerkraut' or cooked as a substitute for apple sauce. From the latex of scratched unripe fruits, papain is prepared; this is used as a tenderizer for meat and for medical and industrial purposes (Foyet, 1972).

Origin, distribution and production

Papaya is native to tropical America and has never been found wild. It probably originated in Central America, thousands of years ago. From there it spread to South America and the West Indian islands; it was taken to the Philippines by the Spaniards and later reached other regions of South-East Asia and Africa. It is now present in every tropical and subtropical country.

The major producers are shown in Table 10.1. There is a sizeable export to the continental United States from Hawaii and Mexico. The import into Europe is growing but has not yet attained a large volume. The Netherlands, for instance, imported only 30 tonnes in 1977 and 73 tonnes in 1982, mainly from Brazil.

Until recently, Tanzania was the main producer of papain and the chief importer is the United States.

Table 10.1   *The major producers of papaya (× 1,000 tonnes)*

	1975	1980	1981	1982	1983	1984
Brazil	129	427	450	460	460	470
Indonesia	220	315	312	300	310	300
India	221	265	270	273	270	272
Mexico	220	221	226	201	230	300
Zaire	165	155	156	156	156	160
World	1,168	1,862	1,957	1,930	1,982	2,097

Source: FAO Production Yearbooks

Growth and development

Under optimum conditions the growth and development of papaya proceed at a fast rate. Seeds germinate in two weeks and from then on two leaves emerge each week. Flowers and fruits are produced in the leaf axils after a juvenile period. How long this period lasts, depends on the cultivar; for instance cv. Betty begins to flower at the node 24, 'Solo' at number 49 and their hybrid at node 32. Were fruits to set in all axils, one could expect a yield of 100 fruits a year (Storey, 1969). At an average weight of 0.5 kg and a density of 2,000, this would result in a production of 100 tonnes/ha/year. However, half this amount is already a respectable figure.

The tree remains in prime condition for two or three years, but its economic life comes to an end when the fruits are hanging so high that they cannot be picked profitably. A small farmer may cut back such a tree, after which some branches will sprout and bear fruit for some time yet; however, this practice is not recommended for orchards. It takes four to six months, depending on climate, for fruit to mature. Sex in papaya is determined by three genetic factors: M_1 is dominant for maleness, M_2 is dominant for herma-phroditism and m is recessive, for femaleness. Each ovule or pollen grain, being haploid, can carry only one of these factors. The diploid zygotes carrying two sex factors with capital letters cannot live; thus, the combinations M_1M_1, M_2M_2 and M_1M_2 are elimi-nated. Of the remaining ones M_1m are male trees, M_2m herma-phrodites and mm females.

If we cross a female with a male we get mm $\times$ M_1m with the result mm + M_1m: half the seeds will produce female trees, half male trees. Similarly the other crosses give predictable issues (Table 10.2). With open, i.e. uncontrolled, pollination a cultivar will lose its identity in a few generations. Table 10.2 indicates that all males must be removed from a bisexual orchard, and that seed from round fruits is to be discarded. Inbreeding causes no loss of vigour (Storey, 1969). The sex of a young papaya seedling cannot be predicted; one has to wait until flowers appear. All methods to separate male from female plants at an early stage have failed so far.

Table 10.2   *The mating system of papaya*

	F	H	M
Female (F) $\times$ male (M)	1	0	1
Hermaphrodite (H) selfed	1	2	0
H $\times$ H	1*	2†	0
H $\times$ M	1*	1†	1
F $\times$ H	1*	1†	0

* Roundfruited   † Pear-shaped or long

Ecology and physiology

Papaya is grown in the tropics up to an elevation of about 1500 m and in the frostless subtropics, from approximately 32° North to 32° South. The minimum temperature for survival is $-1\,°C$, lower temperatures kill the tree. Chandler (1958) puts the maximum at $44\,°C$ and Anon. (1982) states that optimal temperatures are between 25 °C and 38 °C. No mention is made in literature of the minimum temperature for growth; we assume it to be 15 °C. Lassoudière (1968–9) states that 22 °C–26 °C is probably optimal for growth, whereas 35 °C by day and 26 °C at night gave the fastest germination; daylength had no effect. Best quality fruit, which is determined largely by sugar content, develops under full sunlight in the final four to five days to full ripeness on the tree (Storey, 1972). As fruit is harvested almost every week, the tree always needs full sunlight.

Storey stipulates an annual rainfall of 1,000–1,500 mm and Anon. (1982) puts this at an evenly distributed 1,500–2,000 mm (Venezuela). Terra (1949) found best growth of papaya on Java with more than 100 mm rain for every month. The potential evapotranspiration has been determined at 1.3 times the class-A pan evaporation (Anon. 1981a). Irrigation must be provided in climates with a long dry season.

In South Africa papaws of best quality are usually grown in areas low in air humidity (Malan, 1953). On the other hand, it has been claimed (Anon. 1982) that relative air humidity should be at least 60 per cent. It seems likely that the sugar content of the fruit, one of the best marks of quality, will be higher at low humidity. It is probably correct to say that good papayas can be grown under diverging conditions of air humidity, as long as irrigation facilities are available.

A delicate tree like papaya must be protected against strong wind; therefore, windbreaks must be established a year or more before the orchard is planted.

Like banana, papaya needs good soil: a well-drained, permeable, well-aerated, fertile loamy soil, preferably rich in organic matter, with neutral reaction (pH 6–7) should be chosen, if possible on flat land. Trees in water-logged soil will die by drowning in three or four days (Storey, 1972). As Popenoe (1920) stated: 'papaya is one of the most insistent plants in the matter of drainage'.

Cultivars

It is, perhaps, incorrect to speak of cultivars in a crop that is almost entirely propagated by seed. However, in some papaya strains the type is retained by inbreeding and we may regard them as cultivars. Storey (1969) quotes as examples: 'Solo' of Hawaii, possibly 'Hortus Gold' of South Africa, 'Improved Petersen' of Australia and 'Betty' of Florida. 'Solo' was then in its 25th generation of inbreeding since its introduction from Barbados in 1910.

Several improved 'Solo' lines were selected after prolonged selfing, e.g. 'Solo 5' (1948), 'Solo 8' (1953) and 'Solo 10'. All are hermaphrodites (bisexuals) with a pear-shaped fruit that weighs about 400 g; 'Solo 8' has a high sugar content. They were crossed with 'Betty', a dioecious dwarf cv. from Florida, to induce early bearing. After backcrossing and selfing during many generations 'Sunrise' resulted, a red-fleshed type with the desirable 'Solo' flavour (Hamilton and Ito, 1968).

Later work produced 'Waimanalo', an early flowering, vigorous grower with short internodes and short-necked fruits with firm flesh, excellent texture, high sugar content and a low cavity percentage (Nakasone *et al.* 1972). 'Waimanalo' is well accepted locally for its

Fig. 10.3   'Thailand', a papaya cv. with nearly entire leaves, Indonesia

fruit quality, but the fruit is considered too large for export – the normal weight in Hawaii being 450 g.

Two new cvs, 'Higgins' and 'Wilder', were introduced in 1974 after tests on different soil types and rainfall regimes of the four main islands of Hawaii. 'Higgins' performed well in the high rainfall area and 'Wilder' was good in all other areas (Nakasone, 1974).

'Hortus Gold' is dioecious, but only seed from hand-pollinated fruits is used for propagation. It has a golden colour, remains firm and weighs 1.5–2 kg (Malan, 1953). 'Coorg Honey', 'Co. 1' and 'Co. 2' are the best cvs of India. In Indonesia 'Semangka' has big red-fleshed fruits and 'Thailand' (see Fig. 10.3) has practically unlobed leaves. The bisexual 'Guinea Gold' and the dioecious 'Sunnybank' and 'Hybrid 5' are grown successfully in Queensland, Australia (Agnew, 1968).

Bharath (1969) in Trinidad mentions: 'Santa Cruz Giant', a hermaphrodite with fruit weighing over 5 kg; 'Cedros' which is dioecious and resistant to anthracnose with fruit weighing 3 kg; 'Singapore Pink', a hermaphrodite, very sensitive to anthracnose that bears fruits weighing 2 kg. 'Pusa 1–15' is an outstanding Indian cv. with high yield, good quality and 13 per cent sugar (Ram, 1981). Anon. (1982) reports good performance of cvs Maradol roja, Cubana and Paraguanera in Venezuela.

Local selections in Surinam were crossed with Hawaiian lines; results of this research are summarized in Table 10.3. A potential yield of 100 tonnes/ha was calculated for cvs Santo 3 and 4, but actual yield came to 60 tonnes/ha (Soerodimedjo, 1978).

Table 10.3 *Comparison of three Surinam cultivars to three Hawaii cultivars*

Cultivar	1	2	3	4
Santo 3	103	65	919	59.8
Santo 4	101	58	1077	62.4
Santo 7	108	55	771	42.4
Solo 8	137	72	384	27.6
Red Solo[5]	94	75	353	26.5
Waimanalo	94	53	575	30.5

Source: Soerodimedjo (1978)
1 = average height of first production in cm
2 = average number of fruits, first year
3 = average weight of fruit in g
4 = calculated annual yield per tree in kg
5 = probably a local name for 'Sunrise'

Fig. 10.4 Hermaphrodite papaya orchard in the Ivory Coast

The improvement of cvs aims at: high yield, good flesh texture, high sugar content, intermediate fruit size, small fruit cavity, uniformity and resistance to pests and diseases (Yee, 1970); to this low-bearing habit could be added.

Cultivation measures

Propagation

Papaya is almost exclusively multiplied by seed; vegetative propagation is possible but expensive and only used for experimental reasons. For this, the top of a tree is broken off and some 50 shoots develop which can be used for cleft grafting or making cuttings. The same tree can be used many times.

Uncontrolled seed propagation soon leads to inferior types. Inbreeding by hand pollination is required for both bisexuals and dioecious cvs. No ill results have been reported, even after 20 or more generations. Every fruit produces over a thousand seeds, which brings the cost down to an acceptably low level.

The gelatinous covering is removed by rubbing thin layers of seed on cloth or a sheet of rubber. The seed is washed and dried on paper, out of the sun. Such seeds can be stored in an air-tight container for up to three years. Flats with sterilized potting soil, placed in partial shade, are used for sowing. After two weeks the seedlings are moved into full sunlight and a week later they are transferred to pots.

Direct sowing in bottomless pots eliminates some transplantation problems: six seeds are sown per pot and after emergence, they are thinned to four if dioecious. The chance that all will be male is 1 in 16. Hermaphrodite seedlings are thinned to three per pot and the chance that all will be female is 1 in 27, as is shown in Table 10.2. The pots must be 10 cm deep and should not be coated with wax; 300 g of dolomite is added and they are watered once a week with a solution containing 100 g NPK 5–6–5 per 100 litres (Anon. 1981b).

A third and simpler method is direct sowing at stake; this is more suitable for the tropics. It was used for papain production in East Africa (Purseglove, 1968) but also for fruit production elsewhere. In Surinam eight seeds per hill were sown; they had to be protected against ants with diazinon.

Planting

Six weeks after sowing the seedlings are taken to the field. The density varies from 1,000 to 2,000 plants/ha, with spacings of 3 × 3 m to 2.5 × 2 m; when machinery is used, 4.5 × 2 m is more suitable. The planting hole must be deep and wide enough to accommodate the roots; there is no need to make the hole bigger on permeable soil. Half a kg superphosphate is mixed with the topsoil to fill the hole. As with other fruit crops, the plant should be slightly above field level after the soil has settled. The start of the rainy season is the best planting time.

Three or four plants are left until flowers appear. They are then thinned out to one female on most hills; one male in 20 is left as a pollinizer. For Hawaii, Yee (1970) recommended five seeds per hill. The first thinning takes place after a month and the final thinning when the flowers appear; only bisexuals are left.

Fig. 10.5   Papaya plantation in East Java

Soil management

Little space for intercropping remains at a density of 2,000 trees/ha. If it is undertaken, then it should only be for one season and with annuals. Weeds must be controlled, especially during the initial period. Diuron, paraquat or dalapon can be used between the rows and around older trees, provided they are shielded from the spray. Near young trees, hand weeding is necessary. A trash mulch, peat and sawdust are useful (Younge and Plucknett, 1981); in the wet tropics a cover crop of *Pueraria* and *Centrosema* is recommended. Black polythene has been reported to give significant increases in growth and yield (Kasasian, 1971).

Irrigation

Thanks to its dense root-system, papaya is fairly drought-resistant but irrigation is required after two dry months. Malan (1953) advises 50–75 mm water every three to four weeks. It is better to adapt the need for irrigation to rainfall and potential evapotranspiration. Ep equals 1.3 times the class-A pan evaporation (Anon. 1981a). This and the water supply in the rooted zone can be taken as a guide.

Crop fertilization

'In view of its fast growing character, papaya needs readily available plant food at all times in order to ensure good growth and high fruit yields' says de Geus (1973). He continues: 'the best practice seems to be to maintain the nitrogen supply somewhat higher than for other orchard species'. Phosphorus deficiency causes dark green foliage with a reddish purple discoloration of leaf veins and stalks. Next to superphosphate, a basal dressing of farmyard manure is useful; however, applying this to the plant hole would promote damping-off in seedlings.

Soil and leaf analysis must provide the data needed for specific fertilizer recommendations. In Hawaii, samples are taken from fully developed leaves, with a flower bud just visible in the axil. In Cameroon the leaf is small when the flower appears; there the developed leaf is taken for N and S analyses, while P and cation analyses are made from petioles (Anon. 1978).

Nitrogen should be applied early, as canneries do not accept fruit that contains nitrate. Control of pH is important: Younge and Plucknett (1981) report a 33 per cent decline in yield without lime. A maximum yield removed 154 kg N/ha and 375 kg K/ha. According to Cunha and Haag (1980) the removal of nutrients amounted to: N 1770, P 200, K 2120, Ca 350, Mg 180 and S 200 g per tonne of fruit. These figures may serve as as a guide for fertilization.

Pruning

A young papaya tree is never pruned, but an old tree may be rejuvenated by cutting back to 30 cm above ground. Many shoots develop and after a few weeks all but the strongest one are removed. This shoot must be staked to protect it against strong wind. The stump is cut back later on, at a slant from the shoot (Chandler, 1958). However, such trees are usually infested with nematodes and fungi. A three-year rotation seems best; every year one third of the orchard is replanted.

Production of papain

Tapping begins when unripe fruits are 10 cm wide. Three or four vertical cuts, each 3 mm deep, are made with a razor blade mounted on a stick. The latex is caught on trays attached to the tree; it coagulates and is dried quickly in the sun, or in ovens at a temperature of 55 °C. Once a week fruits are tapped between former cuts. They are unfit for export, but can be eaten or processed. The latex yield is 70–130 kg/ha/year (Purseglove, 1968).

Diseases and pests

Numerous fungi cause disease on papaya, e.g. *Glomerella* (the perfect form of *Colletotrichum*) causes anthracnose which makes the fruit unattractive; *Pythium*, *Rhizoctonia* and *Fusarium* cause damping-off in seedlings and *Phytophthora* is partly responsible for the replant problem. Maneb and a warm–water treatment, 20 minutes at a temperature of 45 °C, are used against anthracnose; maneb or zineb against the other fungi.

Mosaic is a very serious disease, caused by a virus. Several areas in the West Indies abandoned papaya growing because of this 'curse of the industry' (Storey, 1972). It is transmitted by aphids, as is distortion ringspot virus. Bunchy top is a mycoplasma disease, transmitted by a hopper (*Empoasca*). Control of the vectors is the only practical measure.

A white scale, *Pseudaulacaspis pentagona*, thickly encrusts young trees from the bottom up. Scrubbing with a stiff brush is effective if done early; otherwise, *Xyleborus* beetles bore into the weakened stem and kill the tree (van Dinther, 1960). Fruit flies cause rot of ripening fruits. Several mites attack papaya; in serious cases spraying with Kelthane is necessary. Anon. (1982) gives recommendations for the control of diseases and pests in Venezuela; they may be useful in neighbouring countries too.

Root-knot nematodes and soil-borne fungi cause replant problems. Soil disinfection with vapam or methyl bromide is an effective, but expensive measure. For most tropical orchards rotation and mulching are recommended.

From harvest to consumption

The potential yield of papaya is very high, perhaps over 100 tonnes/ha during the productive years; however, the average yield is much lower. Younge and Plucknett (1981) report 154 tonnes/ha over four years, with no production in the first, one half of the total in the second and most of the rest in the third year. Anon. (1982) mentions 15.6 tonnes/ha for Venezuela.

The labour requirement was 169 man-days in Surinam, which breaks down into: nursery 22, planting 36, fertilizing 14, weeding 42 and harvest 55 days; weeding between rows was done mechanically (Landb. Proefst., 1975). According to Anon. (1981a), a three-man crew with a machine picked more than double the amount of a three-man hand-picking crew; the percentage culls dropped from 40 per cent (by hand) to 25 per cent.

Fruit for local consumption is picked when the green colour changes into yellow halfway up the fruit; for export it is necessary

Fig. 10.6 Gathering fruits from female trees in Surinam

to pick sooner, after the blossom end has turned colour. The fruit is cut off with a thin-bladed knife and placed stem down in a box padded with paper or wood wool to prevent bruising. Pickers should wear gloves and aprons as the latex may irritate the skin: however, these subtropical rules are not heeded in the tropics (Fig. 10.6). It may be necessary to treat the fruit with hot vapour or ethylene dibromide to kill fruit fly, after which it is sized, graded, packed and shipped.

References

Adsuar, J. (1977) in: *Diseases, pest and weeds in tropical crops* (eds Kranz *et al.*), Parey.

Agnew, G. W. (1968) 'Growing quality papaws in Queensland', *Queensl. Agric. J.* **94**, 24–36.

Anon. (1978) 'Inventaire et état actuel etc., 'Le Papayer'. *Fruits* **33**, 623–7.

Anon. (1981a) *Achievement report July 1976–June 1979*, Hawaiian Agric. Exp. Sta. Misc. Publ. **184**.

Anon. (1981b) 'Raising quality papaw seedlings', *Queensl. Agric. J.* **107**, 317.

Anon. (1982) *Recommendaciones para el cultivo de la lechosa*, Noticias agricolas Fusagri, Venezuela IX, 35.

Bharath, S. (1969) Cultivation of papaw in Trinidad and Tobago, *Crop Bull.* **15**, Trin. and Tob.

Chandler, W. H. (1958) *Evergreen orchards*, Philadelphia.

Cunha, R. J. P. and **Haag. H. P.** (1980) 'Mineral nutrition of papaya', IV, *Anais escuela sup. agric. Luiz de Queiroz* **37**, 169–78.

Dinther, J. B. M. van (1960) *Insect pests of cultivated plants in Surinam*, Landb. Proefst. Suriname, Bull. **76**.

FAO (1983) Production Yearbook 1982, Roma.

Foyet, M. (1972) 'L'extraction de la papaine', *Fruits* **27**, 303–6.

Geus, J. G. de (1973) *Fertilizer guide for the tropics and subtropics*, Zürich.

Hamilton, R. A. and **Ito, P.** (1968) 'Sunrise Solo, a different colored Solo papaya', *Haw. Agric. Exp. Sta.* circ. **69**.

Kasasian, L. (1971) *Weed control in the tropics*, Leonard Hill.

Landb. Proefst. Suriname (1975) *Annual report for 1974*, Paramaribo.

Lassoudière, A. (1968–9) 'Le papayer', *Fruits* **23**, 523–9 and 585–96; *Fruits* **24**, 105–13, 143–51 and 217–21.

Malan, E. F. (1953) 'The production of papaws', Reprint 57, *Farming in South Africa*.

Nakasone, H. Y. *et al.* (1972) 'Evaluation of Waimanalo, a new papaya strain', *H.A.E.S.* Techn. Bull. **79**.

Nakasone, H. Y. *et al.* (1974) 'Evaluation and naming of two new Hawaii papaya lines, Higgins and Wilder', *H.A.E.S.* Res. Bull. **167**.

Nakasone, H. Y. *et al.* (1982) 'Fruit crops' in *Crop improvement in Hawaii* (ed. J. L. Brewbaker), Un. Haw. Misc. Publ. **180**.

Popenoe, W. (1920) *Manual of tropical and subtropical fruits*, New York.

Purseglove, J. W. (1968) *Tropical crops, Dicotyledons*, Longman.

Ram, M. (1981) 'Pusa 1–15, an outstanding papaya', *Indian Hortic.* **26(3)** 21–2.

Soerodimedjo, F. W. (1978) 'Papaya research in Surinam', in *Symp. on maize and peanut*, Proc. Car. Food Crops Soc. XV, 342–52.

Storey, W. B. (1969) 'Papaya' in Ferwerda and Wit (eds) *Outlines of perennial crop breeding in the tropics*, Wageningen.

Storey, W. B. (1972) 'Papaya', *Capita selecta*, Wageningen (stencil).

Teaotia, S. S. and **Singh, R. N.** (1967) 'Seasonal variation in sex expression of papaya', *Indian Agriculturist* **11**, 45–9.

Terra, G. J. A. (1949) *De tuinbouw in Indonesië*, 'sGravenhage.

Yee, W. *et al.* (1970) *Papayas in Hawaii*, Univ. Haw. Coop. Ext. Serv. circular 436.

Younge, O. R. and **Plucknett, D. L.** (1981) *Papaya fruit yield as influenced by crop rotation, cover cropping, liming and soil fumigation in Hawaii*, H.A.E.S. Research Bull. **155**.

Chapter 11

The minor tropical fruits

There are about 3,000 species of edible fruits in the tropics; 250 can be thought of as widespread, 50 as well known, and perhaps a dozen as principal species (Martin, 1976). Counting *Citrus* as four (orange, mandarin, grapefruit, lemon) and banana with plantain as two, we have only dealt with ten so far. Of course, it is impossible to mention the other 2,990 or even 240, in this chapter. A selection was made which includes about 50 of the most well-known ones. I have not put them in alphabetical order of families or genera because their economic value and ecological relationship are of greater importance to the fruit grower.

Guava and two relatives

Guava or guayaba (*Psidium guajava*) is a small tree of the Myrtaceae family. It is easily recognized by its peeling bark, four-angled twigs, opposite leaves and axillary flowers with many stamens. The fruit is a berry, topped by calyx lobes (Fig. 11.1), with many seeds embedded in a white or reddish pulp.

Ripe fruit emits a sweet aroma and has a pleasant sour-sweet taste, but in a later stage produces a penetrating scent. The fruit is eaten fresh or cooked, but is principally used for the preparation of jellies and jams; it contains 82 per cent water, 0.7 per cent protein, 11 per cent carbohydrate and fair to good amounts of vitamins A, B1, B2 and C, plus some minerals. The average weight is 100 to 150 g.

The tree is native to tropical America but is now present in every tropical and subtropical country. In some places it is regarded as a pest, just like its relative *Eugenia cumini*, because the seeds are spread by birds; pastures may be overgrown completely.

Guava tolerates a wide range of climates, provided they are frost-free. In India, Allahabad (25° N) has the reputation of growing the best guava (Singh *et al.* 1963). The optimum rainfall is 1,000 to

Fig. 11.1   Guava fruits and leaves

2,000 mm, but fruit quality is poor in areas of high rainfall or high atmospheric humidity. The optimum temperature is between 23 °C and 28 °C.

The guava tree is fairly salt and drought-resistant and can grow on many different soil types: sand, clay, deep, shallow and acid to alkaline; the pH may range from 4.5 to 8.2 (El Baradi, 1975). The root system has been described as shallow by Singh (1963), but measurements by Moutounet *et al.* (1977) demonstrate that it can also be deep and intensive; this probably depends on the soil.

In India, guava is usually propagated from seed although seedlings are known to be variable. An effort was made in Surinam to exploit natural guava groves by selective thinning, but this failed: production was low and fruit quality poor. In order to attain high-yielding trees with fruit of good quality, selection of superior seedlings is necessary; their properties must be maintained by vegetative propagation.

Except for some clones, guava is difficult to root and cuttings can therefore not be relied on. If used they are taken from strong, succulent green wood. Two 3-node cuttings, with the upper two pairs of leaves halved, can be taken from each branch. The lower pair of leaves is removed and the base is treated with a growth regulator, e.g. Indole-3-butyric acid (IBA). Forkert budding is the best propagation method; a patch of 1×1.5 cm is placed on a seedling rootstock at 15 cm above ground level. No special rootstocks have been selected, so far. Layering, air–layering and inarching are possible. The reader can find more information on the propagation of guava and many of the following crops in Garner *et al.* (1976).

'Safeda' and 'Lucknow 49' are the best known cvs of India. 'Centeno Prolific' of Trinidad produces more than 50 tonnes/ha; it can be multiplied by cuttings and is tolerant to flooding (Bharath, 1969). The best Hawaiian cvs are 'Beaumont' and 'Ka hua kula'. Seedlings of 'Beaumont' differed little from the mother clone and produced 12–22 tonnes/ha in the third year and 40–50 tonnes/ha in the fifth year (Nakasone, 1976, 1982). 'Suprème' and 'Elisabeth' are good clones of Ivory Coast.

The best time for planting is the beginning of the rainy season. The trees are set out at 6×6 m, or 6×8 m, but closer spacings are used on poor soils; Mohammed *et al.* (1984) went up to 27,000 trees/ha. Shigeura and Bullock (1983) advise an equilateral triangle rather than a square system. An intercrop of vegetables may be grown during the first three years and in the wet tropics a cover of *Pueraria* is recommended.

Fertilization, irrigation and weed control needs are similar to those of citrus cultivation. Optimum levels of elements in leaves were determined at N 2, P 0.25, K 1.60 and Mg 0.25 per cent of dry matter (Anon. 1981). Shigeura and Bullock (1983) give slightly lower figures for N and K and 0.18 per cent for S; for Zn, Mn, Cu and B, optimum levels were found to be 20, 60, 8 and 20 ppm respectively.

Good pruning is important. The tree naturally puts out suckers from the stem and rootstock; these should be removed to keep the stem free up to 60 cm above the ground. A framework of four strong branches, radiating to all quarters and as far apart as possible, should be formed. The angle between the branches and the stem must be wide so that sunlight is able to penetrate to the centre of the tree. A good shape should be maintained by light pruning of vertical shoots and dead wood.

Guava can flower in every month of the year near the equator. In India it flowers twice in the north and three times a year in the southern part of the country. The best quality fruit is obtained when it ripens during a dry period. It is therefore useful to give the tree a rest by withholding irrigation periodically. A concentrated bloom facilitates pest control and harvest operations. An extreme example of this practice is root exposure, as done in the Bombay-Deccan

Fig. 11.2 Young guava trees in Surinam

area (Singh *et al.* 1963). Shigeura and Bullock (1983) describe a system for year-round processing.

The tree is susceptible to several fungus diseases and insect pests; for a list of the latter see Butani (1977). Probably the worst pests are the fruit flies. In Surinam and elsewhere in South America and the West Indies *Anastrepha* spp. (Fig. 11.3) infest the fruit so heavily that it may be virtually crawling with larvae ('worms'). Control was achieved by van Brussel and van Vreden (1968) by spraying twice, with an interval of ten days, with Lebaycid or Dipterex and by soil treatment with dieldrin; however, it would be better for the environment to replace dieldrin by diazinon. Bharath (1969) advises the use of Sevin against fruit fly in Trinidad; trees are sprayed once a week from the time the first ripe fruit is picked. In Hawaii a thrips causes much damage.

Fouqué (1979) in Ivory Coast found a time lapse of 102–124 days from bloom to harvest; Singh (1963) puts this interval at nearly five months in India. The fruit should be processed on the day of picking says Bharath (1969), and Wills (1981) sets the storage life of the fruit at one week (at a temperature of 20 °C) or two weeks (below 10 °C).

Fig. 11.3   *Anastrepha striata* laying eggs on guava fruit

The strawberry guava, *Psidium cattleianum*, has recently received much attention. It is an 'almost ideal fruit: slightly acid-sweet, fragrant, nutritious, with highly attractive colour inside and outside' (Barrett, 1928). The fruit is round, reddish, 3–4 cm in diameter and is eaten fresh or as jams and jellies. It can stand some frost, down to −4 °C. Propagation is by seed, it comes true to type. The bush is very attractive as an ornamental or hedge plant. A yellow cv. exists, with larger and sweeter fruit (Maxwell, 1980). However, at the present time, there is no large-scale cultivation of strawberry guava anywhere.

Acca sellowiana, feijoa or pineapple guava, combines in its fruit 'the exquisite flavour of pineapple and strawberry but is not a safe cropper and difficult to propagate' (Barrett, 1928). It is native to southern Brazil and is now cultivated on a fairly large scale in New Zealand. The bush is 2–4 m high, evergreen, with big axillary flowers and fruits that weigh between 25–60 g each. They contain many small seeds and have a pleasant aromatic taste, except when grown on a lime-rich soil; in that case the fruit acquires a medicine-like taste.

Although feijoa will survive ten degrees of frost when dormant, it is damaged by light frost in autumn: −1 °C during one hour can destroy the harvest. The tree is drought and salt-resistant but needs a moist soil for a good yield. An average annual temperature of 15 °C, 1,500 mm rainfall/year and high humidity are required. The soil must be well drained and irrigation should be feasible.

Well known cvs are 'Triumph' and 'Mammoth' in New Zealand and the self-compatible 'Coolidge' in Calfornia. Maxwell (1980)

mentions 'Choiceana' and 'Superba' in Florida. Other cvs than 'Coolidge' have to be planted in mixed stands, for pollination. Propagation is preferably by cuttings, but air-layering and grafting are also used. Fruit is formed on the outside, on first–year wood; yearly pruning assures a regular production. A single stem is kept up to 50 cm high, with three to four main branches. A spacing of 4.5–5 m between rows, with either 3–3.5 m in the row or 2 m in hedges, is commonly used.

Trees in full bearing receive 120 g N, 80 g P_2O_5 and 100 g K_2O/year. The yield amounts to 25 tonnes/ha. Fruit can be picked when the colour changes to light green. It may also be gathered soon after it has fallen (Azam *et al.* 1981).

Soursop and other annonas

Soursop, *Annona muricata*, is called 'guanábana' in Spanish and 'corossol' in French; it belongs to the Annonaceae family, as do:

(*a*) *Annona squamosa*, sweetsop, sugar apple or custard apple;
(*b*) *Annona reticulata*, bullock's heart or custard apple;
(*c*) *Annona cherimolia*, cherimoya;
(*d*) *Annona diversifolia*, ilama.

A hybrid between *A. cherimolia* and *A. squamosa* is known as atemoya. Efforts to cross soursop with other *Annona* spp. had no success. Ilama produces fruit as rich in flavour as a good cherimoya and can grow at low altitudes in the tropics.

The annonas are small trees, up to 7 m tall, with simple, entire leaves. They have bisexual but protogynous flowers, with many pistils and stamens. Pollen is only shed when the stigmas of the same flower are no longer receptive; these crops therefore depend on cross pollination for fruit setting. This can be effected by insects but often fails, in which case hand pollination is required. The fruit is a compound structure, formed by fusion of many single-seeded fruitlets with the receptacle. The seeds are flat, black and very hard; they contain oil, used for paint or insecticide, and are surrounded by a white edible pulp. Sweetsop and soursop contain 60–80 seeds per fruit, cherimoya 10–15.

Cherimoya and sweetsop are best for eating fresh. Soursop pulp clings to the seeds; it makes an excellent drink or ice cream after straining. Soursop pulp contains 80 per cent water, one per cent protein, 18 per cent carbohydrate and fair amounts of vitamins B1, B2 and C. Sweetsop and cherimoya contain somewhat more sugar.

The annonas are native to tropical America but are now widespread in the tropics and frost-free subtropics of other continents. Production mainly takes place in home gardens. India, probably the

Table 11.1   *Properties of four Annona species*

Species	*A. muricata*	*A. cherimolia*	*A. squamosa*	*A. reticulata*
Common name	soursop	cherimoya	sweetsop, sugar apple	bullock's heart custard apple
Habit	evergreen	deciduous	deciduous	deciduous
Growth	upright	spreading	open	spreading
Leaf struct.	leathery*	velvety	thin, soft	downy
Leaf colour	shiny green	light green	dull green	red nerves
Fruit colour	bright green	light green	blue green	dull green
Fruit shape	ovoid	heartshaped	heartshaped	heartshaped
Fruit surface	soft spines	lumpy	lumpy	smooth
Fruit weight (g)	500–2000	200–500	100–200	300–500
Fruit quality	good	excellent	good	poor
Opt. climate	equatorial	trop. highl. subtropical	subhumid trop. warm subtrop.	subhumid trop. warm subtrop.
Opt. rain (mm)	1,800	1,200	800	1,200
Opt. pH	5–6.5	6	7–8	5–7.5

Sources: Chandler (1958), Maxwell and Maxwell (1980), Purseglove (1968), Singh *et al.* (1963)
* with 'domatia' (small cavities), as in coffee (Ostendorf, 1962)

largest producer, had 44,000 ha planted with sweetsop (sitaphal) chiefly in the state of Hyderabad (17–20° N), in 1955.

The annonas are slow growers. With the exception of soursop, they drop their leaves in the cool season and remain bare and dormant for several months. Some of their properties are summarized in Table 11.1. All annonas are sensitive to frost and heat and need partial shade when young. The best growth occurs on moist medium loams, but sweetsop also grows on sandy, stony or rocky soils in India and on lime rock in Florida. A close relative to the four annonas is biribá, *Rollinia pulchrinervis*, found in northern South America; it resembles cherimoya in size, shape and taste.

The annonas generally come true to type when propagated by seed. Some named cvs are known; they can be multiplied by budding or grafting on their own seedlings, in some cases on each other and on two wild relatives: *Annona glabra* and *A. montana*. One-year-old budwood has given the best results with sweetsop. Inarching is the common method in India and cuttings are not difficult to make.

The plants are set out at 6 × 6 m, or somewhat closer on poor soil. A cover crop of *Pueraria* may be grown with soursop. *Crotalaria* is sown as a green manure between sweetsop trees at the start of the rainy season, to be ploughed under later on; mulching is beneficial too. Weed must be controlled, especially during the initial years of the orchard. Regular applications of fertilizer are needed.

Fig. 11.4 Soursop fruit and foliage

A 3-10-10 formula has been recommended in India, but a balanced formula, such as 10-10-10, seems more appropriate for most other regions. Young trees should be pruned until a good framework of branches is acquired; otherwise only light pruning is needed.

A mealy bug attacks sweetsop in India. Soursop is infested by a moth (*Cerconota* sp.) and a wasp (*Bephrata* sp.) in the Caribbean area; larvae of the moth do only superficial damage, but the larva of the wasp leaves a black trail through the fruit, rendering it very unattractive. Prevention is possible by placing muslin bags around young fruits; this is a laborious but effective way to protect the fruit against these pests (van Dinther, 1960).

Sweetsop yields 60–70 fruits per tree, or about 20 tonnes/ha. The fruit is harvested when still firm and plump, the skin between segments turning light yellow. It can be transported, but after a

Fig. 11.5 *Annona squamosa. Source*: Ochse (1931)

week the fruit cracks and splits into segments. If the fruit is harvested immature, the quality will be poor. Cherimoya fruit does not crack and stands transport better. Soursop yields up to 10 tonnes/ha and is also difficult to transport (Nat. Acad. 1978).

Cashew and other nuts

Botanically speaking, a nut is an indehiscent (not opening naturally) dry fruit with one seed. However, this definition does not apply to all crops to be considered here. Brazil and sapucaia nut, for

Table 11.2   *Composition of shelled nuts (percentage)*

Crop	Protein	Oil	Carbohydrate
Almond	20	55	20
Brazilnut	15	66	9
Cashew nut	20	45	26
Macadamia	9	76	13
Walnut	15	65	13

Source: Rehm and Espig (1976)

Table 11.3   *Export of major nuts (× 1,000 tonnes)*

	1974	1975	1976	1977	1978
Hazel nut	161	142	180	75	200
Almond	68	78	98	113	107
Cashew nut	94	96	94	72	65
Walnut	42	52	63	48	42
Brazilnut	23	34	27	25	26
World	387	402	462	332	439

Source: Indian Cashew J. **13**, 1(1980)

instance, have capsules, i.e. dehiscent dry fruits with many seeds. The average world annual nut production, from 1975 to 1979 in 1,000 tonnes was: hazelnut 401, cashew 379, walnut 231, (shelled) almond 210, pecan 119, brazilnut 55 and pistache 51 (*Indian cashew J.* **13**, 1980). They are a very nutritious food, (see Tables 1.2 and 11.2). Data on export of nuts are given in Table 11.3. The reader who would like more information on temperate and subtropical nuts should consult Rebour (1968), Jaynes (1969) and Menninger (1977).

The cashew nut, *Anacardium occidentale* originated in the northern part of South America but is now common in all tropical countries with one or two reliable dry seasons, especially in India and East Africa. The tree is hardy and drought-resistant and grows best on well-drained sandy soils with an annual rainfall of at least 900 mm. It can do with less rain, but then yields decrease, unless it is irrigated. The tree ceases to function as a fruit tree when rainfall drops below 500 mm but is still valuable as it yields timber and firewood. Far more than 900 mm rain is tolerated if drainage is good and a sufficiently long dry period allows bloom and fruit set. According to Ohler (1967): Cashew produces on soils which are too poor and too dry for other crops. The tree grows well at a pH of 4.5 to 6.5.

The real cashew fruit is a kidney-shaped nut about 3 cm long; it is attached to a bigger 'apple' that is sometimes mistaken for the fruit, but that is only the edible receptacle plus fruit stalk. The cashew nut of commerce is the seed. To extract it, the shell must

Fig. 11.6   Cashew, botanical drawing. *Source*: Greshoff (1894)

be removed. This shell contains a skin-blistering oil which was formerly regarded as an unpleasant nuisance, but is now highly valued for its many industrial applications. Therefore, open-air roasting in which the shell oil evaporates is gradually being replaced by roasting in a shell oil bath at 180 °C; in this process the oil is recovered (Ohler, 1966).

Generally, seedling trees are grown. They are either raised in nurseries or sown at stake. However, if high yields of superior quality are desired, then the best clones must be selected and propagated vegetatively. Perhaps the best propagation method is grafting on seedling rootstocks, raised in deep bags (Fig. 11.7). However, it is more convenient to graft or bud seedlings 'on site', as a suitable method to transplant nursery–grown trees is not available (van Eijnatten and Abubaker, 1983).

Fig. 11.7   Cashew grafted on seedling rootstock, Kenya

An initial spacing of 6 × 6 m can be used, with one grafted tree or three seeds to a planting site; this gives a density of 278 trees/ha. As the trees grow, they must be thinned and pruned. Cashew has terminal inflorescences and intermingling branches become unproductive. Other spacings are 5 × 5 m and 8 × 8 m, up to 12 × 12 m, some leading to a final spacing of 18 × 18 m (31 trees/ha). The yield/tree increases, but yield/ha decreases.

An alternative has been proposed by van Eijnatten and Abubaker (1983): hedgerows 9 to 12 m apart and trees at 2 to 3 m in the rows. This, combined with a high-yielding clone, opens perspectives of greatly increased production. Rain during the bloom may cause the rotting of flowers and young fruits; to counteract this clones have been selected with widely spreading inflorescences (Brian Adams, personal communication).

Cashew was formely believed to be almost free from serious pests and diseases; however, Ohler (1979) presents long lists of them. Anthracnose (*Glomerella*) is one of the most harmful. Intini and Sijaoni (1983) found a close correlation between the inoculum, climatic conditions and stages of development of the inflorescence. It is important to spray with sulphur at the critical time.

The harvest is commonly done by collecting fallen fruit which is labour consuming work (Morton and Venning, 1969). Yields fluctuate between one and five tonnes/ha. The principal producers are shown in Table 11.4. Large quantities of cashew nuts were formerly sent from East Africa to India, where labour was abundant and cheap, for processing. This practice is currently diminishing as the African countries have built their own processing plants. The demand for cashew nuts is rising steadily and the prices are good, with the result that cashew has become a popular small-holders crop. Development projects are in progress in India in the states of Kerala, Karnataka and Orissa and in Brazil in the north-east. IRFA (see page 10) workers put the optimum size of a family farm at nine ha (private communication).

Until 1977, no books on cashew were available. Then, almost simultaneously, three books were published: Agnoloni and Giuliani (1977), Nair *et al.* (1979) and Ohler (1979). This shows that cashew is indeed a crop with a future (Morton, 1961).

Table 11.4 *The major producers of cashew nut (1,000 tonnes)*

	1975	1980	1981	1982	1983	1984
India	141	180	190	194	200	180
Mozambique	180	95	95	75	70	20
Brazil	37	75	74	76	90	65
Tanzania	122	41	61	45	35	49
Nigeria	30	33	34	35	36	37
Kenya	16	15	15	15	12	15
World	654	457	465	472	467	397

Source: FAO Production Yearbooks

Another delicious and nutritious nut is macadamia. In the opinion of Hamilton and Fukunaga (1959) the botanical name *Macadamia ternifolia*, which is often seen in early literature, is only valid for a wild relative. In culture we find two species that differ as follows:

	M. integrifolia	*M. tetraphylla*
fruit	smooth	rough-skinned
leaves	3 per node	4 per node, spiny
flower	white	pink

Practically all commercial cultivars belong to *M. integrifolia*.

Macadamia is native to eastern Australia and is at home in the same type of climate as *Coffea arabica*, i.e. from equatorial highlands at 1,500 m altitude to frost-free subtropical areas at sea level. The crop requires about 1,000 mm of rain and is fairly drought-resistant but needs windbreaks in exposed locations. Cool nights induce flowering.

The highest distribution of dry matter was found at 20 °C to 25 °C by Trochoulias and Lahav (1983); no growth occurred at 10 °C and the flush was chlorotic and necrotic at 30 °C.

For propagation of rootstocks, husked seed is sown; the seedlings are grafted when their diameter at ground level is at least 10 mm. Slightly thinner scion-wood is girdled six weeks before grafting. The wedge graft has given good results. After two years in the nursery the roots are pruned and ten weeks later the plants can be set out in the field, usually at a spacing of 8 × 10 m, or 125 trees/ha. Weeds must be controlled constantly, e.g. with paraquat, glyphosate or atrazine.

The object of pruning in macadamia is to form a tree with a single main stem and a framework of horizontal branches, starting at 1 m above the ground and from there at intervals of 0.5 m. In order to achieve this, one must know how the tree is constructed. There are three buds in a vertical row in each of the three leaf axils of a node. When a stem is topped, all three upper buds will grow straight up. Only one of these must be allowed to remain and to continue the main stem, the other two being clipped off to a stub of 1 cm. Now the buds below those two stubs will grow out in a more or less horizontal direction; only these branches flower and fruit. This process is repeated until a good framework has been established.

The tree begins to bear fruit in the sixth year and continues to do so for nearly 50 years. The best Hawaiian cvs were, in 1959, 'Keauhou', 'Ikaika' and 'Kakea'. Hamilton and Ito (1982) have summed up the criteria for selection: a good cv. may produce four times as much as a seedling and may have nuts with 10 per cent more kernel. On average a fruit weighs 7 g, of which 40 per cent is kernel. It is wise to plant at least two cvs, allthough self-pollination usually occurs. During the initial years intercropping may be practised.

Hamilton and Fukunaga (1959) advise fertilizing with ammonium sulphate before the bloom and with NPK twice afterwards; the amounts to be increased yearly. Few serious pests have been reported. The nutborer *Cryptophlebia* in Queensland, Australia is hard to control, because it has many alternative hosts (Ironside, 1982). Macadamia, like avocado and mango, is also sensitive to *Phytophthora cinnamomi* (Pegg and Forsberg, 1982).

The harvest takes place by gathering fallen fruit at least once every two weeks. This is husked and then carefully dried to a moisture content of 3.5 per cent. Uneven drying will cause cracking and rot. The commercial product is the vacuum-packed roasted nut. The actual production, about 10,000 tonnes/year, is far below the estimated world market of 30,000 tonnes (Anon. 1981). A yield of 8–10 tonnes/ha is considered normal and the economic life of a well-run orchard is 40–50 years.

Fig. 11.8   Macadamia nuts in a Dutch market

The Brazil nut, *Bertholletia excelsa*, is the leading nut of the humid tropics. It is a very large (up to 40 m high) forest tree with roundish, woody fruits 12–15 cm in diameter, containing 12–24 angular seeds; it takes the fruit over a year to ripen. Nearly all the world's supply of 50,000 tonnes/year is obtained from wild trees in South America (Purseglove, 1968). However, Müller (1981) describes ways to cultivate this crop: plant grafted trees on clay soils at a spacing of 10 × 15 m or 20 × 20 m.

Sapucaia or paradise nuts come from related large forest trees of Brazil and the Guianas: *Lecythis zabucajo* and other *Lecythis* spp. The seeds are more difficult to gather as they do not remain in the fruit, but are scattered.

The souari nut, *Caryocar nuciferum*, occurs wild in the forests of Surinam and Guyana. The large fleshy fruit contains one to four seeds that are covered by an extremely hard layer. The seed is 3 cm long and has a delicious flavour. About 100 trees were planted in the botanical garden of Paramaribo, Surinam, but their production was low and uncertain (Ostendorf, 1962).

Important nuts of the subtropics are almond (*Prunus amygdalus*), pecan (*Carya illinoensis*), Turkish hazel (*Corylus colurna*) and pistachio (*Pistacia vera*). All these need chilling and cannot therefore be grown in tropical lowland. However, they deserve trial in subhumid tropical mountain areas; stripping of leaves and hormonal treatments might produce good results. The interested reader is referred to Rebour (1968), Jaynes (1969), Rehm and Espig (1976) and Menninger (1977) for additional information.

The date and other palms

The date is one of the main food crops of the arid regions, from Morocco to Pakistan. The major producers are shown in Table 11.5. *Phoenix dactylifera* is a dioecious palm that can be propagated vegetatively by shoots. There are three groups of cultivars: dry, half dry and soft; 'Deglet-Noor' (half dry) and 'Sayir' (soft) are among the best known cvs.

Table 11.5   *The major producers of dates (× 1,000 tonnes)*

	1980	1981	1982	1983	1984
Saudi Arabia	342	350	400	440	450
Iraq	395	400	400	400	115
Egypt	446	391	393	440	450
Iran	300	301	301	302	330
Algeria	201	195	207	210	207
Pakistan	202	205	205	218	450
World	2,573	2,530	2,630	2,776	2,379

Source: FAO Production Yearbooks
NB: FAO indications for estimates have been omitted

Date cultivation requires high temperature, much sunlight, low humidity, low rainfall and much irrigation. An Arab proverb states: the date must have its feet in heaven, but its head in hell. The date is very salt-tolerant and irrigation water may contain up to 1 per cent salt, provided drainage is good. The soil must be permeable and well-aerated. From all this we may conclude that the date prefers a climate with winter rains. However, south of the Sahara cultivars tolerant of summer rains are grown. The reader who desires more information can find this in El Baradi (1968), Munier

(1973), Nixon and Carpenter (1978). For diseases, see Djerbi (1983).

The date can also be tapped for sugar. The yield is about 2 kg/year; tapped trees have a zig-zag appearance (van Heurn, personal communication). In 1982 a periodical, *The Date Palm Journal* was founded to deal specifically with this crop.

Salak, *Salacca edulis,* is cultivated in Indonesia and Malaysia in regions with at least 1,700 mm of rain and only a short dry season, from sea level to 300 m altitude. It is a stemless, spiny, tillering, dioecious palm, that only grows well in shade. Propagation is usually by seed, as suckers and layers often do not survive. Generally two to five seeds are sown together under heavy shade at a

Fig. 11.9   Salak. *Source*: Ochse (1931)

spacing of 2 × 2 m. After the first bloom the seedlings are thinned out, leaving 10 per cent males (Ochse 1931, 1961; Brückmann, 1938). Polprasid (1982) points out that breeding is possible with several other wild *Salacca* spp. called Rakam, Sala, Sakam and Som Lumphi.

Tropical America is particularly rich in palms with edible fruits. The leading one is pejibaye, *Guilielma gasipaes,* but this is not eaten fresh. It is one of the most balanced tropical foods and contains twice as much protein as the banana. Thirteen bunches of fruit, each about 12 kg, can be harvested twice a year and the tree can live for 75 years (Nat. Acad. 1978). Blaak (1980) has described vegetative propagation: the growing point of a young seedling or offshoot is removed and side shoots will emerge within six weeks.

Other well-known palms are *Astrocaryum* and *Maximiliana* spp. When the Amerindians are clearing land for their villages they do not cut down these palms and veritable palm forests may thus ensue. According to a physician who knew them well (Schuite-maker, personal communication) the health of these people is evidently correlated with the seasonal fluctuations of the palm fruits. In Africa the fruits of *Borassus aethiopum* and *Hyphaene thebaica* are eaten.

The vine crops

The word 'vine' specifically refers to the grape but can also be used for other trailing and climbing plants; here grape, kiwi and the passion fruits will be considered. All are grown along wire on stakes, trellises or pergolas, which involves a high capital invest-ment. The grape can be pruned to become self-supporting.

Grape is a deciduous crop of warm temperate regions and of the subtropics with winter rains. A cool, but not very cold winter and a dry, hot summer are required for the best results. The crop is preponderantly grown from 20° to 50° North and from 20° to 40° South. However, grape culture is also possible in the tropics under certain conditions. Suitable cultivars and cultural practices must be adopted.

Grape cultivars are generally grouped according to the use made of them: for wine, raisins, juice or fresh fruit; we shall only consider the last type here. Furthermore they are classified according to origin in four groups: *Vitis vinifera* or European grape, *V. labrusca* and other American grapes, the French hybrids between those two and *V. rotundifolia* or muscadine, which is tolerant to hot condi-tions. Well known European cvs are 'Thompson seedless' and 'Muscat of Alexandria'. They are generally grafted on American rootstocks, which are tolerant to phylloxera, a louse that attacks European grape roots.

The grape requires a deep, loamy soil with good structure. It must contain organic matter and should be very well drained and aerated. A pH of 6 is preferred and the soil must be practically salt-free.

Propagation is by budding or grafting, but cuttings and layers are also used. The rootstock must be tolerant for, or resistant against, phylloxera and nematodes. Young plants are set in the field at a spacing of 2 × 4 m, or somewhat closer, at densities of 1,250 to 2,000/ha. They must be supported, but how depends on the pruning system. In 'head pruning' stakes are used for a few years, after which the plants have become self-supporting. In the 'cordon' system the vines are trained along wire.

Grape growing in northern India differs little from European cultivation. In the South and West of India, however, grape becomes an evergreen. It is pruned twice a year. In October fruiting canes are cut back to 4–10 buds, depending on the cv.; growth starts within a few days and the harvest takes place in February–March, which are relatively cool and dry months. The quality is good, thanks to slow ripening. Typical centres are Poona (18° N), Hyderabad (17° N) and Bangalore (13° N), at elevations of 500–900 m. The second pruning is in April, when vines are cut back to one bud; all flower clusters are pinched off to let the wood mature (a second harvest would be of low quality and would affect the next crop). 'Thompson seedless' and local cvs such as 'Anab-e-shahi' and 'Bangalore blue' are grown on their own roots (Bammi and Randhawa, 1968).

Descriptions of grape growing in other regions can be found in Pansiot and Libert (1971), Aubert (1972) and Chadha (1977). Tropical countries with large grape-growing areas are India, Peru, Bolivia, Mexico, Thailand and Taiwan.

An interesting pruning system for Kenya, altitude 1,500 m, has been designed by Shalitin (1973). Conventional practices fail here, so close to the equator, as there is practically no chilling and therefore no well-defined growth cycle. Young plants must be pruned in such a way that a strong framework of branches is formed; then during a dry period (June to August) irrigation is withheld to induce water stress. As a consequence the plant drops its leaves and goes dormant, the canes (young branches) meanwhile hardening.

In August the vines are pruned. To understand the effect of pruning we must know that two buds are present in the axil of each leaf: a big and a small one; the big bud forms leaves and flowers, the other only leafy shoots. Now, if we 'pinch' the top of a cane, the small buds are stimulated to grow out. This helps to break the dormancy of the big buds and bloom occurs. The fruit

Fig. 11.10   Grape growing at Probolinggo, north coast of East Java

can now develop and ripen in a sunny and rather warm season.

Readers wishing to know more about pruning, girdling, thinning and other cultural practices in grape growing, and associated pests and diseases, should turn to Winkler (1962), or Shoemaker (1975).

'Kiwi' is the New Zealand name now generally used for yang tao, *Actinidia chinensis*. Another name, chinese gooseberry, is best forgotten since it neither looks like, nor is related to the real gooseberry. It was known for centuries in China but has only become important in the Bay of Plenty (northern New Zealand) since 1960. A fast development is now taking place in California (Sommer *et al.* 1983), Italy and other suitable areas. New Zealand produced 12,000 tonnes in 1981, California 4,000 tonnes and France 1,700 tonnes; Italy's production was 5,800 tonnes in 1982.

Kiwi is deciduous, has woolly-haired leaves and branches that climb by tendrils; they need support. Every plant is either male or female and it takes a seedling five to six years to flower, but cuttings bear fruit in the third year. There are vegetative and mixed buds; bloom occurs on two-year old wood. The fruit is a hairy brown berry with greenish pulp and small black seeds; it is sweet, sub-acid, with good flavour and contains 1.6 per cent protein, 11 per cent carbohydrate and 300 mg vitamin C per 100 g. It is eaten fresh and is used in salads or pastry for its decorative appearance.

When fully dormant and leafless, kiwi tolerates low temperature, down to $-15\,°C$, but otherwise it is harmed by even light frost. Hail, strong wind and salt are also harmful. Kiwi needs a warm and moist summer. The average temperature of the Bay of Plenty area

Fig. 11.11   Kiwi growing along wire in Israel

is 24 °C and the average minimum is 5 °C; the rainfall is well-distributed, 1,300 mm/year and there are 2,300 hours of sunshine (Sale, 1980). The soil must be well-aerated, deep, light and rich in organic matter; the optimum pH is 6–7. Kiwi does not tolerate a water-logged soil and even a short dry spell in the growing season can be deleterious.

Several cultivars have been selected in New Zealand, their properties are set out in Table 11.6.

Far more male than female seedlings are formed and they do not come true to type; it is therefore necessary to propagate asexually. Softwood cuttings, hardwood cuttings and grafting are used. The spacing is 6–7 m between rows and 4–5 m in the row, giving a density of 300–400 plants/ha. Many training systems exist, of which we mention counter espalier, double pergola and T-bar. The double

Table 11.6   *Properties of four kiwi cultivars*

Cultivar	'Bruno'	'Abott'	'Monty'	'Hayward'
Season	early	mid-	mid-	late
Plant vigour	VG	G	VG	M
Production	G	G	VG	M
Flowers/node	1	2	2–3	1
Fruit shape	long	round	round	round
Fruit weight (g)	60–70	65–70	35–40	90–95

Source: Youssef and Bergamini (1981)
G = good, M = medium, VG = very good

pergola is best adapted to the natural growth of the plant, which makes picking and spraying easier (Youssef and Bergamini, 1981) but Sale (1978) states that the pergola system costs more with no advantages: one man can only manage 2 ha of plants grown as pergolas, but 4 ha of plants grown on the T-bar.

In the even rows only female plants are set out, in the odd rows one plant in every four is male. As in other fruit crops, the root-crown should be slightly above the ground. Pruning is needed to regulate growth and flowering. Multiple stems require less work to achieve a good frame. Side branches are cut back to two to four buds in summer. Sale (1978) favours one stem with one leader to each side and fruiting arms every 50 cm.

Kiwi needs a regular but moderate water supply. Irrigation is applied frequently in small amounts, e.g. 2–3 mm. Much farmyard manure is given, e.g. 80 tonnes/ha, and is supplemented with phosphate and potassium. Bearing starts in the third year and reaches a level of 20 tonnes/ha in the tenth; 40-year-old trees are still productive in New Zealand. Foot-rot is caused by *Phytophthora cinnamomi* and fruit-rot by a *Botrytis*. Root-knot nematodes do much damage if not controlled.

The fruit can be stored at 0 °C and 96 per cent humidity but must be kept away from ethylene producing fruits, such as citrus and banana. The reader is referred to Beutel *et al.* (1976), Nihoul (1976), Youssef and Bergamini (1981) and Hudson (1982) for more information.

Table 11.7 *Importation of kiwis into the Netherlands (tonnes)*

	1980	1981	1982	1983
New Zealand	2,111	3,636	2,306	4,679
USA	361	528	983	1,296
France	243	322	380	531
Italy	6	52	53	193
Total import	2,742	4,547	3,859	7,463

Source: Produktschap voor groenten en fruit, The Hague

Passiflora is a large genus of perennial climbing plants, many with edible fruits. The best known is the purple passion fruit, *Passiflora edulis*, var. *edulis*, which thrives in the subtropics and in tropical highlands; e.g. in Kenya at 2,000 m and even higher. A yellow form, *P. edulis*, var. *flavicarpa*, with very aromatic and rather acid juice, is better adapted to the tropical lowlands. Their properties are compared in Table 11.8.

The fruit contains 100–150 seeds and is as big as a tennis ball. The juice contains 1.2 per cent protein, 18 per cent sugars, much vitamin A and fair amounts of vitamins B1 and C. There are no

Table 11.8   *Properties of two varieties of Passiflora edulis*

Variety	edulis	flavicarpa
Plant vigour	medium	high
Yield (tonnes/ha/year)	8–10	10–20
Juice content	medium	high
Acidity	medium	high
Tolerance to cold	yes	no
id. to nematodes	no	yes
id. to woodiness	no	yes
id. to *Fusarium*	no	yes

reliable statistics on production and planted areas but it is apparent that production of passion fruit juice in Brazil is a rapidly growing export industry: exports increased from 1,800 tonnes in 1979 to 2,600 tonnes in 1980 and 4,800 tonnes in 1981 (*Fruits* **37**, 425).

Both varieties are self-compatible but cross pollination gave better results than selfing: more fruits and seeds, higher weight. It took 24 days to reach the maximum size and 88 days to reach maturity in var. *edulis* (Gachanja and Gurnah, 1978), whereas the fruits were mature 51–64 days after the flower had opened in var. *flavicarpa* (Fouqué, A. and R., 1980). Pollination is by wild bees (*Xylopa* sp.) and honey bees; where they are lacking hand pollination is required.

Purple passion fruit needs a moist climate with at least 1,000 mm rain; the yellow variety needs considerably more rainfall. Both need a fertile, well-drained soil with the pH around 6. For every ha of passion fruit 2 ha must be available for rotation, otherwise nematodes and soil-borne fungi will do much damage.

Propagation is by seed, cuttings or grafts. Seedlings are variable but initially virus-free; cuttings are easily made but suffer from soil-borne diseases in the purple variety. Grafts of var. *edulis* are best made on seedlings of the yellow variety. Seed is taken from fully mature fruit and fermented in glass for four days, then washed and dried out of the sun. It is sown in polythene bags, seven seeds per bag of 10 cm diameter and 20 cm depth, containing 1.5 litres of soil. After germination they are thinned to three seedlings per bag. When the seedlings are about 20 cm high, they are taken to the field for planting (GTZ, 1978).

The usual spacing is 4 × 4 m, but GTZ advises 3 × 3 m; perhaps this has to do with the kind of support used. There are three main systems of wiring: vertical (one wire), T (two wires on a short cross-bar) and double T (four wires with double support). In all cases the wire is 2–2.4 m above the ground. Strong poles must be used – teak if available – cut 2.60–3 m long and set 60 cm deep; the end poles must be well anchored. On sloping land the rows follow the contours, while on flat land the rows are in line with the prevailing

Fig. 11.12 Yellow passion fruit growing along wire, Surinam

wind, if possible. As a windbreak for higher altitudes, *Grevillea robusta* may be used.

At 4 × 4 m, 625 poles/ha are needed, excluding the anchors, and at 3 × 3 m more than 1,100 poles are needed. After the bark has been stripped off, the poles are dried and treated with a preservative such as carbolineum. Living poles of *Gliricidia* have been used with good results in Fiji.

The start of the rainy season is the best time for planting. The hole must be big enough to accommodate the three plants in the bag. Superphosphate ($\frac{1}{2}$ kg) is deposited at the bottom of each hole. The plants are wetted and planted after the plastic has been removed. Long stakes are placed near the plants, to allow them to climb up to the wire. The best two are selected for further growth and tied to the stakes. Later on, they may be grafted. When they reach the wire, they are wound several times around it in opposite directions. The third plant is now cut off near the ground and side shoots are also removed. For more vivid information, the reader should look at the drawings in GTZ (1978) or Aubert (1974). An intercrop of vegetables may be grown between the rows for the first year. Melon and other Cucurbitaceae should be avoided, as they carry a virus that is harmful to passion fruit. A cover crop is unsuitable as fallen fruit would be hard to find.

Pruning must be done frequently during the first year, e.g. every two weeks. The object is to have secondary branches hanging down freely. A shoot flowers only once; after its fruit has been gathered, it must be cut back so that new shoots can appear. The shoots have to be disentangled regularly and are cut off at 15 cm above the ground.

Fertilizer practices vary with soil and climate. A mixture of urea, superphosphate and potassium sulphate is given in Sri Lanka: 500 g/plant in the first year in two portions; this is increased to 1,800 g in the fourth year (Bertin, 1976). GTZ recommends 300 g calcium ammonium nitrate per plant in two portions and trace elements every three months. One leaf analysis contained, in dry matter/100 g 3.47 g N, 0.21 g P, 2.36 g K, 1.49 g Ca and 0.22 g Mg (Landb. Proefst., 1974); this is an indication for the proportion of elements to be used in the fertilizer. Irrigation is needed when annual rainfall is less than 1,600 mm for the yellow and 1,000 mm for the purple var., or when a dry period coincides with bloom and fruit setting.

A yield of 60 tonnes/ha is possible in Kenya on T-wires, against 35 tonnes/ha on single wire (Aubert, 1974). Salazar and Torres (1978) achieved a yield of 22.4 tonnes/ha on overhead wires, 10 tonnes/ha on T and 8.5 tonnes/ha on single wire. On the other hand, Anon. (1978), found T-wires and double T-wires less satisfactory than single wire. An average yield is 10–20 tonnes/ha/year.

There are two crops annually in the subtropics, but in the tropics cropping is almost continuous. Fruit for the local market in South Africa is picked when the colour is light purple. Fruit for the factory is picked, when the purple colour is well defined, or is gathered from the ground (Malan, 1953). Yellow passion fruit is never picked, but gathered every other day.

Root nematode is the most important pest and its best remedy is rotation. Woodiness is caused by the cucumber mosaic virus; it is spread by asexual propagation, by aphids and on tools. The symptoms, such as misshapen fruit, disappear in summer (Peasley and Fitzell, 1982). Infested plants must be destroyed immediately. Several fungi, fruit flies, mites and scales also attack passion fruit (Beal and Farlow, 1982).

Passion fruit juice is becoming ever more popular for mixture with other fruit juices. Pectin and an oil with high linolic acid

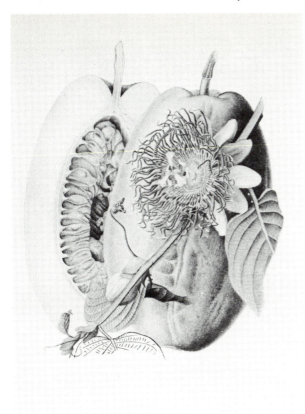

Fig. 11.13 *Passiflora quadrangularis. Source:* Ochse (1931)

content are obtained from seeds and waste peel (Lopez, 1980; Prasad, 1980).

Other cultivated passion fruits are: the giant granadilla, *P. quadrangularis* that is grown extensively in Venezuela, *P. ligularis* and *P. laurifolia*. Fruits from many other *Passiflora* spp. are gathered from plants growing wild. The reader should consult Martin and Nakasone (1970), Akamine *et al.* (1974), Chandra (1976) and Fouqué (1982) for more information.

Fruits of the Asian rain forest zone

Four crops are considered here: mangosteen (*Garcinia mangos-*

Fig. 11.14   Mangosteen tree in Bogor, West Java

tana), durian (*Durio zibethinus*), rambutan (*Nephelium lappaceum*) and pulasan (*Nephelium mutabile*). They all need a humid, equatorial climate with no dry season, or only a short one. Furthermore, they all thrive in different kinds of soils, if moist and rich in organic matter. They also prefer to grow in the shade when young.

Mangosteen (Fig. 11.14) was called 'queen of fruits' by Fairchild, who travelled all over the world to collect new crop plants. The fruit is 4–7 cm in diameter and contains 4–8 fleshy segments. Practically everybody who has tasted it agrees that it is one of the finest fruits in the world. It is grown extensively in its native area, Indonesia and Malaysia, but is difficult to establish elsewhere. Economically speaking, it is not an attractive crop as eight to fifteen years pass before it starts to bear and the yield is low and uncertain. Only female trees are known. However, 'seeds' are formed from nucellar tissue in the parthenocarpic fruits. The seedlings are set out at a spacing of 10 × 10 m but Almeyda and Martin (1976) advise an initial 6 × 6 m for Puerto Rico. Bourdeaut and Moreuil (1970) have reported on mangosteen planted in Ivory Coast. I have seen these trees in 1978; they were growing well, but were still unproductive. A yield of 200–800 fruits/tree in good years is reported by Nat. Acad. (1978) and a yield of 500–1,500 fruits by Almeyda and Martin (1976).

Fig. 11.15 Durian fruit

Opinions on the durian (Fig. 11.15) differ greatly. This is expressed in the following (pseudo-)limerick (*Hortscience 9*, 1973):

The durian – neither Wallace nor Darwin agreed on it.
Darwin said: 'may your worst enemies be forced to feed on it'.
      Wallace cried, 'It's delicious.'
      Darwin replied, 'I'm suspicious,
For the flavour is scented like papaya fermented, after a fruit-eating bat has pee'd on it.'

Darwin never saw or smelt a durian and his supposed lack of appreciation was undoubtedly caused by various sulphur compounds; their odours have been described by Stanton and Howard (1969) as 'garlic, fruity, rancid, pungent and sooty'. The fruit is spiny, some 20 cm wide and usually weighs about 4 kg.

A seedling durian tree can be very tall, but clonal trees are not more than 10 m high (Chua and Chuo, 1981). Most trees are seedlings although it is easy to make cuttings, grafts and buddings. Bagakalie and Anwarudin (1980) report 40–50 per cent success from budding in the dry season and a 60–90 per cent take in the wet season. The take in approach grafts is 100 per cent (H. Sumaryono, personal communication, 1981).

More than one clone must be planted for cross pollination. Trees are set 12 m apart or at 10×10 m to be thinned later to 50/ha (Watson, 1982). The fruit is allowed to ripen on the tree and is gathered after it has fallen, some 16 weeks after bloom. A good orchard yields 10–18 tonnes/ha. Thailand with 470,000 tonnes/year and Indonesia with 160,000 tonnes/year are the largest producers.

A patch canker caused by *Phytophthora* and defoliation by a hawk moth (*Daphnusa*) have been reported (Watson, 1982; Ramasamy, 1980). Chua and Chuo (1981) mention a stem borer, a fruit borer and three fungi as pests in the Singapore area.

The rambutan was said to be monoecious in the first edition of this book because of Chandler's remark: 'Separate male and female flowers are usually found in the same inflorescence but some seed-

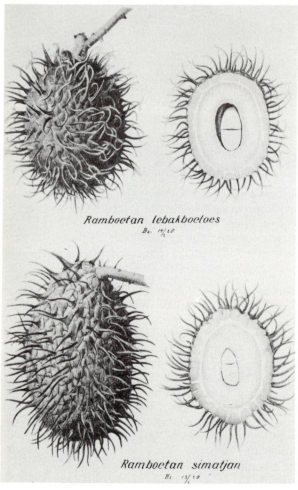

Fig. 11.16   Two rambutan cultivars *Source*: Ochse (1931)

lings may bear female and a few hermaphrodite flowers.' Ochse (1931) claimed that seedling trees may have bisexual or exclusively male flowers, but according to Purseglove (1968) it is a dioecious tree. F. W. Soerodimedjo in Surinam (personal communication) supports Purseglove; he adds that male flowers may have a rudimentary pistil and female flowers may have very small staminodes whereas the style is usually split. It is likely, considering these diverging opinions, that seedlings vary in sex expression but clones are fixed in this respect.

Rambut means hair in Malay or Indonesian and the name rambutan refers to the soft spines surrounding the fruit. It is a bushy and wide-crowned tree; grafted and layered plants are smaller and

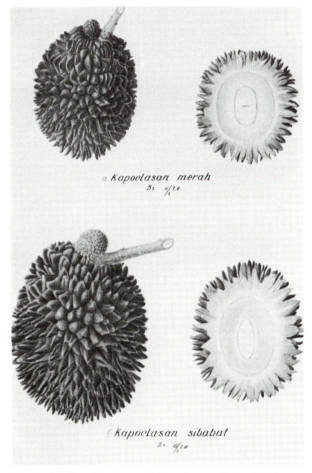

a Kapoelasan merah

b Kapoelasan sibabat

Fig. 11.17 Two kapulasan cultivars *Source*: Ochse (1931)

bear fruit earlier. The fruit is oblong to round, 5–6 cm long and its colour changes from green to red and yellow. The fleshy aril of good cultivars easily parts from the seed; it is translucent, firm and juicy, containing 16 per cent carbohydrate, 0.7 per cent protein and some vitamin B1 and C.

Seedling trees are planted at 12 × 12 m but grafted or layered trees can be set out at 9 m intervals (Whitehead, 1959). Lai Kwong-Kwong (1974) states that cultivation of rambutan and durian can be as profitable as that of rubber and oil palm.

Pulasan, or kapulasan, resembles rambutan but has stouter spines on the fruit and smaller fruit bunches; the aril is yellow. On Java, the best fruits come from an altitude of 200–400 m. Budding is advised and the trees should be planted eight m apart.

For other fruit trees of the rain forest zone, see *Antidesma*, *Baccaurea*, *Bouea*, *Lansium*, *Pourouma* and *Spondias mombin*.

Litchi and longan

Litchi (*Litchi chinensis*) and longan (*Euphoria longana*) are closely related to rambutan taxonomically but differ much from it ecologically. Unlike rambutan, they tolerate cool weather and even need chilling temperatures for flowering; the critical temperature is variously given as 12 °C, 10 °C and 7 °C and apparently depends on the cultivar. They thrive in a climate with a warm, humid summer and at least 1,200 mm rain. These plants, native to south China, feel at home in the subtropics with summer rains and a dry, frostless winter. Longan is hardier and less sensitive to wind than litchi. The best soil for these crops is a deep loam with good drainage (Joubert, 1970).

The litchi tree is evergreen and has a dense growth of compound leaves with two to four pairs of leaflets. The flowers are small and greenish; both male and female flowers are present in the terminal panicles; they are highly dependent on bees for pollination (Rao, 1983). The fruit is 4 cm long, covered by tubercles and red when ripe. A white, translucent, juicy aril with excellent flavour lies over the seed; it contains up to 24 per cent soluble solids and is eaten either fresh or dried.

Early maturing cvs e.g. 'Tai So' (Mauritius) and 'Bengal' (or Brewster) are better suited to warmer areas. Late maturing cvs such as 'No Mai Chee' and 'Wai Chee' need more chilling. 'Haak Yip' is known for its outstanding quality. All cultivars are propagated as marcots from outside branches. Grafting is more difficult but succeeds if the scion is ringed three weeks ahead at 60 cm from the top. Cv. Tai So is a good rootstock.

The trees are set out at an initial spacing of 8 × 8 m

Fig. 11.18 Litchi tree in Israel

(156 trees/ha); when 15 years old, they are thinned out diagonally to reduce the density to 78 trees/ha. The holes should be filled with soil mixed with compost or mulch but no manure is added. Young trees must not be carried by the stem, otherwise roots will be lost (Cull and Paxton, 1983). They are tied to a stake and protected against wind by a hessian or plastic cover and by windbreaks of e.g. *Grevillea robusta*. Litchi roots live in symbiosis with mycorrhiza fungi; it is therefore necessary to inoculate young trees with soil from old trees.

Pruning is needed to train trees into a good shape. Branches break easily if the angle with the trunk is too acute. Fertilizer and irrigation are applied sparingly in winter: three months of dormancy are required prior to flowering. During that period, available soil moisture may fall to 10 per cent but it should be at least 50 per cent for the rest of the year (Menzel, 1983).

The *Dysdercus* bug, which also infests cotton, is a major pest of litchi; other pests are the macadamia nut borer, a mite, fruit flies, scales and nematodes. No major diseases have been described.

The harvest in South Africa takes place 98–106 days after the bloom. The fruit must be fully ripe as it contains no starch, like citrus and grape. Whole clusters are picked. One fruit weighs 21–22 g for litchi and 11–20 g for longan. The aril comprises 75 per cent of the weight, the skin 14 per cent and the seed 11 per cent. Death of the embryo results in 'chicken-tongue' seeds, with bigger and tastier fruits; this occurs in 15 per cent of 'Tai So' fruit.

Northern Thailand is the major producer of longan, the yield is 6 tonnes/ha or 120 kg per tree. The grower receives about 0.50 to

1.10 US dollar per kg, the consumer in Singapore pays 2–4 $ per kg. Much more is paid to the farmer for litchi: $3.50 per kg (*Fruits* **35**, 265).

Sapodilla and other 'sapotes'

Sapodilla, *Manilkara achras,* is also called chiku, nispero and naseberry. The now invalid name *Achras sapota* is often found in literature. It occurs wild in the forests of Central America, where the tree is tapped for its latex; this is concentrated to chicle, the base of chewing gum. The tree is now grown throughout the tropics for its

Fig. 11.19   Two sapodilla cultivars *Source*: Ochse (1931)

delicious fruit or as an ornamental. It is a handsome, slow-growing tree with dense foliage. The round or ovoid fruit is rust-brown in colour, 5–10 cm across and contains up to 12 hard, black seeds. The pulp is brown and very sweet, containing 14 per cent sugar and very little acid.

This crop is at home in the lowland tropics and grows best in drained locations at a pH of 6–7. It is resistant to drought, wind and salt sprays. A 6-2-6 NPK fertilizer is recommended in southern Florida (Campbell *et al.* 1975). Sapodilla requires 1.16 kg N, 0.17 kg P_2O_5, 1.69 kg K_2O, 1.12 kg Ca and 0.14 kg Mg to produce 1,000 kg fresh fruit (Avilan *et al.* 1980); this leads to a 7:1:10:7:1 formula for bearing trees.

Practically no pruning is needed as the branches sit at nearly right angles to the stem. Propagation is usually by seed, but inarching and marcottage are practised in India. As with rubber and other latex-bearing trees, modified Forkert budding has given good results.

There are some named cvs, e.g. 'Prolific'. The island of Tobago is well known for its tasty sapodillas but, as far as I know, this is more a matter of climate and soil than of special cvs. A good tree will yield 3,000 fruits/year or about 300 kg. Larvae of fruit flies may enter the fruit and feed on it, making it unfit for consumption. A rust fungus does some damage too.

Mamey sapote (*Calocarpum sapota*) is in the same family but the fruit is bigger, up to 12 cm, and pear-shaped. A poetical description:

> Or a fruit, *mamey*,
> cased in rough brown peel, the flesh
> rose-amber and the seed:
> the seed a stone of wood, carved and
>
> polished, walnut-colored, formed
> like a brazilnut, but large,
> large enough to fill
> the hungry palm of a hand.

(from 'Pleasures' by Denise Levertov in The new Oxford book of American verse, 1976).

Less lyrical but more exact botanical and horticultural descriptions are found in Almeyda and Martin (1976), Maxwell and Maxwell (1980) and Campbell (1983). To sum up: the leaves are clustered at branch tips in groups of 8–10; below them, on leafless parts, are the flowers. The ovoid fruit weighs 1 to 3 kg and contains one large seed.

The plants are easily killed by frost and poor drainage. A pH of 5.5–7 is preferred and irrigation must be available at all times because all leaves are lost during even a short dry spell. Asexual propagation by Forkert budding, side-veneer graft and air-layering is possible. Initial shade, mulch and a cover crop are beneficial. The trees are planted at 8 × 8 m. The sugarcane root-borer and scales are the prevalent pests; anthracnose also occurs, The fruit is harvested for the market when it begins to redden and for home use when completely red.

Star apple or caimito (*Chrysophyllum cainito*) is a strikingly beautiful large tree, with green or purple round fruits; because of the latex in the skin, the juicy white pulp should be eaten with a spoon. Clonal selections can be air-layered or grafted.

Egg fruit, *Lucuma nervosa* (also known as *Pouteria campechiana*) has egg-shaped, orange yellow fruits with a mealy pulp. These two species also belong to the Sapotaceae family.

However, some fruit trees belonging to quite different families are also called 'sapotes', for no good reason. In terms of clarity it would be preferable to avoid those names, but they seem to have found general acceptance. The white 'sapote' (*Casimiroa edulis*) belongs to the Rutaceae family, has palmately compound leaves and is adapted to the subtropics or tropical highlands; the fruit tastes somewhat similar to an apple. Propagation is best done by air-layers or grafts. The black 'sapotes', *Diospyros ebenaster*, *D. digyna* and perhaps other *Diospyros* spp., have brown fruits with chocolate–brown flesh and many seeds. These are tropical lowland crops.

'Cherries' and 'apples'

Diverse tropical fruits are called 'cherry' or 'apple', without being related to their namesakes of temperate climes. Foremost is the West Indian or Barbados cherry, *Malpighia glabra*; 'acerola' is its common name in Spanish (Fig. 11.20). Its chief claim to fame rests on the incredibly high vitamin C content: 2–3 per cent! One cherry of 20 g, half of which is pulp, thus contains 200–300 mg ascorbic acid, some five times the recommended daily dose. It is a small tree that is easily propagated by cuttings. Making new cultivars is therefore very easy.

The fruit has shallow ridges and contains three triangular seeds. Under favourable conditions, as in Surinam with its warm and humid climate, five blooms and harvests per year are possible, permitting a yield of 60 tonnes/ha. A rosy future was predicted for this crop, but was not realized. It is too perishable to succeed as a fresh fruit and as processed fruit has to face stiff competition from the much cheaper synthetic ascorbic acid (Ostendorf, 1963).

The Surinam cherry or pitanga, *Eugenia uniflora*, is more valuable as an ornamental than as a fruit tree. The cherry is deeply ribbed and the seed is spherical or splits into two half-globes.

Another 'cherry', *Flacourtia jangomas*, is recognizable by enormous doubly branched thorns that fall off after a few years. The fruit

Fig. 11.20 West Indian cherry, fruit and leaves

Fig. 11.21 Fruit of the Surinam cherry

Fig. 11.22   *Flacourtia rukam. Source*: Ochse (1931)

contains several membranous seeds from which either male or female plants arise. To *Flacourtia* also belong: lovi-lovi (*F. inermis*), rukam (*F. rukam*) and governor's plum or ciruela forastera, *F. indica.*

We have already encountered the cashew 'apple', the swollen fruit stalk. Most other 'apples' belong to *Syzygium*, a genus that is closely related to and sometimes included in *Eugenia*. As a rule, *Eugenia* is limited to the New World species, while *Syzygium* comprises the species of the eastern hemisphere. The fruit of *Syzygium malaccense*, pomerac or Malay apple, resembles a pear rather than an apple. The tree is beautiful, especially when the bright red

Fig 307

Fig. 11.23 Flowers of *Syzygium malaccense*

flowers (Fig. 11.23) shed their many stamens on the ground. The fruit is somewhat astringent when eaten raw, but improves much on stewing. The rose apple, *S. jambos*, has rather insipid spherical white fruits with a rose scent; the seed is polyembryonic.

Perhaps the most valuable member of the group is jamun or jambolan, *S. cumini*, with its dark purple fruits from which wine and all kinds of preserves are made. Birds scatter the seeds widely and the tree thus becomes a serious pest in pastures. The seedlings are extremely variable but good cultivars have been selected. Other *Eugenia* spp. with edibile fruits (Brazil) are pitomba, grumichama and lipote.

Fig

Ficus carica is grown almost exclusively in the Mediterranean area; Spain and Italy alone produce two thirds of the world's harvest. Another area of significance is California. Cultivation in the arid and semi-arid tropics is possible. When the crop cannot be dried, it has to be consumed fresh or processed.

Fig is a large shrub with big, deeply-lobed leaves. The inflorescence consists of a cup-like receptacle with only a small opening and numerous unisexual flowers inside. There are several types: in the well-known 'Smyrna' fig all flowers are female and have long pistils.

In caprifig on the other hand, there are female flowers with short pistils at the base and male flowers around the opening.

Pollination is brought about by a wasp, *Blastophaga psenes*, that lives in the caprifig. Consequently, 'Smyrna' figs must be interplanted with about 5 per cent caprifigs. The caprifig blooms three times a year and one of these periods coincides with the bloom of 'Smyrna'. The common figs, such as 'Kadota', and the San Pedro figs do not need caprification as their fruits are formed by parthenocarpy.

The fig needs a certain amount of chilling and can tolerate frost, down to −8 °C. It is grown up to an altitude of 1,200 m. About 700 mm rainfall is required and where less falls, irrigation is necessary. Some dry months, particularly when the fruit is ripening, are also essential. The crop is not exacting as regards soils; they must be well drained and contain some free lime. Fig is one of the most salt and drought-resistant crops.

Propagation is generally by cuttings. Ordinarily a spacing of 8 × 8 m is used, to be increased to 10 or even 15 m square if water is scarce. Close plantings, of 5 × 5 m, have also been tried. A good harvest is 12 tonnes/ha fresh figs. These are soaked for a half minute in boiling salt water, dried for a few hours in the sun and for eight days under shade; at the end of the process the weight is reduced to a little over one third of the fresh weight. More detail on fig growing and breeding can be found in Rehm and Espig (1976), Storey (1974), Rebour (1968) and Condit (1947).

Pome and stone fruits

Most cold-requiring fruit crops are deciduous members of the Rosaceae family. They can grow in tropical lowlands, but never set fruit there. For that, they must have a certain number of hours with temperatures below 7 °C, to replace the dormant period of the temperature climates. This is necessary to break down the growth inhibitors which accumulate in leaves and bud-scales during the season of active growth. With insufficient chilling, the development is hindered by disorders known as prolonged rest and delayed foliation. Thin, long branches with only a few leaves near the top are formed and fruit spurs are lacking.

Close to the equator, at altitudes where temperatures below 7 °C regularly occur, the day temperature is too low for normal growth and fruit ripening. Therefore, an intermediate altitude of about 1,000 m is chosen in combination with cultivars with low chilling requirement. Furthermore, certain cultivation and chemical methods also help to solve the problem: light pruning, bending of branches, spraying with oil/DNOC emulsions and stripping of leaves to break the rest and to stimulate flowering.

In general the chilling requirement increases in the following order: almond, apricot, peach, Japanese plum, pear, apple, European plum and cherry. However, there are also large differences between cultivars of the same species, e.g. from 200 to 1,400 hours in apple.

More than two million apple trees grow in Indonesia. They are on wild rootstock, from 700 to 1,200 m above sea level; 'Rome Beauty' and 'Princess Noble' are the most popular cvs. The leaves are stripped every six months to induce flowering and there are two harvests a year (Kusomo, 1978; Yusuf, 1983). In Queensland, Australia, 'Granny Smith', 'Delicious' and 'Jonathan' take up 90 per cent of the apple area (Ledger, 1979).

More information on these fruits, especially the apple, and names of suitable cultivars can be found in Giesberger (1972), Janick (1974), Ruck (1975) and van Epenhuijsen (1976). A description of apricot growing in the tropics and subtropics has been given by Larue (1976).

The loquat, *Eriobotrya japonica*, is an evergreen Rosacea with a rather low chilling requirement. It grows best in the subtropics, but is also suitable for tropical highlands. Propagation is mainly by budding or grafting (Rivals and Assaf, 1977). Among the non-rosaceous fruits requiring chilling we mention: fig, grape, kiwi, kaki, litchi, longan, olive, pecan, pistachio and walnut.

Cucurbits

Few annual crops are considered in this book. Foremost among them are watermelon and melon; they creep and bear male and female flowers on the same plant. Watermelon, *Citrullus lanatus*, is grown throughout the tropics and subtropics, and also in temperate regions with a warm summer. Its chief requirements are a long dry season and much sunlight. It generally grows on light, permeable soils. There are many cultivars, even seedless ones. World production amounts to about 25 million tonnes, Turkey and China being the largest producers.

Usually six seeds are sown together on mounds 2.5 m apart; the seedlings are thinned out to one or two plants/mound. It takes four to five months to ripen the crop.

Melon, *Cucumis melo*, exists in various forms such as cantaloupe and musk melon. Well known cultivars are 'Honeydew' and 'Ogen'. Arid regions are most suitable. Seed is sown 1.5 m apart and the fruit can be harvested three to four months later. An

Fig. 11.24   Watermelon, fruit and leaves

elegant system for growing (water) melons on salty soil was described by Mondal (1974): pitchers filled with good soil are buried at 3 × 3 m, seed is sown in them and three litres of water/pitcher is given every day.

World production is six to seven million tonnes. There is an extensive literature on cucurbits, of which we only mention Whitaker and Davis (1962).

Miscellaneous fruits

A selection of the remaining fruit crops has been arranged in alphabetical order, according to their botanical names. Climatic preferences and chief methods of propagation are recorded.

Antidesma bunius, bignay, is a dioecious small tree native to South-East Asia. Fruit is produced in clusters of 20 to 40 near branch tips; it is 1 cm in diameter, juicy, sub-acid, with a rather large seed and is eaten fresh or converted to excellent jams and wine. Propagation is by seed, cuttings, air layers or grafting. It is a crop for the humid tropics.

Artocarpus altilis occurs in two forms: seedless, breadfruit and seeded, breadnut. Breadfruit is an important food crop in Polynesia, the West Indies and San Tomé, in the Gulf of Guinea. It is more used as a vegetable than as a fruit. The roasted seeds of breadnut taste like chestnut. This species grows well in humid tropical lowlands; it has large, deeply-lobed leaves and separate

Fig. 11.25   Breadfruit tree, Ivory Coast

male and female inflorescences. The compound fruit is formed
by the whole inflorescence and may be up to 30 cm in size
(Fig. 11.25). Propagation is by root cuttings.
Artocarpus heterophylla, jackfruit or nangka, is one of the most
useful trees (Thomas, 1980), especially in South-East Asia. The
fruit may weigh up to 30 kg (Fig. 11.26) and a good tree may
bear up to 200 fruits/year. Air-layering is the common way of
propagation but it is also grafted on seedling rootstock. Seedlings
do not withstand transplantation. The pests and diseases have
been described by Butani (1978).
Averrhoa bilimbi and *A. carambola* need shade. The seeds are
covered by a fatty layer; washing them with soap improves
germination. Bilimbi fruit contains up to 6 per cent oxalic acid
and is pickled, while carambola fruit contains 1 per cent oxalic
acid or less. Bilimbi is round, like a gherkin, and carambola or
star fruit is 5-angled (Fig. 11.27). Both belong in the humid
tropics and are propagated by seed, but carambola can be
extended to the frost-free subtropics and is also propagated by
budding, grafting or layers; its fruit, of which sweet and acid
forms exist, is used for salads and drinks.

Fig. 11.26   Jackfruit, for sale in Indonesia

Baccaurea motleyana, rambai, typical of Malaysia, and *B. racemosa*, menteng, grown throughout Indonesia, are monoecious plants flowering on older wood; they are cultivated from sea level to 500 m altitude. The fruit is small, 2–3 cm, and well-liked by the local people, but too perishable for export.

Blighia sapida, akee, is popular in Jamaica. The aril is edible when fully ripe; it resembles scrambled eggs when fried. However, when the aril is eaten in an unripe or overripe state, it is poisonous. Akee grows well in the humid tropics and is propagated by seed and budding.

Bouea macrophylla, gandaria, is well-liked in West-Java. The fruit is 3 × 5 cm large and has one seed. Cultivars with very sweet and very sour flesh exist. They are mainly propagated by air-layering and grow well in the humid tropics.

Carissa grandiflora, natal plum, is used as an ornamental shrub and for hedge planting; the branches are very spiny. The plum-like fruit is small and contains many seeds, it is eaten fresh or in salads. Lower branches are layered, marcotts and cuttings are also used. *C. carandas* is used for fruit drinks and jellies in India but is too acid for eating fresh. Both crops do well in the warm subtropics and subhumid tropics.

Chrysobálanus icaco is native to northern South America. It is a shrub or small tree with plum-like white to yellow, sometimes red-tinged fruits (Ostendorf, 1962). It is canned in Venezuela.

Clausena lansium, wampi, is native to South China and Indo-China. The small fruits are eaten fresh or used for marmalade. Propa-

Fig. 11.27   Flowers, leaves and a fruit of carambola

Fig. 11.28   *Bouea macrophylla*, gandaria. *Source*: Ochse (1931)

gation is usually by seed but grafting is possible. It grows well in the subhumid tropics.

Coccoloba uvifera, sea grape, grows on sandy sea shores of the humid tropics; it is propagated by seeds and cuttings.

Cyphomandra betacea, tree tomato or tamarillo, is eaten raw or stewed. It is typical of tropical highlands, e.g. the Andes region, and is propagated by seeds and cuttings.

Diospyros kaki, persimmon or kaki, is an important crop in Japan and China and is becoming prominent in Mediterranean countries. It can also be grown in tropical highlands, e.g. on Java

Fig. 11.29 Persimmon, *Diospyros kaki. Source*: Ochse (1931)

(Terra, 1936). The tree is deciduous, dioecious and resistant to cold and drought. Propagation is by root suckers and by grafting on *D. lotus*. There are astringent and non-astringent cultivars, with many or no seeds and in various sizes and shapes. 'Hana-fuyu', e.g., has non-astringent reddish-orange fruits, with excellent flavour (Maxwell and Maxwell, 1980). *D. discolor* (velvet apple) and *D. ebenaster* (black sapote) are evergreen and much less resistant to cold and drought.

Dovyalis caffra, kei-apple and *D. hebecarpa*, kitembilla, are spiny shrubs well suited for hedge planting. The acid fruit is used for jellies and preserves. *D. abyssinica* is eaten fresh. All three are dioecious and are propagated by seed, grafting, budding, layers and cuttings. They do well in the subtropics and tropical highlands. A hybrid, *D. abyssinica* × *D. hebecarpa*, is called 'tropical apricot.'

Fragaria ananassa and hybrids, strawberry, is a very important creeping herbaceous perennial in temperate climates. Bloom occurs in short days and is practically continuous in tropical mountain areas; on the other hand, the runners (creeping stems) are only formed in long days. Rotation is necessary, but tomato and its relatives must be avoided. The planting density is around 20,000/ha. It is successfully grown in temperate, subtropical and tropical highland climates; propagation is by runners.

Grewia asiatica, phalsa, is a drought resistant crop of India. It can be grown in the subtropics to arid tropics and is propagated by seed or layers.

Inga spp. are leguminous shade trees with big pods, containing a sweet edible pulp around the seeds. They are grown in the humid tropics ranging to the highlands; propagation is by seed.

Lansium domesticum takes two forms: duku and langsat. Langsat is a slender tree, duku is shorter and more spreading. The racemes grow on the stem; langsat fruit contains latex in the skin. The flesh separates into five sub-acid segments. A fruit weighs 20 g and contains 13–17 per cent sugars. Propagation is by seed, which is true to type, or by budding. Young plants are grown under *Leucaena* shade for six years and are set out at 5 × 6 m (Ochse, 1931; Watson, 1983). *Lansium* is at home in the same climate as rambutan but is more tolerant to lower temperature, humidity and rainfall.

Mammea americana, mamey, is a large tree of the West Indies with big fruit. It is cultivated in South and Central America and is propagated by seed.

Melicocca bijuga, genip, is called mamoncillo in Cuba. It is a big, spreading tree, native to South and Central America. The berry contains one rather big seed, surrounded by a sweet and sub-acid pulp. The thin green skin is easily removed. It grows well in the humid tropics and is propagated by seed or grafting.

Monstera deliciosa, ceriman, has compound cone-like fruit containing needle-sharp crystals when not fully ripe. It tastes like a cross between pineapple and banana. This crop belongs to the humid tropics and is multiplied by cuttings.

Myrciaria cauliflora, jaboticaba, is very popular in Brazil. The fruit contains much vitamin C. It can be grown in the humid tropics to subtropics and is propagated by cuttings.

Olea europaea, olive, is mostly grown for oil production but special cultivars with big fruit for pickling have been selected. Total

Fig. 11.30 Duku, *Lansium domesticum*

exports of table olives amounted to 160,000 tonnes in 1970–71, half of which came from Spain (Loussert and Brousse, 1978). This crop has a high chilling and heat requirement and is therefore limited to subtropical areas with winter rains; it is propagated by cuttings, grafting and seed.

Opuntia ficus-indica, prickly pear, is a cactus. Spineless cultivars have been selected. It is grown in the subtropics with winter rains and in the arid tropics; propagation is by cuttings.

Pandanus odoratissimus and other *Pandanus* spp. are native to Micronesia. They can also be grown in coastal belts of the humid tropics and are propagated by cuttings.

Phyllanthus emblica, aonla or myrobolan, a monoecious tree of

India, has very acid fruit containing 1.5 per cent vitamin C. It grows well in the humid to semi-arid tropics and is propagated by seed, budding and root sprouts. *P. acidus*, Otaheite gooseberry, also has acid fruits, but with little vitamin C.

Physalis peruviana, cape gooseberry (a doubly misleading name) is an herbaceous plant. The fruit is covered by the calyx and contains much vitamin A. It is native to tropical highlands, but will also perform well in a subtropical climate; propagation is by seed.

Pourouma cecropiaefolia, uvilla, is recommended as a home garden crop for the humid tropics (Nat. Acad. 1978). It has large racemes of purple, grape-like fruits. The fruit of *P. mollis* in Surinam has an inedible skin which, however, peels off easily.

Punica granatum, pomegranate, is a deciduous ornamental shrub with beautiful orange to red flowers. The fruit is up to 10 cm large, spherical, with a leathery skin; the sub-acid pulp contains many seeds. The fruit is eaten fresh and used for decorative purposes (LaRue, 1969) or is medicinally used (Ostendorf, 1962). The tree is propagated by hardwood cuttings or by root suckers.

Rubus spp., raspberry, is characterized by canes growing from underground parts that set fruit in the second year and then die. Two cultivars of low chilling requirement have been bred in Florida; they are worthy of trial in other areas (Sherman and Sharpe, 1971). Raspberry can be grown in tropical highlands and is multiplied by suckers, cuttings, layers or seed.

Solanum quitoense, naranjilla, is a large shrub with yellow-orange fruit covered with white hairs (they are easily removed). It grows best on fertile well-drained slopes of humid valleys at an elevation of 1,500 near the equator. It is just as susceptible to nematodes, viruses and fungi as the tomato and must be rotated every two years. *S. macranthum* can be used as a nematode-resistant rootstock (Nat. Acad. 1978).

Solanum topiro, cocona and *S. muricatum*, pepino dulce, grow at lower altitudes and have bigger fruit. Pepino dulce is said to taste like a mixture of pear and muskmelon. All three *Solanum* spp. are native to the Andes region and are multiplied by seed, cuttings and grafting.

Spondias cytherea, Otaheite apple or pomme de cythère, is a small tree with horizontal branches that suffers much from *Phytophthora* in humid climates. The seed is enclosed in a stone with woody fibres, which makes it difficult to remove the rather acid pulp. The tree is propagated by cuttings or seed and is at home in the semi-arid tropics and frost-free subtropics.

Spondias mombin, yellow mombin or hog plum, is a very large forest tree of the American tropics. The fruit measures 2.5 × 4 cm at maturity and contains a woody seed surrounded by a

Fig. 11.31 Pomme de cythère, *Spondias cytherea*. *Source*: Ochse (1931)

juicy sour pulp. A tree may produce 200 kg fruit, which rots rapidly after it has fallen; it should be gathered daily. The juice content is 50–55 per cent. The fruit is eaten fresh or is converted into jam, juice or sorbet (Joas, 1982). Propagation is by seed.

Spondias purpurea, red mombin, ranges from the humid tropics to the frost-free subtropical zone and is propagated by cuttings and seed.

Spondias tuberosa, umbú, is a tree of semi-arid North-East Brazil with refreshing yellow-green fruits.

Tamarindus indica, tamarind, is a large and very beautiful tree. It is often planted in lanes, 15–20 m apart. From fruit-set to harvest takes 245 days (Hernandez-Unson, 1982) and a tree may yield 200 kg of fruit. A pod weighs 10–15 g, 40 per cent of which is pulp. India, the largest producer, harvests 250,000 tonnes/year (Jansen, 1981). A bibliography has been compiled by Lefèvre (1971).

Zizyphus jujuba, Chinese jujube, is grown in temperate to subtropical climates and is propagated by seed and grafting.

Zizyphus mauritiana, Indian jujube, is adapted to subtropical and arid tropical conditions; it is multiplied by seed and grafting.

The arid tropics are perhaps under-represented in the list above. It stands to reason, however, that most crops can be grown there under irrigation. The reader who desires more information on these fruits, and possibly on others not mentioned here, should consult any or all of the following books: Chandler (1958), Kennard and Winters (1960), Ochse *et al.* (1961), Mowry *et al.* (1967), Rebour (1968), Purseglove (1968–72) and Fouqué (1977). And, if he can lay his hands on it – for it is out of print – Bailey's *Standard Cyclopedia of Horticulture* in six or three volumes (1914, reprinted 1925 and 1944). After all these years this is still one of the richest sources of horticultural knowledge. Finally we have to mention once again Popenoe's book of 1920, recently published in facsimile, Barrett (1928) and Ochse (1931).

If the reader of this book has become interested in establishing a fruit industry in his region, he may like to have more information on how to go about it. He must resist an inclination to do it alone. The closest possible cooperation is needed between three specialists, to wit:

1. *A horticulturist* who surveys existing fruit crops and cultivars, assembles data on climate and soils, availability of water, nursery facilities etc. An experimental fruit farm, with seedlings as well as cvs, should be started at an early stage. Plant quarantine measures must be maintained.
2. *A technologist* who, after trials in a small experimental plant, can give his opinion on the suitability of diverse fruits and their cvs for processing.
3. *An economist* who advises on choice of products, marketing, shipping, financing, labour etc.

For more particulars the reader is referred to Samson (1963), Samson *et al.* (1964) and Aylward (1969).

References

Agnoloni, M. and **Giuliani, F.** (1977) *Cashew cultivation*, Florence.
Akamine, E. K. *et al.* (1974) *Passion fruit culture in Hawaii*, Coop. Ext. Service, Univ. Hawaii, circular 345.

Almeyda, A. N. and **Martin, F. W.** (1976) *Cultivation of neglected tropical fruits with promise*: 1. *The mangosteen*; 2. *Mamey sapote*, ARS, USDA.

Anon. (1978) diverse fruits in: 'Inventaire et état actuel' etc., *Fruits* **33**, 610–15 (passion fruit), 627–8 (guava), 628–30 (cashew), 630–2 (date) and 634–6 (grape).

Anon. (1981) *Achievement report* July 1976–June 1979, Hawaii Agric. Exp. Sta., Misc. Publ. 184.

Aubert, B. (1972) 'Viticulture en region tropicale pour la production de raisins de table. Aspects et possibilités, *Fruits* **27**, 513–37.

Aubert, B. (1974) 'La culture de la grenadille au Kenya', *Fruits* **29**, 323–8.

Aubert, B. (1975) 'Précocité de production de la grenadille violette, *Passiflora edulis*, à la Réunion', *Fruits* **30**, 535–40.

Avilan-R., L. *et al.* (1980) 'Absorción de nutrimentos por una cosecha de nispero', *Agronomia tropical* **30**, 7–16.

Aylward, F. (1969) 'Requirements for the establishment of a tropical fruit industry', *Proc. Conf. trop. and subtr. fruits*, London, TPI, 15–22.

Azam, B. *et al.* (1981) 'Le feijoa en Nouvelle Zélande', *Fruits* **36**, 361–84.

Bagakalie, M. and **Anwarudin, J.** (1980) 'The vegetative propagation of durian', *Bull, Pen. Hortik. Indonesia* **8**(9), 37–42.

Bammi, R. K. and **Randhawa, G. S.** (1968) 'Viticulture in the tropical region of India', *Vitis* **7**, 124–9.

Barrett, O. W. (1928) *The tropical crops*, Macmillan, NY.

Beal, P. R. and **Farlow, P. J.** (1982) 'Passion fruit', *Austr. Hortic.* **80**(2), 57–65.

Bertin, Y. (1976) 'La culture de la grenadille au Sri Lanka', *Fruits* **31**, 171–6.

Beutel, J. A. *et al.* (1976) 'A new crop for California: kiwifruit', *Calif. Agric.* **30**(10), 5–7.

Bharath, S. (1969) *Producing guavas for the factory*, Min. ALF Crop Bull. 13, Trinidad.

Blaak, G. (1980) 'Vegetative propagation of pejibaye', *Turrialba* **30**, 258–61.

Bourdeaut, J. and **Moreuil, C.** (1970) 'Le mangoustanier, ses possibilités de culture en Côte d'Ivoire et à Madagascar', *Fruits* **25**, 223–45.

Brückmann, J. H. (1983) 'Salakcultuur en handel in Noord Jogjakarta', *Landbouw* (Bogor) **14**, 436–45.

van Brussel, E. W. and **van Vreden, G.** (1968) 'Studies on the biology, damage and control of the guava fruit fly *Anastrepha striata* Schiner in Surinam', *Surin. Landb.* **16**, 110–22.

Butani, D. K. (1977) 'Insect pests of guava in India and their control', *Fruits* **32**, 61–6.

Butani, D. K. (1978) 'Pests and diseases of jackfruit in India and their control', *Fruits* **33**, 351–7.

Campbell, C. W. *et al.* (1975) *The sapodilla*, Florida Fruit Crops Fact Sheet FC-1.

Campbell, C. W. (1983) 'The mamey sapote, a new tropical fruit crop in Florida', *Hortsci.* **18**, 581.

Chadha, K. L. *et al.* (ed. 1977) *Viticulture in the tropics*, Bangalore.

Chandler, W. H. (1958) *Evergreen orchards*, Philadelphia.

Chandra, S. (1976) 'A review of recent research on the yellow passion fruit', *Fiji Agric. J.* **38**, 41–8.

Chin, H. F. and **Young, H. S.** (1981) *Malaysian fruits in colour*, Trop. Press, Kuala Lumpur.

Chua, S. E. and **Chuo, S. K.** (1981) *A guide to tropical fruit tree cultivation*, Agric. Handbk. 5, Singapore.

Condit, I. J. (1947) *The fig*, Waltham, Ma.

Cull, B. W. and **Paxton, B. F.** (1983) 'Growing the lychee in Queensland', *Qld. Agric. J.* **109**, 53–9.

van Dinther, J. B. M. (1960) *Insect pests of cultivated plants in Surinam*, Landb. Proefst. Bull. 76.

Djerbi, M. (1983) *Diseases of the date palm*, Region. proj. palm and dates Res. Centre, Baghdad.

van Eijnatten, C. L. M. and **Abubaker, A. S.** (1983) 'New cultivation techniques for cashew', *Neth, J. agric. Sci.* **31**, 13–25.

El Baradi, T. A. (1968) 'Date growing', *Trop. Abstr.* **23**, 473–9.

El Baradi, T. A. (1975) 'Guava', *Abstr. on Trop. Agr.* **1**(3), 9–16.

van Epenhuijsen, C. W. (1976) *Deciduous fruits in Tanzania*, Gorssel.

Fouqué, A. (1977) *Espèces fruitières d'Amérique tropicale*, IRFA, Paris.

Fouqué, A. (1982) 'Quelques passiflores de Guyane', *Fruits* **37**, 599–608.

Fouqué, A. and **R.** (1980) 'Quelques notes sur la grenadille jaune', *Fruits* **35**, 309–12.

Gachanja, S. P. and **Gurnah, A. M.** (1978) 'Flowering and fruiting of purple passion fruit at Thika', *E. Afr. Agric. J.* **44**, 47–51.

Garner, R. J. *et al.* (1976) *The propagation of tropical fruit trees*, FAO/CAB, Farnham Royal.

Giesberger, G. (1972) 'Climate problems in growing deciduous fruit trees in the tropics and subtropics', *Trop. Abstr.* **27**, 1–8.

Greshoff, M. (1894) *Nuttige Indische planten*, Kol. Mus. Amsterdam.

GTZ (1978) *Passion fruit growing in Kenya; a recommendation for smallholders*, Eschborn.

Hamilton, R. A. and **Fukunaga, E. T.** (1959) *Growing macadamia nuts in Hawaii*, Hawaii Agric. Exp. Sta. Bull. 121.

Hamilton, R. A. and **Ito, P. J.** (1982) 'Macadamia nut' in: *Crop improvement in Hawaii* (ed. J. L. Brewbaker), Univ. Haw. Misc. Publ. 180.

Hernandez-Unson, H. Y. and **Lakshminarayana, S.** (1982) 'Developmental physiology of tamarind fruit', *Hortsci.* **17**, 938–40.

Hudson, J. P. (1982) 'The kiwifruit', *Span* **25**, 28–30.

Intini, M. and **Sijaona, M. E. R.** (1983) 'Calendar of disease control with reference to phenological phases of cashew in Tanzania', *Rivista di agric. subtrop. e trop.* **77**, 419–22.

Ironside, D. A. (1982) 'Macadamia nutborer and the new macadamia orchard', *Qld. Agric. J.* **108**, 263–5.

Janick, J. (1974) 'The apple in Java', *Hortsci.* **9**, 13–15.

Jansen, P. C. M. (1981) *Spices, condiments and medicinal plants in Ethiopia, their taxonomy and agricultural significance*, PUDOC, Wageningen.

Jaynes, R. A. (ed., 1969) *Handbook of North American nut trees*, Northern Nut Growers Assoc., Knoxville, Tenn.

Joas, J. (1982) 'Les mombins', *Fruits* **37**, 727–9.

Joubert, A. J. (1970) 'The litchi', *S. Afr. Bull.* 389.

Kennard, W. C. and **Winters, H. F.** (1960) *Some fruits and nuts for the tropics*, USDA Misc. Publ. 801.

Kheill, A and **Kellal, A.** (1980) 'Possibilités de culture et délimitation des zones à vocation pistachier en Algérie', *Fruits* **35**, 177–85.

Kusomo, S. (1978) 'More than 2 million apple trees grow in Indonesia', *Indon. Res. and Developm. J.* **1 & 2**, 11–12.

Lai Kwong-Kwong, A. (1974) 'The economics of establishing a 500 acre fruit orchard in Peninsular Malaysia', *Mal. Agric. J.* **49**, 421–32.

Landb. Proefst. Surinam (1974) *Annual report for 1971*, Paramaribo.

LaRue, J. H. (1969) *Growing pomegranates in California*, UC AXT-305.

Larue, M. (1976) 'La culture de l'abricotier outre-mer', *Fruits* **31**, 157–70

Lefèvre, J. C. (1971) 'Revue de la littérature sur le tamarinier', *Fruits* **26**, 687–95.

Ledger, S. N. (1979) 'Apple growing in Queensland', *Qld. Agric. J.* **105**, 241–7.

Little, E. L. and **Wadsworth, F. H.** (1964) *Common trees of Puerto Rico and the Virgin Islands*, USDA Agr. Handbk. 249, Wash. DC.

Lopez, A. S. (1980) 'Lipids from the seeds of passion fruit', *Revista Theobroma* **10**(1), 47–50.

Loussert, R. and **Brousse, G.** (1978) *L'olivier*, Maisonrouge et Larose, Paris.

Malan, E. F. (1953) *The production of granadillas*, S. Afr. repr. 91.

Martin, F. W. (1976) 'Introduction and evaluation of new fruits in Puerto Rico', *Acta Hortic.* **57**, 105–10.

Martin, F. W. and **Nakasone, H. Y.** (1970) 'The edible species of *Passiflora*', *Econ. Botany* **24**, 333–43.

Maxwell, L. S. and **B. M.** (1980) *Florida fruit*, Tampa.

Menninger, E. A. (1977) *Edible nuts of the world*, Stuart, Florida.

Menzel, C. M. (1983) 'The control of floral initiation in lychee: a review', *Scientia Hortic.* **21**, 201–15.

Mohammad, S. *et al.* (1984) 'Guava orchard meadow', *Trop. Agric.* (Trin.) **61**, 297–301.

Mondal, R. C. (1974) 'Farming with a pitcher', *World Crops* **26**, 94–7.

Morton, J. F. (1961) 'The cashew's brighter future', *Econ. Bot.* **15**, 57–78.

Morton, J. F. and **Venning, F. D.** (1969) 'Eviten fracasos y pérdidas en el cultivo del marañon', *Proc. Trop. region ASHS* **13**, 235–50.

Moutounet, S. E. *et al.* (1977) 'Étude de l'enracinement de quelques arbres fruitiers sur sol ferrallitique brun profond', *Fruits* **32**, 321–33.

Mowry, H. *et al.* (1967) *Miscellaneous tropical and subtropical Florida fruits*, Univ. Florida Bull. 156A.

Müller, C. H. (1981) *Castanho-do-Brasil; estudos agronomicos*, Embrapa, Belem.

Munier, P. (1973) *Le palmier dattier*, Paris.

Nair, M. K. *et al.* (1979) *Cashew*, Kasaragod, India.

Nakasone, H. Y. *et al.* (1976) *Fruit and yield evaluation of ten clones of guava*, Hawaii Agric. Exp. Sta. Rpt. 218.

Nakasone, H. Y. (1982) 'Fruit crops' in : *Crop improvement in Hawaii*, (ed J. L. Brewbaker), Un. Hawaii Misc. Publ. 180.

Nagy, S. *et al.* (1980) *Tropical and subtropical fruits; composition, properties and uses*, AVI, Westport (Conn.).

National Acad. of Sciences (1978) *Underexploited tropical plants with promising economic value*, Wash. DC.

Nihoul, E. (1976) 'Le yang-tao', *Fruits* **31**, 97–109.

Nixon, R. W. and **Carpenter, J. B.** (1978) *Growing dates in the United States*, Agric. Inf. Bull. 207, USDA, Wash. DC.

Ochse, J. J. and **Bakhuizen van den Brink, R. C.** (1931) *Vruchten en vruchtenteelt in Nederlandsch-Oost-Indië*, Batavia.

Ochse, J. J. *et al.* (1961) *Tropical and subtropical agriculture*, Macmillan.

Ohler, J. G. (1966) 'Cashew nut processing', *Trop. Abstr.* **21**, 549–54.

Ohler, J. G. (1967) 'Cashew growing', *Trop. Abstr.* **22**, 1–9.

Ohler, J. G. (1979) *Cashew*, Comm. **71**, Royal Inst. Tropics, Amsterdam.

Ostendorf, F. W. (1962) *Nuttige planten en sierplanten in Suriname*, Landb. Proefst. Bull. 79.

Ostendorf, F. W. (1963) 'The West Indian cherry', *Trop. Abstr.* **18**, 145–50.

Pansiot, F. P. and **Libert, J. K.** (1971) 'Culture de la vigne en pays tropicaux', *Bull. OIVV* **44**(485–6) 595–661.

Peasley, D. and **Fitzell, R.** (1982) 'Virus-free scionwood for passionfruit', *Austral. Hortic.* **80**(5) **109**, 112–13.

Pegg, K. G. and **Forsberg, L. I.** (1982) 'Avocado root rot', *Qld. Agric. J.* **108**, 162–8.

Polprasid, P. (1982) 'Genetic resources of *Salacca* palms in Thailand', IBPGR **6**(4), 11.

Popenoe, W. (1920) *Manual of tropical and subtropical fruits*, Hafner.

Prasad, J. (1980) 'Pectin and oil from passion fruit waste', *Fiji Agric. J.* **42**, 45–8.

Purseglove, J. W. (1968–72) *Tropical crops*, Longman.

Ramasamy, S. (1980) 'An outbreak and some aspects of the biology of the hawk moth *Daphnusa ocellaris* Walk. on durian in Perak', *Malaysian Agric. J.* **52**, 213–18.

Rao, G. M. (1983) 'Litchi and bee-keeping', *Indian Hortic.* **28**(2) 19–20.

Rebour, H. (1968) *Fruits méditerranéens autres que les agrumes*, Paris.
Rehm, S. and **Espig, G.** (1976) *Die Kulturpflanzen der Tropen und Subtropen*, Ulmer.
Rivals, P. and **Assaf, R.** (1977) 'Modalités de croissance et système de reproduction du néflier de Japon (*Eriobotrya*), *Fruits* **32**, 105–13, 237–51.
Ruck, H. C. (1975) *Deciduous fruit tree cultivars for tropical and subtropical regions*, Commonw. Agr. Bur., Farnham Royal.
Salazar, C. R. and **Torres, M. R.** (1978) 'Determinación de soportes para maracuyá', *Rev. Inst. Colomb. Agropec.* **13**, 281–9.
Sale, P. R. (1978) 'The commercial production of kiwifruit in the Bay of Plenty', *Cal. Avocado Soc. Yearbk.* **62**, 95–9.
Sale, P. R. (1980) 'The New Zealand avocado industry', *Cal. Avocado Soc. Yearbk.* **64**, 71–4.
Samson, J. A. (1963) 'Remarks on new fruit crops for Surinam', *FAO Conf. on fruit propagation*, Jamaica.
Samson, J. A. *et al.* (1964) 'Nieuwe vruchtencultures in Suriname', *Surin. Landb.* **12**, 62–7.
Scott, F. S. and **Marutani, H. K.** (1982) *Economic viability of small macadamia nut farms in Kona*, HITAHR, Hawaii, Res. ser. 009.
Shalitin, G. (1973) *New approaches to grape-growing in the tropics; grape-vine training and pruning studies in Kenya*, Nairobi. Also published in French: *Fruits* **29**, 375–83.
Sherman, W. B. and **Sharpe, R. H.** (1971) 'Breeding *Rubus* for warm climates', *Hortsci.* **6**, 147–9
Shigeura, G. T. and **Bullock, R. M.** (1983) *Guava in Hawaii – history and production*, Res. Ext. Series 035, Univ. Hawaii.
Shigeura, G. T. and **Ooka, H.** (1984) *Macadamia nuts in Hawaii: history and production*, HITAHR (Un. Haw.) res. extension series 039.
Shoemaker, J. S. (1975) *Small fruit culture*, AVI.
Singh, S. *et al.* (1963) *Fruit culture in India*. ICAR, New Delhi.
Sommer, N. F. *et al.* (1983) 'Minimizing postharvest diseases of kiwi fruit', *Cal. Agric.* **37**(1 + 2) 16–18.
Stanton, W. R. and **Howard, G. E.** (1969) 'Fruits of South-East Asia', *Proc. Conf. trop. and subtrop. fruits*, London, 237–44.
Storey, W. B. (1974) 'Figs' in *Advances in fruit breeding* (eds Janick and Moore), Purdue.
Terra, G. J. A. (1936) 'De stand van de cultuur van kesemek (*Diospyros kaki*) in het regentschap Garoet', *Landbouw* (Bogor) **11**, 326–36.
Thomas, C. A. (1980) 'Jackfruit as source of food and income', *Econ. Botany* **34**, 154–9.
Trochoulias, T. and **Lahav, E.** (1983) 'The effect of temperature on growth and dry-matter production of macadamia', *Scientia Hortic.* **19**, 167–76.
Watson, B. (1982) 'The durian', *Austral. Hortic.* **80**(7), 109.
Watson, B (1983) 'Langsat and duku (*Lansium domesticum*)', *Austral. Hortic.* **81**(3), 99–104.
Whiley, A. W. and **Saranah, J. B.** (1984) 'Kiwifruit: a new fruit crop for Queensland', *Qld. Agric. J.* **110**, 167–76.
Whitaker, T. W. and **Davis, G. N.** (1962) *Cucurbits*, Leonard Hill.
Whitehead, C. (1959) 'The rambutan, a description of the characteristics and potential of the more important varieties', *Malayan Agric. J.* **42**, 53–75.
Wills, R. H. H. *et al.* (1981) *Postharvest*, Granada Publ. Ltd.
Winkler, A. J. (1962) *General viticulture*, Berkeley.
Youssef, J. and **Bergamini, A.** (1981) *L'actinidia (kiwi – yangtao), sa culture*, Paris.
Yusuf, R. (1983) 'Fruit production in Indonesia', *Chronica Hortic.* **23**(1) 4–5.

Appendices

Appendix 1 – List of families and genera of fruit crops

Dicotyledons

Actinidiaceae	*Actinidia*
Anacardiaceae	*Anacardium, Bouea, Mangifera, Pistacia, Spondias*
Annonaceae	*Annona, Rollinia*
Apocynaceae	*Carissa*
Betulaceae	*Corylus*
Bombacaceae	*Durio*
Cactaceae	*Opuntia*
Caricaceae	*Carica*
Caryocaraceae	*Caryocar*
Cucurbitaceae	*Citrullus, Cucumis*
Ebenaceae	*Diospyros*
Euphorbiaceae	*Antidesma, Baccaurea, Phyllanthus*
Flacourtiaceae	*Dovyalis, Flacourtia*
Grossulariaceae	*Ribes*
Guttiferae	*Garcinia, Mammea*
Juglandaceae	*Carya*
Lauraceae	*Persea*
Lecythidaceae	*Bertholletia, Lecythis*
Leguminosae	*Inga, Tamarindus*
Malpighiaceae	*Malpighia*
Meliaceae	*Lansium*
Moraceae	*Artocarpus, Ficus, Pourouma*
Myrtaceae	*Acca, Eugenia, Myrciaria, Psidium, Syzygium*
Oleaceae	*Olea*
Oxalidaceae	*Averrhoa*
Passifloraceae	*Passiflora*
Polygonaceae	*Coccoloba*

Proteaceae	*Macadamia*
Punicaceae	*Punica*
Rhamnaceae	*Zizyphus*
Rosaceae	*Chrysobalanus, Eriobotrya, Fragaria, Malus, Prunus, Pyrus, Rubus*
Rutaceae	*Casimiroa, Citrus, Clausena, Fortunella, Poncirus*
Sapindaceae	*Blighia, Euphoria, Litchi, Melicocca, Nephelium*
Sapotaceae	*Calocarpum, Chrysophyllum, Lucuma, Manilkara*
Solanaceae	*Cyphomandra, Physalis, Solanum*
Tiliaceae	*Grewia*
Vitaceae	*Vitis*

Monocotyledons

Araceae	*Monstera*
Bromeliaceae	*Ananas*
Musaceae	*Musa*
Palmae	*Astrocaryum, Borassus, Guilielma, Hyphaene, Maximiliana, Phoenix, Salacca*
Pandanaceae	*Pandanus*

Appendix 2 – Common names of fruit crops and their botanical equivalents

acerola	*Malpighia glabra* L.
African fan palm	*Borassus aethiopum* Mart.
akee	*Blighia sapida* Koenig
almond	*Prunus amygdalus* Batsch.
aonla	*Phyllanthus emblica* L.
apple	*Malus sylvestris* Mill.
apricot	*Prunus armeniaca* L.
atemoya	*Annona* hybrid
avocado	*Persea americana* Mill.
bael	*Aegle marmelos* (L.) Correa
banana	*Musa* cvs
Barbados cherry	*Malpighia glabra* L.
bergamot	*Citrus aurantium* L., subsp. *bergamia*
bignay	*Antidesma bunius* (L.) Spreng.
bilimbi	*Averrhoa bilimbi* L.
biribá	*Rollinia pulchrinervis* A.DC.
black sapote	*Diospyros ebenaster* Retz.
Brazilnut	*Bertholletia excelsa* HB

breadfruit, breadnut	*Artocarpus altilis* (Park) Fosberg
bullock's heart	*Annona reticulata* L.
cantaloupe	*Cucumis melo* L.
Cape gooseberry	*Physalis peruviana* L.
caprifig	*Ficus carica* L.
carambola	*Averrhoa carambola* L.
cashew nut	*Anacardium occidentale* L.
ceriman	*Monstera deliciosa* Liebm.
cherimoya	*Annona cherimolia* Mill.
Chinese gooseberry	*Actinidia chinensis* Planch.
ciruela forastera	*Flacourtia indica* Merr.
citrange	*Citrus* × *Poncirus* hybrid
citron	*Citrus medica* L.
cocona	*Solanum topiro* HBK
custard apple	*Annona reticulata* L.
date palm	*Phoenix dactylifera* L.
duku	*Lansium domesticum* Correa
dum palm	*Hyphaene thebaica* (L.) Mart.
durian	*Durio zibethinus* Murr.
egg fruit	*Lucuma nervosa* A.DC.
emblic	*Phyllanthus emblica* L.
ethrog	*Citrus medica* L.
feijoa	*Acca sellowiana* Berg
fig	*Ficus carica* L.
gandaria	*Bouea macrophylla* Griff.
genip	*Melicocca bijuga* L.
governor's plum	*Flacourtia indica* Merr.
granadilla	*Passiflora quadrangularis* L.
grape	*Vitis vinifera* L.
grapefruit	*Citrus paradisi* Macf.
guava	*Psidium guajava* L.
hog plum	*Spondias mombin* L.
icaco	*Chrysobálanus icaco* L.
ilama	*Annona diversifolia* Saff.
jaboticaba	*Myrciaria cauliflora* (Mart.) Berg
jackfruit	*Artocarpus heterophylla* Lam.
jambolan, jamun	*Syzygium cumini* (L.) Skeels
jujube, chinese	*Zizyphus jujuba* Mill.
jujube, indian	*Zizyphus mauritiana* Lam.
kaki	*Diospyros kaki* Thunb.
kapulasan	*Nephelium mutabile* Bl.
kei apple	*Dovyalis caffra* Warb.
King mandarin	*Citrus reticulata* Blanco
kitembilla	*Dovyalis hebecarpa* Warb.
kiwi	*Actinidia chinensis* Planch.
kumquat, oval	*Fortunella margarita* (Lour.) Swing.

kumquat, round	*Fortunella japonica* (Thunb.) Swing.
langsat	*Lansium domesticum* Correa
lemon	*Citrus limon* (L.) Burm. f.
lime	*Citrus aurantifolia* (Christm.) Swing.
litchi, lychee	*Litchi chinensis* Sonn.
longan	*Euphoria longana* Lam.
lovi lovi	*Flacourtia inermis* Roxb.
loquat	*Eriobotrya japonica* (Thunb.) Lindl.
macadamia, smooth	*Macadamia integrifolia* Maiden et Betch.
macadamia, rough	*Macadamia tetraphylla* L. Johns
Malay apple	*Syzygium malaccense* (L.) Merr. et Perry
mamey, mammey apple	*Mammea americana* L.
mamey sapote	*Calocarpum sapota* (Jacq.) Merr.
mamoncillo	*Melicocca bijuga* L.
mandarin	*Citrus reticulata* Blanco
mango	*Mangifera indica* L.
mangosteen	*Garcinia mangostana* L.
melon	*Cucumis melo* L.
menteng	*Baccaurea racemosa* Muell. Arg.
mombin, red	*Spondias purpurea* L.
mombin, yellow	*Spondias mombin* L.
muscadine	*Vitis rotundifolia* Michx.
myrobalan	*Phyllanthus emblica* L.
nangka	*Artocarpus heterophylla* Lam.
naranjillo	*Solanum quitoense* Lam.
natal plum	*Carissa grandiflora* A.DC.
olive	*Olea europaea* L.
orange, sour	*Citrus aurantium* L.
orange, sweet	*Citrus sinensis* (L.) Osbeck
Otaheite apple	*Spondias cytherea* Sonn.
Otaheite gooseberry	*Phyllanthus acidus* (L.) Skeels
pandan	*Pandanus tinctorius* Soland. ex Parkins
papaw, papaya	*Carica papaya* L.
paradise nut	*Lecythis zabucajo* Aubl.
passion fruit, giant	*Passiflora quadrangularis* L.
passion fruit, purple	*P. edulis* Sims, var. *edulis*
passion fruit, yellow	*P. edulis* Sims, var. *flavicarpa*
peach	*Prunus persica* (L.) Batsch
peach palm, pejibaye	*Guilielma gasipaes* (HBK) Bailey
pear	*Pyrus communis* L.
pecan	*Carya illinoensis* (Wangenh.) K. Koch
pepino dulce	*Solanum muricatum* L'Hérit. ex Ait.
persimmon	*Diospyros kaki* Thunb.
phalsa	*Grewia asiatica* L.
pineapple	*Ananas comosus* (L.) Merr.
pistachio	*Pistacia vera* L.

pitanga cherry	*Eugenia uniflora* L.
plantain	*Musa* cvs
plum	*Prunus domestica* L.
pomegranate	*Punica granatum* L.
pomelo	*Citrus paradisi* Macf.
pomerac	*Syzygium malaccense* Merr. et Perry
pomme de cythère	*Spondias cytherea* Sonn.
prickly pear	*Opuntia ficus-indica* (L.) Mill.
pulasan	*Nephelium mutabile* Bl.
pummelo	*Citrus grandis* (L.) Osbeck
Queensland nut	*Macadamia* spp.
rambai	*Baccaurea motleyana* Muell. Arg.
rambutan	*Nephelium lappaceum* L.
raspberry	*Rubus* spp.
rose apple	*Syzygium jambos* (L.) Alston
rough lemon	*Citrus jambhiri* Lush.
rukam	*Flacourtia rukam* Z. & M.
salak palm	*Salacca edulis* Reinw.
sapodilla	*Manilkara achras* (Mill.) Fosb.
sapote, mammey	*Calocarpum sapota* (Jacq.) Merr.
sapote, white	*Casimiroa edulis* La Llave
sapucaia nut	*Lecythis zabucajo* Aubl.
screw pine	*Pandanus odoratissimus* L.f.
sea grape	*Coccoloba uvifera* Jacq.
Seville orange	*Citrus aurantium* L.
shaddock	*Citrus grandis* (L.) Osbeck
souari nut	*Caryocar nuciferum* L.
sour orange	*Citrus aurantium* L.
soursop	*Annona muricata* L.
star apple	*Chrysophyllum cainito* L.
star fruit	*Averrhoa carambola* L.
strawberry	*Fragaria ananassa* Duch. and *F.* hybrids
strawberry guava	*Psidium cattleianum* Sabine
sugar apple, sweetsop	*Annona squamosa* L.
Surinam cherry	*Eugenia uniflora* L.
sweet orange	*Citrus sinensis* (L.) Osbeck
tamarind	*Tamarindus indica* L. .
tangelo	*Citrus* hybrid (mandarin × grapefruit)
tangerine	*Citrus reticulata* Blanco
tangor	*Citrus* hybrid (mandarin × sw. orange)
tree tomato	*Cyphomandra betacea* (Cav.) Sendt
trifoliate orange	*Poncirus trifoliata* (L.) Raf.
Turkish hazelnut	*Corylus colurna* L.
umbú	*Spondias tuberosa* Arruda
uvilla	*Pourouma cecropiaefolia* Mart.
wampi	*Clausena lansium* (Lour.) Skeels

watermelon	*Citrullus lanatus* (Thunb.) Mansf.
watery rose apple	*Syzygium aqueum* (Burm.f.) Merr. et Perry
West Indian cherry	*Malpighia glabra* L.
white sapota	*Casimiroa edulis* La Llave
yang-tao	*Actinidia chinensis* Planch.
yellow mombin	*Spondias mombin* L.

Appendix 3 – Conversion of some non-metric to metric units and vice versa

Length

1 inch = 25.4 mm   1 cm = 0.394 in
1 foot = 30.5 cm   1 m = 3.28 ft
1 mile = 1,609 m   1 km = 0.621 mile

Area

1 sq. foot = 929 cm²   1 m² = 10.8 sq ft
1 acre   = 4,047 m²   1 ha = 2.47 acre

Volume

1 gallon (Imp) = 4.546 dm³   1 dm³ = 0.22 gallon (Imp)
1 gallon (US) = 3.785 dm³   1 dm³ = 0.264 gallon (US)

Mass

1 ounce   = 28.35 g   1 g = 0.035 oz
1 pound (lb) = 0.454 kg   1 kg = 2.205 lb

lb/acre to kg/ha multiply by 1.12 kg/ha to lb/acre multiply by 0.89

Temperature

°C	0	5	10	15	20	25	30	35	40	45	50
°F	32	41	50	59	68	77	86	95	104	113	122

°F to °C Subtract 32, then multiply by 5/9
°C to °F Multiply by 9/5, then add 32

Crop Index

Geographical Index

General Index